Computer-Aided Design and Manufacturing

Computer-Aided Design and Manufacturing

Edited by **Justin Riggs**

C WILLFORD PRESS

New York

Published by Willford Press,
118-35 Queens Blvd., Suite 400,
Forest Hills, NY 11375, USA
www.willfordpress.com

Computer-Aided Design and Manufacturing
Edited by Justin Riggs

© 2016 Willford Press

International Standard Book Number: 978-1-68285-009-1 (Hardback)

Printed in the United States of America.

Contents

Preface

Computer Aided Design (CAD) and Computer Aided Manufacturing (CAM) has revolutionised the process of designing and manufacturing of machinery and electronic equipment with precision and efficiency. Computer aided softwares have led to the creation of products with precise dimensions and have increased the rate of production. This book explains the innovative aspects of computer-aided design and manufacturing with the help of core subjects like technical and engineering drawings, geometric configuration for solid modeling, user and system interfaces, etc. With state-of-the-art inputs by acclaimed experts of this field, this book targets students and professionals alike.

The information contained in this book is the result of intensive hard work done by researchers in this field. All due efforts have been made to make this book serve as a complete guiding source for students and researchers. The topics in this book have been comprehensively explained to help readers understand the growing trends in the field.

I would like to thank the entire group of writers who made sincere efforts in this book and my family who supported me in my efforts of working on this book. I take this opportunity to thank all those who have been a guiding force throughout my life.

Editor

Operational matrix approach for the nonlinear Volterra-Fredholm integral equations: Arising in physics and engineering

B. Basirat[1]*, K. Maleknejad[2] and E. Hashemizadeh[2]

[1]Department of Mathematics, Birjand Branch, Islamic Azad University, Birjand, Iran.
[2]Department of Mathematics, Karaj Branch, Islamic Azad University, Karaj, Iran.

An approximation method based on hybrid Legendre polynomials and Block-Pulse functions used for the solution of nonlinear Volterra-Fredholm integral equations (NV-FIEs). The operational matrices of these functions are utilized to reduce a nonlinear Volterra-Fredholm integral equation to a system of nonlinear algebraic equations. In addition, convergence analysis and numerical examples that illustrate the pertinent features of the method are presented.

Key words: Hybrid functions, nonlinear integral equations, operational matrix, product matrix, coefficient matrix.

INTRODUCTION

Integral equation has been one of the principal tools in various areas of applied mathematics, physics and engineering. Integral equation is encountered in a variety of applications in many fields, including continuum mechanics, potential theory, geophysics, electricity and magnetism, antenna synthesis problem, communication theory, mathematical economics, population genetics and radiation, the particle transport problems of astrophysics and reactor theory, fluid mechanics etc. Many of these integral equations are nonlinear (Voitovich and Reshnyak, 1999; Jaswon and Symm, 1977; Schiavane et al., 2002; Abdou, 2003; Bloom, 1980; Jiang and Rokhlin, 2004; Semetanian, 2007).

In this paper we deal with nonlinear Volterra-Fredholm integral equations as follows:

$$u(x)=f(x)+\lambda_1\int_0^x k_1(x,s)\psi_1(s,u(s))ds+\lambda_2\int_0^1 k_2(x,s)\psi_2(s,u(s))ds,$$

(1)

where the parameters λ_1, λ_2 and

*Corresponding author. E-mail: behrooz.basirat@kiau.ac.ir.

Mathematics subject classification (2010): 45G10, 45D05, 45B05, 40C05.

functions $f(x)$, $\psi_1(s,u(s))$, $\psi_2(s,u(s))$, $k_1(x,s)$ and $k_2(x,s)$ are known while $L^2[0,1)$ and $u(x)$ is an unknown function. In this work we suppose $\psi_1(s,u(s))=u(s)^\alpha$ and $\psi_2(s,u(s))=u(s)^\beta$ where α, β are positive integers (Maleknejad et al., 2011; Maleknejad et al., 2010; Yalcinbas, 2002; Yousefi and Razzaghi, 2005; Ordokhani, 2006; Babolian et al., 2008; Sepehrian and Razzaghi, 2005; Ordokhani and Razzaghi, 2008)

Maleknejad, Hashemizadeh and Basirat solved Equation 1 by Bernstein operational matrices method (Maleknejad et al., 2011). Maleknejad, Almaslh and Roodaki reduced these kinds of equations to algebraic equations by triangular functions (Maleknejad et al., 2010). Yalcinbas (2002) applied Taylor series to solve Equation 1. Yousefi and Razzaghi (2005) used Legendre wavelets for the numerical solution of these equations. Ordokhani (2006) used rationalized Haar functions for solving these equations and Babolian et al. (2008) used Block–Pulse functions to encounter these kinds of NV-FIEs.

In this paper we use the hybrid Legendre polynomials and Block-Pulse functions as basis for reducing NV-FIEs to a system of nonlinear algebraic equations. We present hybrid Legendre polynomials and Block-Pulse useful properties such as operational matrix of integration, product matrix, integration of the cross product and coefficient matrix, and use them to transform our NV-FIE to an algebraic system. As shown in our examples, our method works better in comparison to the existing methods.

This paper introduces hybrid functions and their properties. Application of these set of hybrid functions for approximating the solution of NV-FIEs was done, convergence analysis given and the proposed method tested with some examples; thereafter the results were compared with some existing methods results. Thus a conclusion was drawn.

HYBRID FUNCTIONS OF BLOCK-PULSE AND LEGENDRE POLYNOMIALS

The orthogonal set of hybrid functions is $h_{ij}(x), i = 1, 2, ..., n$, $j = 0, 1, ..., m-1$ in which i is the order for Block-Pulse functions, j is the order for Legendre polynomials, and x is the normalized time and is defined on the interval $[0,1)$ (Hsiao, 2009).

$$h_{ij}(x) = \begin{cases} L_j(2nx - 2i + 1), & \frac{i-1}{n} \le x < \frac{i}{n}, \\ 0, & \text{otherwise.} \end{cases}$$

(2)

Here, the Legendre polynomials $L_m(x)$ on the interval $[-1,1]$:

$L_0(x) = 1, \ L_1(x) = x,$

$(m+1)L_{m+1}(x) = (2m+1)xL_m(x) - mL_{m-1}(x), \ m = 1, 2, 3, ...$

The set $\{L_m(x) : m = 0, 1, ...\}$ in Hilbert space $L^2[-1,1]$ is a complete orthogonal set.

A set of Block-Pulse functions $b_i(x), i = 1, 2, .., n$ on the interval $[0,1)$ is defined as follows:

$$b_i(x) = \begin{cases} 1, & \frac{i-1}{n} \le x < \frac{i}{n}, \\ 0, & \text{otherwise.} \end{cases}$$

(3)

The Block-Pulse functions on $[0,1)$ are disjoint, that results for $i, j = 1, 2, ..., n$, we have: $b_i(x)b_j(x) = \delta_{ij}b_i(x)$, also these functions have the property of orthogonality on $[0,1)$. Since $h_{ij}(x)$ is the combination of Legendre polynomials and Block-Pulse functions which are both complete and orthogonal, thus

the set of hybrid functions is complete orthogonal set too.

PROPERTIES OF HYBRID FUNCTIONS

Function approximation

Any function $u(x) \in L^2[0,1]$ can be expanded in hybrid functions (Hsiao, 2009).

$$u(x) = \sum_{i=1}^{\infty} \sum_{j=0}^{\infty} c_{ij} h_{ij}(x),$$

(4)

where the hybrid coefficients are given by

$$c_{ij} = \frac{(u(x), h_{ij}(x))}{(h_{ij}(x), h_{ij}(x))}$$ for $i = 1, 2, ..., \infty$, $j = 0, 1, ..., \infty$, so

that, $(\cdot, \cdot)$ denotes the inner product.

Usually, the series expansion Equation 4 contains an infinite number of terms for a smooth $u(x)$. If $u(x)$ is piecewise constant or may be approximated as piecewise constant, then the sum in Equation 4 may be terminated after nm terms, that is,

$$u(x) \simeq \sum_{i=1}^{n} \sum_{j=0}^{m-1} c_{ij} h_{ij}(x) = C^T \mathbf{h}(x),$$

(5)

where

$$C = [c_{10}, ..., c_{1,m-1}, c_{20}, ..., c_{2,m-1}, ..., c_{n0}, ..., c_{n,m-1}]^T,$$

(6)

$$\mathbf{h}(x) = [h_{10}(x), ..., h_{1,m-1}(x), h_{20}(x), ..., h_{2,m-1}(x), ..., h_{n,m-1}(x)]^T.$$

(7)

We can also approximate the function $k(x,s) \in L^2([0,1) \times [0,1))$ as follows:

$$k(x,s) \simeq \mathbf{h}^T(x) K \mathbf{h}(s),$$

(8)

where K is an $nm \times nm$ matrix that

$$K_{ij} = \frac{(\mathbf{h}_{(i)}(x), (k(x,s), \mathbf{h}_{(j)}(s)))}{(\mathbf{h}_{(i)}(x), \mathbf{h}_{(i)}(x))(\mathbf{h}_{(j)}(s), \mathbf{h}_{(j)}(s))}$$ for $i, j = 1, 2, ..., nm$.

Operational matrix of integration

The integration of the vector $\mathbf{h}(x)$ defined in Equation 7 is given by

$$\int_0^x \mathbf{h}(x')\,dx' \simeq P\mathbf{h}(x),\tag{9}$$

where P is the $nm \times nm$ operational matrix for integration and is given (Hsiao, 2009; Chang and Wang, 1983) as:

$$P = \begin{bmatrix} E & H & H & \cdots & H \\ O & E & H & \cdots & H \\ O & O & E & \cdots & H \\ \vdots & \vdots & \vdots & \ddots & \vdots \\ O & O & O & \cdots & E \end{bmatrix},$$

that E and H are $m \times m$ matrices that have the following shapes, respectively.

$$H = \frac{1}{n}\begin{bmatrix} 1 & 0 & 0 & \cdots & 0 \\ 0 & 0 & 0 & \cdots & 0 \\ 0 & 0 & 0 & \cdots & 0 \\ \vdots & \vdots & \vdots & \ddots & \vdots \\ 0 & 0 & 0 & \cdots & 0 \end{bmatrix},$$

$$E = \frac{1}{2n}\begin{bmatrix} 1 & 1 & 0 & 0 & 0 & \cdots & 0 & 0 & 0 & 0 & 0 \\ \frac{-1}{3} & 0 & \frac{1}{3} & 0 & 0 & \cdots & 0 & 0 & 0 & 0 & 0 \\ 0 & \frac{-1}{5} & 0 & \frac{1}{5} & 0 & \cdots & 0 & 0 & 0 & 0 & 0 \\ 0 & 0 & \frac{-1}{7} & 0 & \frac{1}{7} & \vdots & 0 & 0 & 0 & 0 & 0 \\ 0 & 0 & 0 & \frac{-1}{9} & 0 & \cdots & 0 & 0 & 0 & 0 & 0 \\ \vdots & \vdots & \vdots & \vdots & \vdots & \ddots & \vdots & \vdots & \vdots & \vdots & \vdots \\ 0 & 0 & 0 & 0 & 0 & \cdots & 0 & \frac{1}{2m-9} & 0 & 0 & 0 \\ 0 & 0 & 0 & 0 & 0 & \cdots & \frac{-1}{2m-7} & 0 & \frac{1}{2m-7} & 0 & 0 \\ 0 & 0 & 0 & 0 & 0 & \cdots & 0 & \frac{-1}{2m-5} & 0 & \frac{1}{2m-5} & 0 \\ 0 & 0 & 0 & 0 & 0 & \cdots & 0 & 0 & \frac{-1}{2m-3} & 0 & \frac{1}{2m-3} \\ 0 & 0 & 0 & 0 & 0 & \cdots & 0 & 0 & 0 & \frac{-1}{2m-1} & 0 \end{bmatrix}.$$

The integration of the cross product

The integration of the cross product of two hybrid function vectors $\mathbf{h}(x)$ in Equation 7 can be obtained as

$$D = \int_0^1 \mathbf{h}(x)\mathbf{h}^T(x)\,dx = \begin{bmatrix} L & O & \cdots & O \\ O & L & \cdots & O \\ \vdots & \vdots & \ddots & \vdots \\ O & O & \cdots & L \end{bmatrix},\tag{10}$$

where L is an $m \times m$ diagonal matrix that is given by

$$L = \frac{1}{n}\begin{bmatrix} 1 & 0 & \cdots & 0 \\ 0 & \frac{1}{3} & \cdots & 0 \\ \vdots & \vdots & \ddots & \vdots \\ 0 & 0 & \cdots & \frac{1}{2m-1} \end{bmatrix}.\tag{11}$$

The efficacy of matrix D is used for converting the Fredholm part of NV-FIEs to an algebraic equation. Because of its diagonal shape, it can increase the calculating speed.

Product operational matrix

It is always necessary to evaluate the product of $\mathbf{h}(x)$ and $\mathbf{h}^T(x)$, that is called the product matrix of hybrid functions. Let

$$\mathbf{H}(x) = \mathbf{h}(x)\mathbf{h}^T(x),\tag{12}$$

where $\mathbf{H}(x)$ is an $nm \times nm$ matrix. Multiplying the matrix $\mathbf{H}(x)$ in vector C that was defined in Equation 6, we obtain

$$\mathbf{H}(x)C = \tilde{C}\mathbf{h}(x),\tag{13}$$

where $\tilde{C}$ is an $nm \times nm$ matrix and is called the coefficient matrix. To illustrate the calculation procedure in Equation 13 we consider that $n = 4$ and $m = 3$ (Hsiao, 2009; Chang and Wang, 1983; Marzban and Razzaghi, 2003). We have

$$h_{ij}(x)h_{kl}(x) = 0 \;\text{ if }\; i \neq k,$$

$$h_{i0}(x)h_{ij}(x) = h_{ij}(x),$$

$$h_{i1}(x)h_{i1}(x) = \frac{1}{3}h_{i0}(x) + \frac{2}{3}h_{i2}(x),$$

$$h_{i1}(x)h_{i2}(x) = \frac{2}{5}h_{i1}(x) + \frac{3}{5}h_{i3}(x),$$

$$h_{i2}(x)h_{i2}(x) = \frac{1}{5}h_{i0}(x) + \frac{2}{7}h_{i2}(x) + \frac{18}{35}h_{i4}(x).$$

Then we get

$$\mathbf{H}(x) = \begin{bmatrix} h_{10}(x) & h_{11}(x) & h_{12}(x) & & & & & \\ h_{11}(x) & \dfrac{1}{3}h_{10}(x) \\ +h_{12}(x) & \dfrac{2}{5}h_{11}(x) & & & & O & \\ h_{12}(x) & \dfrac{2}{5}h_{11}(x) & \dfrac{1}{5}h_{10}(x) \\ +\dfrac{2}{7}h_{12}(x) & & & \ddots \\ & & & & h_{40}(x) & h_{41}(x) & h_{42}(x) \\ & & & & h_{41}(x) & \dfrac{1}{3}h_{40}(x) \\ +h_{42}(x) & \dfrac{2}{5}h_{41}(x) \\ & O & & & h_{42}(x) & \dfrac{2}{5}h_{41}(x) & \dfrac{1}{5}h_{40}(x) \\ +\dfrac{2}{7}h_{42}(x) \end{bmatrix}.$$

The matrix $\tilde{C}_{12\times12}$ in Equation 13 is given by

$$\tilde{C} = \begin{bmatrix} \widetilde{C_1} & O & O & O \\ O & \widetilde{C_2} & O & O \\ O & O & \widetilde{C_3} & O \\ O & O & O & \widetilde{C_4} \end{bmatrix},$$

where $\widetilde{C_i}$, $i = 1,\dots,4$ are given by

$$\widetilde{C_i} = \begin{bmatrix} c_{i0} & c_{i1} & c_{i2} \\ \dfrac{1}{3}c_{i1} & c_{i0}+\dfrac{2}{5}c_{i2} & \dfrac{2}{3}c_{i1} \\ \dfrac{1}{5}c_{i2} & \dfrac{2}{5}c_{i1} & c_{i0}+\dfrac{2}{7}c_{i2} \end{bmatrix}.$$

With the powerful properties of Equation 13, we can convert the Volterra part of NV-FIEs to an algebraic equation.

OUTLINE OF THE METHOD FOR NV-FIES VIA HYBRID FUNCTIONS

Consider the nonlinear Volterra–Fredholm integral Equation 1. We put

$$u(x) \simeq U^T\mathbf{h}(x), \tag{14}$$

where U is an unknown nm-vector and $\mathbf{h}(x)$ is given by Equation 7.

Likewise $k_1(x,s)$, $k_2(x,s)$ and $f(x)$ are expanded into the hybrid functions as follows:

$$k_1(x,s) \simeq \mathbf{h}^T(x)K_1\mathbf{h}(s), \quad k_2(x,s) \simeq \mathbf{h}^T(x)K_2\mathbf{h}(s), \tag{15}$$

$$f(x) \simeq F^T\mathbf{h}(x), \tag{16}$$

where K_1, K_2 are known $nm \times nm$–matrices and F is an nm-vector.

After substituting the approximate Equations 14, 15 and 16 in Equation 1, we get

$$U^T\mathbf{h}(x) \simeq F^T\mathbf{h}(x) + \lambda_1\mathbf{h}^T(x)K_1\int_0^x \mathbf{h}(s)\psi_1(s,U^T\mathbf{h}(s))ds$$
$$+\lambda_2\mathbf{h}^T(x)K_2\int_0^1 \mathbf{h}(s)\psi_2(s,U^T\mathbf{h}(s))ds. \tag{17}$$

Functions $\psi_1(s,U^T\mathbf{h}(s)) = (U^T\mathbf{h}(s))^\alpha$ and $\psi_2(s,U^T\mathbf{h}(s)) = (U^T\mathbf{h}(s))^\beta$ are known which can be expanded into the hybrid functions as

$$(u(s))^\alpha \simeq U_\alpha^T\mathbf{h}(s), \quad (u(s))^\beta \simeq U_\beta^T\mathbf{h}(s). \tag{18}$$

where U_α, U_β are nm–vectors whose elements are nonlinear combination of the elements of the vector U and are produced as follows:

From Equations 13 and 14, we have

$$(u(x))^2 \simeq (U^T\mathbf{h}(x))(U^T\mathbf{h}(x)) = U^T\mathbf{h}(x)\mathbf{h}^T(x)U$$
$$= U^T\tilde{U}\mathbf{h}(x) = U_2\mathbf{h}(x), \tag{19}$$

where the vector $U_2 = U^T\tilde{U}$ is an nm–row vector, then for $(u(s))^3$ we get

$$(u(x))^3 \simeq (U^T\mathbf{h}(x))(U_2\mathbf{h}(x)) = U^T\mathbf{h}(x)\mathbf{h}^T(x)U_2^T$$
$$= U^T\tilde{U}_2^T\mathbf{h}(x) = U_3\mathbf{h}(x). \tag{20}$$

Therefore with this method we can approximate $(u(s))^\alpha$ and $(u(s))^\beta$ for arbitrary α and β. Suppose that this method holds for $\alpha-1$ where $(u(x))^{\alpha-1} = U_{\alpha-1}\mathbf{h}(x)$, we shall obtain it for α as follows:

$$(u(x))^\alpha = u(x)u(x)^{\alpha-1} \simeq (U^T \mathbf{h}(x))(U_{\alpha-1}\mathbf{h}(x))$$
$$= U^T \mathbf{h}(x)\mathbf{h}^T(x)U_{\alpha-1}^T$$
$$= U^T \tilde{U}_{\alpha-1}^T \mathbf{h}(x) = U_\alpha \mathbf{h}(x), \tag{21}$$

We have a similar relation for β. So, the components of U_α and U_β can be computed in terms of components of unknown vector U.

Substituting Equation 18 in Equation 17 produces

$$U^T\mathbf{h}(x) \simeq F^T\mathbf{h}(x) + \lambda_1 \mathbf{h}^T(x)K_1 \int_0^x \mathbf{h}(s)\mathbf{h}^T(s)U_\alpha ds + \lambda_2 \mathbf{h}^T(x)K_2 \int_0^1 \mathbf{h}(s)\mathbf{h}^T(s)U_\beta ds \tag{19}$$

Note that by use of Equations 9 and 13 we have
$$\int_0^x \mathbf{h}(s)\mathbf{h}^T(s)U_\alpha ds = \int_0^x \tilde{U}_\alpha \mathbf{h}(s)ds = \tilde{U}_\alpha P\mathbf{h}(x)$$
, by this relation and Equation 10, we get

$$U^T\mathbf{h}(x) \simeq F^T\mathbf{h}(x) + \lambda_1 \mathbf{h}^T(x)K_1 \tilde{U}_\alpha P\mathbf{h}(x) + \lambda_2 \mathbf{h}^T(x)(K_2 D U_\beta). \tag{22}$$

In order to find U we collocate Equation 22 in nm nodal points of Newton-Cotes as,

$$x_p = \frac{2p-1}{2nm}, \quad p = 1,2,\dots,nm, \tag{23}$$

then we have the following system of nonlinear equations

$$U^T\mathbf{h}(x_p) \simeq F^T\mathbf{h}(x_p) + \lambda_1 \mathbf{h}^T(x_p)K_1\tilde{U}_\alpha P\mathbf{h}(x_p) + \lambda_2 \mathbf{h}^T(x_p)(K_2 D U_\beta), \quad p=1,2\dots,nm. \tag{24}$$

This nonlinear system of equations can be solved by Newton's method. We used the "*Mathematica 7*" software to solve this nonlinear system. After solving nonlinear system (Equation 24) we can achieve U, then we will have our unknown $u(x)$ as $U^T\mathbf{h}(x)$, that is the approximate solution of NV-FIE (1).

CONVERGENCE ANALYSIS

We assume $(C(J)\|\ \|)$ the Banach space of all continuous functions on $J = [0,1]$ with norm $\| f(x)\| = \max_{0\le x\le 1} | f(x)|$ and the following conditions on k_1, k_2 and ψ_1, ψ_2 for Equation 1. We define $k_x \equiv k(x,s)$ for $x,s \in [0,1]$:

1. $\lim_{x\to\tau} \| k_x - k_\tau\| = 0$, $\tau \in [0,1]$.

2. $M_1 \equiv \sup_{0\le x,s\le 1} | k_1(x,s)| < \infty$, $M_2 \equiv \sup_{0\le x,s\le 1} | k_2(x,s)| < \infty$.

3. $\psi_1(s,x)$ and $\psi_2(s,x)$ are continuous in $s \in [0,1]$ and Lipschitz continuous in $x \in R$, that is, there exist constants $C_1, C_2 > 0$ for which
$$|\psi_1(s,x_1) - \psi_1(s,x_2)| \le C_1 | x_1 - x_2| \quad \text{for all } x_1,x_2 \in R,$$
$$|\psi_2(s,x_1) - \psi_2(s,x_2)| \le C_2 | x_1 - x_2| \quad \text{for all } x_1,x_2 \in R.$$

Theorem 1

The solution of nonlinear Volterra-Fredholm integral equation by hybrid functions converges if $0 < \gamma < 1$.

Proof: (Maleknejad et al., 2010)□.

NUMERICAL EXAMPLES

Here, we implemented the proposed method on 3 different examples. The results achieved by a proper value for m and different values for n. All the results are compared with some existing method results. As in Tables 1 to 5, the error tends to zero when n becomes greater and in analogy to another methods' results the proposed method have better answers in lower n. Although, we do not claim that this method shows superiority over the other methods from the viewpoint of accuracy, but we can say this method is more practical, quite accurate with lower calculation. The matrices P and D are sparse, hence are much faster than other functions' operational matrices and they reduce the CPU time and at the same time keeping the accuracy of the solution. In our examples we get the results by $m = 8$ and $n = 2,4,8,16$. In all of them we compared our answers with some existing methods. We consider the L^2-norm of errors for Examples 1 to 3 which are shown in Tables 4 and 5 by

$$E_2 = \left(\int_0^1 (u(x) - u_{nm}(x))^2 dx \right)^{\frac{1}{2}}$$

For implementation of proposed method we used *Mathematica 7*.

Example 1

Consider the nonlinear Volterra–Fredholm equation given by

$$u(x) = \frac{1}{30}x^6 + \frac{1}{3}x^4 - x^2 + \frac{5}{3}x - \frac{5}{4} + \int_0^x (x-s)u^2(s)ds + \int_0^1 (x+s)u(s)ds,$$

Table 1. Approximate and exact solutions for Example 1.

x	Solution with $n=2$	Solution with $n=4$	Solution with $n=8$	Solution with $n=16$	Method (Ordokhani, 2006) with $k=16$	Exact
0.0	-2.004858	-2.001333	-2.000340	-2.000085	-1.995	-2
0.1	-1.996105	-1.991671	-1.990420	-1.990104	-1.989	-1.99
0.2	-1.967892	-1.962063	-1.960511	-1.960125	-1.965	-1.96
0.3	-1.919818	-1.912286	-1.910599	-1.910152	-1.912	-1.91
0.4	-1.851187	-1.842838	-1.840683	-1.840174	-1.841	-1.84
0.5	-1.761040	-1.753034	-1.750775	-1.750195	-1.752	-1.75
0.6	-1.650618	-1.643166	-1.640844	-1.640208	-1.643	-1.64
0.7	-1.521064	-1.513388	-1.510834	-1.510204	-1.498	-1.51
0.8	-1.371005	-1.362753	-1.360742	-1.360192	-1.359	-1.36
0.9	-1.198995	-1.192383	-1.190576	-1.190149	-1.185	-1.19
1.0	-1.003603	-1.001267	-1.000339	-1.000086	-0.994	-1

Table 2. Approximate and exact solutions for Example 2.

x	Solution with $n=2$	Solution with $n=4$	Solution with $n=8$	Solution with $n=16$	Method (Sepehrian and Razzaghi, 2005) with $M=20$	Exact
0.0	0.00000	0.00000	0.00000	0.00000	0.00000	0
0.1	-0.09007	-0.09002	-0.09000	-0.09000	-0.09000	-0.09
0.2	-0.16032	-0.16008	-0.16002	-0.16000	-0.16000	-0.16
0.3	-0.21066	-0.21018	-0.21005	-0.21001	-0.20999	-0.21
0.4	-0.24094	-0.24027	-0.24005	-0.24001	-0.23999	-0.24
0.5	-0.25111	-0.25022	-0.25005	-0.25001	-0.24999	-0.25
0.6	-0.24127	-0.24013	-0.24004	-0.24000	-0.23999	-0.24
0.7	-0.21167	-0.21025	-0.21002	-0.21001	-0.20999	-0.21
0.8	-0.16252	-0.16047	-0.16008	-0.16002	-0.15998	-0.16
0.9	-0.09377	-0.09075	-0.09020	-0.09004	-0.08997	-0.09
1.0	-0.00503	-0.00113	-0.00027	-0.00006	0.00003	0

with the exact solution $u(x)=x^2-2$ [13]. The comparison among the hybrid solutions beside the exact solutions are shown in Table 1.

Example 2

Consider the nonlinear Volterra integral equation considered in (Sepehrian and Razzaghi, 2005)

$$u(x)=\frac{15}{56}x^8+\frac{13}{14}x^7-\frac{11}{10}x^6+\frac{9}{20}x^5+x^2-x+\int_0^x (x+s)u^3(s)ds,$$

with the exact solution $u(x)=x^2-x$. The comparison

among the hybrid solutions with $m=8$ and $n=2$, $n=4$, $n=8$ and $n=16$ beside the exact solutions are shown in Table 2.

Example 3

Consider the nonlinear Fredholm integral equation given in (Babolian et al., 2008) by

$$u(x)=x^2-\frac{1}{12}+\frac{1}{2}\int_0^1 s\,u^2(s)ds,$$

with the exact solution $u(x)=x^2$. The comparison among the hybrid solutions beside the exact solutions are

Table 3. Approximate and exact solutions for Example 3.

x	Solution with $n=2$	Solution with $n=4$	Solution with $n=8$	Solution with $n=16$	Method (Babolian, 2008) with $m=64$	Exact
0.0	0.020629	0.005630	0.001436	0.000361	0.000054	0
0.1	0.030629	0.015630	0.011436	0.010361	0.010308	0.01
0.2	0.060629	0.045630	0.041436	0.040361	0.038140	0.04
0.3	0.110629	0.095630	0.091436	0.090361	0.092828	0.09
0.4	0.180629	0.165630	0.161436	0.160361	0.158746	0.16
0.5	0.270629	0.255630	0.251436	0.250361	0.257867	0.25
0.6	0.380629	0.365630	0.361436	0.360361	0.361871	0.36
0.7	0.510629	0.495630	0.491436	0.490361	0.483453	0.49
0.8	0.660629	0.645630	0.641436	0.640361	0.647515	0.64
0.9	0.830629	0.815630	0.811436	0.810361	0.807183	0.81
1.0	1.020629	1.005630	1.001436	1.000361	-	1

Table 4. Errors E_2 for Examples 1.

m	Method in [10]	n	present method $m=8$
4	0.011410886610	2	0.009508596380
8	0.002852721653	4	0.002578626395
16	0.000713180414	8	0.000656705013
32	0.003779018306	16	0.000164921678

Table 5. Errors E_2 for Examples 2 to 3.

n	Example 2	Example 3
2	0.001986101366	0.020629858826
4	0.000407433748	0.005630249634
8	0.000098383419	0.001363230304
16	0.000024414251	0.000361078218

shown in Table 3. Our achieved errors for Example 1 is comparable with exhibited errors (Maleknejad et al., 2010), an example is shown in Table 4. As obvious in present method with lower n gets better results.

Conclusion

This work presents a numerical approach for solving NV-FIEs based on the hybrid Legendre polynomials and Block–Pulse functions. These hybrid functions operational matrices of integration D, operational matrix P, product matrix H and coefficient matrix $\tilde{C}$ have been created to convert NV-FIEs to an algebraic equation and then by collocating this equation in Newton-cuts nodes, a nonlinear system of equations that can be solved by Newton method, was produced. Illustrative examples are given to demonstrate the validity and applicability of proposed method. Our compared results show that this method works better and faster than some existing methods.

Acknowledgment

The authors would like to thank Islamic Azad University of Birjand Branch for partially financially supporting this research and providing facilities and encouraging this work.

REFERENCES

Abdou MA (2003). On asymptotic methods for Fredholm–Volterra integral equation of the second kind in contact problems. J. Comput. Appl. Math., 154: 431-446.

Babolian E, Masouri Z, Hatamzadeh-Varmazyar S (2008). New direct method to solve nonlinear Volterra–Fredholm Integral and Integro-Differential equations using operational matrix with Block–Pulse functions. Prog. Electromagn. Res. B., 8: 59-76.

Bloom F (1980). Asymptotic bounds for solutions to a system of damped integro–differential equations of electromagnetic theory. J. Math. Anal. Appl., 73: 524-542.

Chang RY, Wang ML (1983). Shifted Legendre direct method for variational problems. J. Opim. Theory Appl., 39: 299-307.

Hsiao CH (2009). Hybrid function method for solving Fredholm and Volterra integral equations of the second kind. J. Comput. Appl. Math., 230: 59-68.

Jaswon MA, Symm GT (1977). Integral Equation Methods in Potential Theory and Elastostatics. London: Academic Press. 275(23).

Jiang S, Rokhlin V (2004). Second kind integral equations for the calssical potential theory on open surface II. J. Comput. Phys., 195: 1-16.

Maleknejad K, Almasih H, Roodaki M (2010). Triangular functions method for the solution of nonlinear Volterra–Fredholm integral equations. Commun. Nonlinear Sci. Numer. Simul., 15: 3293-3298.

Maleknejad K, Hashemizadeh E, Basirat B (2011). Computational method based on Bernstein operational matrices for nonlinear Volterra–Fredholm–Hammerstein integral equations. Commun. Nonlinear. Sci. Numer. Simulat. In Press. doi:10.1016/j.cnsns.04.023.

Marzban HR, Razzaghi M (2003). Hybrid functions approach for linearly constrained quadratic optimal control problems. Appl. Math. Model, 27: 471-485.

Ordokhani Y (2006). Solution of nonlinear Volterra–Fredholm–Hammerstein integral equations via rationlized Haar functions. Appl. Math. Comput., 180: 436-443.

Ordokhani Y, Razzaghi M (2008). Solution of nonlinear Volterra–Fredholm–Hammerstein integral equations via a collocation method and rationalized Haar functions. Appl. Math. Lett., 21: 4-9.

Schiavane P, Constanda C, Mioduchowski A (2002). Integral Methods in Science and Engineering. Birkhauser, Boston, p. 264 ISBN 0-8176-4213-7.

Schiavane P, Constanda C, Mioduchowski A (2002). Integral Methods in Science and Engineering. Birkhauser, Boston, pp. 250-258.

Semetanian BJ (2007). On an integral equation for axially symmetric problem in the case of an elastic body containing an inclusion. J. Comp. Appl. Math., 200: 12-20.

Sepehrian B, Razzaghi M (2005). Solution of nonlinear Volterra–Hammerstein integral equations via single-term walsh series method. Math. Probl. Eng., 5: 547-554.

Voitovich NN, Reshnyak OO (1999). Solutions of nonlinear integral equation of synthesis of the linear antenna arrays. BSUAE J. App. Electr., 2(1): 43-52.

Yalcinbas S (2002). Taylor polynomial solution of nonlinear Volterra–Fredholm integral equations. Appl. Math. Comput., 127: 195-206.

Yousefi S, Razzaghi M (2005). Legendre wavelets method for the nonlinear Volterra–Fredholm integral equations. Math. Comp. Simul., 70: 1-8.

Modeling flow regime transition in intermittent water supply networks using the interface tracking method

Stephen Nyende-Byakika[1]*, Gaddi Ngirane-Katashaya[2] and Julius M. Ndambuki[1]

[1]Department of Civil Engineering, Tshwane University of Technology, Pretoria, South Africa.
[2]Department of Civil Engineering, Makerere University, Kampala, Uganda.

For several ailing water distribution networks in the world, during conditions of excessive withdrawals or insufficient water production, pressures fall to very low or even negligible values and consequently, no water can be supplied. Usually, most pipes have water which either fills or nearly fills their cross sectional areas but the pressure to push it out is absent. With conditions changing from pressurized to no pressure (free surface flow), existing water supply models are unable to simulate either free surface flow or the transition between free surface to pressurized flow. In this study, transient "low pressure-open-channel flow" (LPOCF) conditions were analyzed. The interest of this research lies in the coexistence of free surface and pressurized flow regimes with the aim of understanding the pressurization process of pipes. This was represented by a flow regime transition from free surface to pressurized flow through a moving interface along the pipeline. Results revealed the merits of applying full dynamic wave equations in the solution of transient LPOCF conditions in water distribution networks.

Key words: Free surface flow, full dynamic equations, low pressures, open channel flow, pressurization, water supply, flow regime transition, mixed flow.

INTRODUCTION

Growing demand for water as a result of increasing urban populations, industrialization and rising water consuming lifestyles puts stress on existing water supply systems. In order to cater for additional demand, distribution networks are expanded often beyond their design capacities, which creates bottlenecks such as development of transient flow conditions ranging from excessive pressures and fluctuating pressures to open-channel flow situations (Nyende-Byakika et al., 2010). This culminates into low pressures with low flows and sometimes no flow at all, thereby compromising service levels and giving planners and engineers a complicated task of supplying the additional resource in sufficient and reliable quantities in the

most feasible way possible. Such problems would best be solved through infrastructural upgrades; however, this is an expensive option not easily affordable in many developing countries that are often faced with this problem.

In order to meet regulatory requirements and customer expectations, water utilities are feeling a growing need to explain better the movement and transformations undergone by water introduced into their distribution systems (Rossman, 2000). If understanding of network behaviour under adverse conditions could be obtained and the impact of these conditions established, networks would be managed better, and more satisfactory customer service would be offered (Nyende-Byakika, 2011). However, modeling intermittent water supply systems of pipeline networks is a challenging task because these systems are not fully pressurized but networks with high

*Corresponding author. E-mail: stenbyak@gmail.com.

Water demands, very low pressures and sometimes restricted by water supply hours per day. Many systems exhibit open channel flow behaviour due to excessive low transient pressure conditions when some sections operate as gravity flow systems under low reservoir conditions and get pressurized under high reservoir level conditions as force mains with low pressures. The alternate emptying and refilling followed by pressurization and depressurization of water pipelines make it problematic to apply standard hydraulic models because of low transient pressures and pipes flowing partially full (Ingeduld et al., 2006). This type of situation is difficult to analyze using conventional approaches and may require special treatment different from that of fully pressurized systems, with more sophisticated/complex algorithms and robust scenario management to model. Ingeduld et al. (2006) notes that hydraulic models of intermittent water supply need to simulate the "charging" process in pipes and this requires integration of continuity and motion equations to indicate the positions of the water front in the network at any time.

Due to the fact that most water supply models operate on the assumption that pressure is sufficient to deliver adequate flows, in situations of low transient pressures, the models do not give reliable results. Thus, in order to address this issue, a tool that treats transient pressures and flows which lie along the continuum between open and closed systems in both time and space has got to be developed. This research was aimed at augmenting existing knowledge on supply of water during transient low pressure open-channel flow (LPOCF) conditions. Knowledge obtained would aid the provision of water supply services even under extreme situations of low flows and low pressures. This will not only improve the understanding of piped water supply systems but also ensure sustainable supply of the basic need for survival.

In this paper, the authors studied co-existence of pressurized and free surface flow regimes in a network; a situation sometimes referred to as mixed flow. Simulation of flow regime changes between pressurized and free surface flow was done; a condition which current water distribution models do not tackle. This aids in understanding the development and propagation of pressure surges in pipelines so that during the transient state while supply is low, pressures can be determined simultaneously with discharges that can be availed to consumers.

Transient low pressure – open channel flow conditions

It is worth noting that in water supply situations where intermittent flow is manifested, transient LPOCF

conditions can best be described as unsteady flow since flow depth and velocity vary with time and as gradually varied flow since the rate of change of flow depth and velocity along the channel is very low. Gradually varied flow is a non-uniform flow whose spatial rate of flow is sufficiently low to imply translatory wave motion of long wavelength and low amplitude (Chadwick et al., 2004) such that the assumption of parallel streamlines and hydrostatic pressure distributions is reasonable. In this research therefore, LPOCF conditions were modeled as unsteady gradually varied flow.

In unsteady non-uniform flow, the discharge Q, varies as a function of time and length along the pipe and all the hydraulic parameters of a cross-section change as a function of time and length that is, water depth, cross-sectional area and water surface width. Thus, unsteady flow equations are key to the understanding of the unsteady flows in pipelines. The (full dynamic wave) equations that are used to solve unsteady gradually varied flow are the Saint Venant / shallow water equations which were derived in 1871 by A.J.C Barre de Saint Venant based upon the following assumptions (Chadwick et al., 2004):

(1) Flow is one-dimensional, that is, velocity is uniform over a cross section and the water level across the section is horizontal.
(2) The streamline curvature is small and vertical accelerations are negligible, hence the pressure is hydrostatic. Gradually varied unsteady flow implies translatory wave motion of long wave length and low amplitude in which case the assumption of parallel streamlines and hydrostatic pressure distributions is reasonable.
(3) Effects of boundary friction and turbulence can be accounted for through resistance laws analogous to those used for steady state flow.
(4) The average channel bed slope is small so that the cosine of the angle it makes with the horizontal may be replaced by unity.

During the transition from free-surface flow to pressurized flow, a moving water interface advances into the free-surface region. There is need to track the interface in order to explain the development of pressures in a pipeline. This enables us to make a contribution in understanding the co-existence of pressurized and free-surface flows in a water supply network. The study of the flow regime transition also greatly enables understanding of the pressurization process in pipelines. However, a major problem with mixed flow analysis is the difficulty involved in treating the moving interface similar to surges (Song et al., 1983) and requires a considerable amount of computational effort to detect its generation and trace its movement.

DEVELOPMENT OF MATHEMATICAL FORMULATIONS AND MODELS USED

Analysis of unsteady flow in this study was carried out using conservation of mass and conservation of momentum principles which yielded two governing algebraic equations because the flow and depth of the water surface were both unknown. Each computational element in the governing equations was written in terms of elevations and flows at the ends of the element. A computational element with respect to time was also considered and due to this, the algebraic governing equations involved not only the unknown flow and depth at two points along the channel but also at two points in time (Chadwick et al., 2004; Franz and Melching, 1997).

Conservation of mass

The continuity equation for one-dimensional unsteady open channel flow can be expressed as (Chadwick et al., 2004; Franz and Melching, 1997):

$$\frac{\partial A}{\partial t} + \frac{\partial Q}{\partial x} = 0 \tag{1}$$

where A is cross sectional area of flow, Q is flow rate; t is the time interval being considered and x is the length of the reach being considered. All quantities in the equation are algebraic expressions and can be positive or negative therefore, a negative outflow is an inflow. The equation is a statement of the conservation of mass principle on a per-unit-length basis and can also be stated as:

$$\frac{\partial y}{\partial t} + V \frac{\partial y}{\partial x} + y \frac{\partial V}{\partial x} = 0 \tag{2}$$

where V and y are the velocity and depth of flow, respectively.

Conservation of momentum

The momentum conservation equation (Equation of motion) can be written in the form (Chadwick et al., 2004; Franz and Melching, 1997):

$$\frac{\partial Q}{\partial t} + \frac{\alpha \partial (Q^2 / A)}{\partial x} + gA \frac{\partial h}{\partial x} - gA(S_0 - S_f) = 0 \tag{3}$$

where Q is discharge, A is the flow cross sectional area, h is depth of flow, g is gravitational acceleration, S_o and S_f are channel slope and friction slope, respectively and α is an energy coefficient normally equated to unity for SI units.

Equations 2 and 3 are called Saint Venant equations. These governing equations for gradually varied unsteady flow in open channel can also be expressed as:

$$\frac{\partial y}{\partial t} + \left(\frac{A}{T}\right)\frac{\partial V}{\partial x} + V \frac{\partial y}{\partial x} = 0 \tag{4}$$

$$\frac{\partial V}{\partial t} + V \frac{\partial V}{\partial x} + g \frac{\partial y}{\partial x} = g(S_0 - S_f) \tag{5}$$

where T is the top width of flow. Equations 4 and 5 represent the continuity and dynamic equation, in non-conservation form, respectively. The friction slope is given by $S_f = \frac{fV|V|}{2D}$ for pressurized flows and $S_f = \frac{n^2 V|V|}{R_h^{4/3}}$ for free-surface flows, with f the Darcy-Weisbach friction factor, n the Manning's Coefficient, D the pipe diameter and R_h the hydraulic radius.

Different types of formulations can be given for the Saint Venant equations depending on the problem. Equations of continuity and motion for a one-dimensional unsteady flow in an open channel can be restated as (Song et al., 1983; Trajkovic et al., 1999; Leon, 2007; Gomez and Achiaga, 2008):

$$\frac{\partial y}{\partial t} + v \frac{\partial y}{\partial x} + \frac{c^2}{g}\frac{\partial v}{\partial x} = 0 \tag{6}$$

and

$$g \frac{\partial y}{\partial x} + \frac{\partial v}{\partial t} + v \frac{\partial v}{\partial x} + g(S_f - S_0) = 0 \tag{7}$$

in which c is the gravity wave celerity given by:

$$c = \sqrt{\frac{gA}{T}} \tag{8}$$

Corresponding equations for pressurized or closed conduit flow can be written as (Song et al., 1983):

$$\frac{\partial y}{\partial t} + v \frac{\partial y}{\partial x} + \frac{a^2}{g}\frac{\partial v}{\partial x} = 0 \tag{9}$$

$$g \frac{\partial y}{\partial x} + \frac{\partial v}{\partial t} + v \frac{\partial v}{\partial x} + g(S_f - S_0) = 0 \tag{10}$$

in which a, the speed of the water hammer wave and y should be regarded as the piezometric head measured from the pipe invert rather than the flow depth.

Solution of unsteady flow equations

An important family of equations that is often encountered in hydraulics is based on the following equation (Chadwick et al., 2004):

$$a \frac{\partial^2 f}{\partial x^2} + b \frac{\partial^2 f}{\partial y \partial x} + c \frac{\partial^2 f}{\partial y^2} = 0 \tag{11}$$

f, some variable/function such as velocity. If $b^2 - 4ac > 0$ then a typical form is:

$$\frac{\partial^2 f}{\partial x^2} - c^2 \frac{\partial^2 f}{\partial y^2} = 0 \qquad (12)$$

This is the hyperbolic equation which can be applied to unsteady flows. The Saint Venant/shallow wave equations are classified as partial differential equations of the hyperbolic type. All flow variables are functions of both time and distance along the channel. In other words, at a given location, the flow depth, discharge and the other flow variables vary with time. Likewise, at a fixed time, the flow variables change along the channel. For a given channel of known properties (cross sectional geometry, roughness factor and longitudinal slope), the unknowns are the discharge Q and flow depth y. The other flow variables such as the area A and the friction slope S_f can be expressed in terms of Q and y. The independent variables are time t and distance along the channel.

Method of characteristics

The differential equations of Saint Venant cannot be solved analytically unless certain simplifications are carried out such as the neglect of certain terms and simplification of boundary conditions as is the case in the kinematic approximation and diffusion analogy and this can lead to serious errors (Tucciarelli, 2003). With the advent of the digital computer, numerical solutions can be obtained and thus, no simplifying assumptions to the basic equations need to be made. Thus, unsteady gravity flows have been traditionally modeled by numerically solving the one-dimensional equations of continuity and momentum. A number of different numerical methods are available to solve hyperbolic differential equations. The best known is the method of characteristics (MOC) (Chadwick et al., 2004; Chou, 2009). The method is widely used for transient flow in a closed conduit because it is simple and also provides good insight into behaviour of hyperbolic equations. In the method, the original set of partial differential equations (Equations 6 and 7) are transformed into two sets of simultaneous ordinary differential equations.

$$\frac{dx}{dt} = v \pm c \text{ and } \frac{dy}{dt} \pm \frac{c}{g} \frac{dv}{dt} \pm c(S_f - S_0) = 0 \qquad (13)$$

Equation 13 is known as characteristic equations and is valid along two different characteristic lines. The first one is called a positive characteristic line and the second one is a negative characteristic line (Chadwick et al., 2004; Gomez and Achiaga, 2008). By discrediting the transformed equations we can obtain the velocity v_p and depth of flow y_p at point P as:

$$y_p = \frac{1}{C_R + C_S}\left[y_R C_S + y_S C_R \left(\frac{v_R - v_S}{g} - \Delta t.(S_{fR} - S_{fS}) \right) \right] \qquad (14)$$

$$v_p = v_R - \frac{g}{C_R}(y_p - y_R) + g.\Delta t.(S_o - S_{fR}) \qquad (15)$$

where $\left(\frac{g}{c}\right)_R = C_R$, $\left(\frac{g}{c}\right)_S = C_S$ and R and S are subscripts for locations defining positive and negative characteristic lines

drawn from P, respectively. They are known quantities (velocity or depth) at the beginning of time step Δt and can be obtained by linear interpolation. If we are at a boundary then one of the characteristic equations is outside of the problem boundary, so we need a boundary condition that will be the pressure head at this boundary or a relationship between pressure and head. When converted into finite difference equations, the characteristic equations lead to a set of simultaneous algebraic equations for the unknowns.

Interface tracking method of analyzing mixed flow

This class of flow regime transition models is exemplified by the works of Wiggert (1983), Hamam and McCorquodale (1982) and Li and McCorquodale (1999) by introducing a moving interface between the free surface and pressurised flow regimes. The method treats the two flow regimes separately but joined together by an interface which is regarded as a shock wave. Song et al. (1983) used the characteristic method for both open channel flow and closed conduit flow regimes. Identical equations and solution techniques were used throughout the system except for a special treatment at the interface. These models solve the ordinary differential equations (ODE) based on a momentum balance in a rigid column represented by the pressurised portion of the flow. In each time step, the ODE is solved and the velocity of the rigid column, speed, location and intensity of the shock wave is updated. The location of the pressurisation front is obtained using the continuity equation across the moving interface. The flow conditions near the interface are thus, calculated using a mass and momentum balance in a control volume. The free surface portion of the flow is solved by the method of characteristics. This model, also called a shock-fitting model is appropriate when the energy contained in the flow is sufficient to pressurize the flow through a hydraulic jump. The water depths and velocities near the interface are obtained using two shock-boundary conditions plus three characteristic equations (Politano et al., 2005).

If velocity changes are more gradual, acceleration of flow between two adjacent sections can be neglected and the flow near the interface can be simulated using momentum and mass balance in a moving control volume. This method facilitates accurate tracking of the interface conserving mass and is the approach that was used in this study.

Modeling flow discontinuities in the transition region

While the MOC is a valuable approach in the sense that it provides a deep understanding of the nature of shallow water fronts, the approach is limited by its inability to handle flow discontinuities (Figure 1). The hyperbolic nature of mass and momentum partial differential equations allows discontinuities in the solution in form of hydraulic bores (Vasconcelos, 2005). Figure 1 shows a smooth interface for illustration purposes only. Actually, the interface is modeled as a steep, near-vertical shock wave (Song et al., 1983) since it behaves as a moving internal hydraulic jump of extremely large magnitude representing abrupt flow change because of $c \to \infty$ as $T \to 0$ (Equation 8) for both pressurization and depressurization.

Solution of free surface side of the interface

For flow downstream of the interface (free-surface flow), we need to

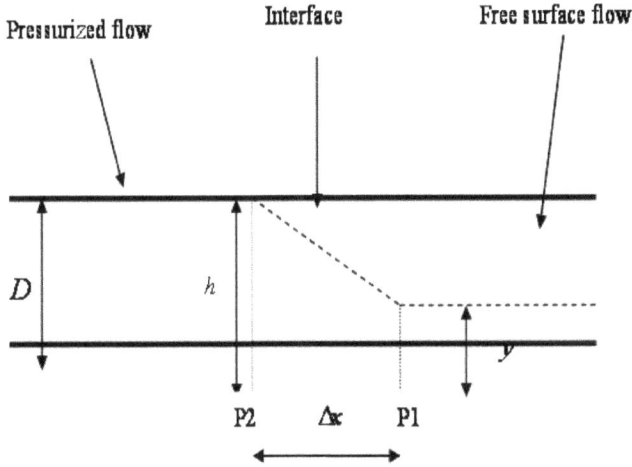

Figure 1. Control volume for the interface.

determine the flow depth and velocity. In the case of a surge that advances from upstream to downstream; the free-surface side of the interface can be solved separately to find the velocity v_1 and the depth y_1 at P1 (Figure 1), by using the characteristic flow equations in the free surface region.

Solution of moving interface

Analysis of the moving interface or surge front requires determination of the interface location x and velocity w. In this region, the equations that should be applied are the conservation of mass and momentum quantities given in Equations 17 and 20, respectively (Fuamba, 2003; Politano et al., 2005; Gomez and Achiaga, 2008).

Continuity equation

The conservation of mass equation for the surge is:

$$\frac{d(A_1 - A_2)}{dt}\Delta x = A_1(v_1 - w) - A_2(v_2 - w) \tag{16}$$

where A_1, v_1 are respectively, the cross sectional area of flow and velocity at the upstream (pressurized) end of the interface; A_2, v_2 are respectively, the cross sectional area of flow and velocity at the downstream (free-surface) end of the interface; Δx, the length of the control volume that contains the interface; w, surge velocity; subscripts 1 and 2, locations at the free-surface and pressurized zones, respectively. If $\dfrac{d(A_1 - A_2)}{dt}$ is taken as zero as is practically expected during pressurization, then:

$$A_1(v_1 - w) = A_2(v_2 - w) \tag{17}$$

Motion equation

In the derivation of the equation of motion, the momentum equation $F = ma$ was used where F is the net force causing acceleration, a, while m is the mass of the fluid. Also, $F = h\rho gA$ where h is the depth of fluid; ρ is the fluid density; g is gravitational acceleration and A is cross-sectional area of fluid. Therefore:

$$F = \frac{mv_2 - mv_1}{t} = \frac{m}{t}(v_2 - v_1) = \rho\frac{Vol}{t}(v_2 - v_1) = \rho Q(v_2 - v_1) = \rho A_1 v_1(v_2 - v_1) \tag{18}$$

where $Q = A_1 v_1$ that is, mass is constant at $A_1 v_1$. The velocity at point 2, velocity at point 1 and the time between these velocities are represented as v_2, v_1 and t, respectively. Vol is volume and Q is discharge. This implies that:

$$(\overline{h}A_2 - \overline{y}A_1)\rho g = \rho A_1 v_1(v_2 - v_1) \tag{19}$$

in which ρ cancels out giving

$$\overline{h}A_2 - \overline{y}A_1 = (V_1 - w)\frac{A_1}{g}(v_1 - v_2) \tag{20}$$

after incorporating the interface velocity. $\overline{h}$ and $\overline{y}$ are depths from the water surface to the centre of gravity of the flow cross-sectional areas of the pressurized and free surface ends of the interface, respectively. The new interface position is found from the kinematic condition $\Delta x = w.\Delta t$.

Solution of the pressurized side of the interface

For flow upstream of the interface (pressurized flow), we need to determine the velocity v_2 and pressure head h_2. The velocity of water at point P2 (Figure 1) is obtained from continuity Equation 17 as:

$$v_2 = w + \frac{A_1}{A_2}(v_1 - w) \tag{21}$$

The pressure at P2 is obtained from the characteristic equation for pressurized flow that does not cross the interface trajectory Equation 22 derived from Equation 15 for the reason that, in the transition region of two different flows (Figure 1), there are no valid characteristic equations that cross the interface trajectory

$$h_2 = h_{L2} + a\left(\frac{v_2 - v_{L2}}{g} - \Delta t(S_0 - S_{fL2})\right) \tag{22}$$

Table 1. Model input parameters.

Input parameter	Value
Flow depth at the start of the pressurized characteristic equation (m)	0.08
Pressure at pressurized interface side (m)	1.1
Velocity at the start of the pressurized characteristic equation (m/s)	2
Friction slope	0.0008
Pipe length (m)	1000
Pipe diameter (m)	0.1

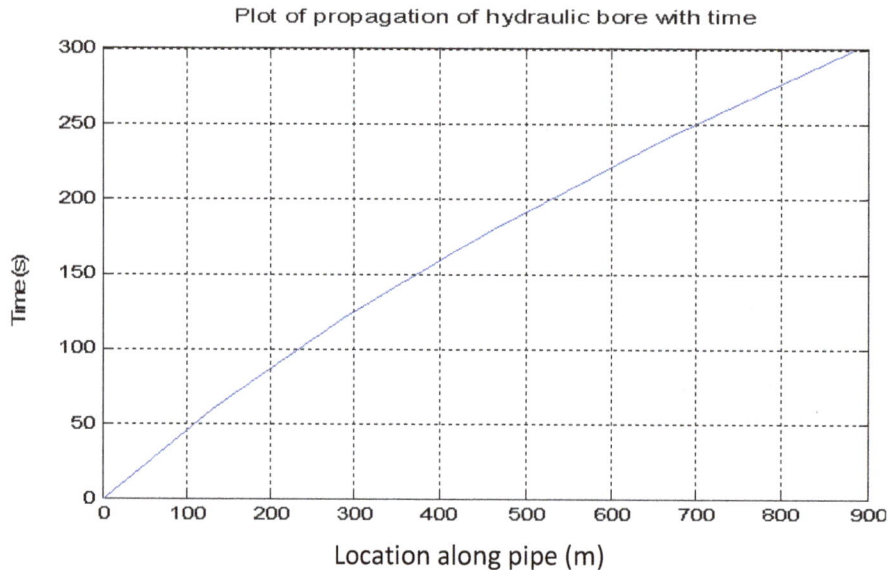

Figure 2. Behaviour of pressure surge.

where a is the celerity of the pressure wave; h_{L2}, v_{L2} and S_{fL2} are the head, velocity and friction slope at the beginning of the characteristic curve, respectively.

There is need to ensure that the size of the time step conforms to Courant's stability condition that ensures convergence of the finite difference equations (Chadwick et al., 2004). As we want information along the pipe to travel along the characteristic lines, we select the time interval Δt such that $\Delta t \leq \dfrac{\Delta x}{|v \pm c|}$ for free surface flow and the one for pressurized flow has c replaced with a. Thus, the size of Δx and the wave celerity c determine the size of the time interval. The parameter a is much greater than c so $\Delta t < \dfrac{\Delta x}{v + a}$ is the stability condition that is applied to the whole grid.

Programming

The aforementioned system of equations developed was programmed in MATLAB to simulate the pressurization of a pipeline. The model equations were discretised using a fixed-grid method with a first-order finite difference approximation. The resulting nonlinear equations were solved using the Newton-Raphson method.

RESULTS

This section presents the results of the simulation of the pipe pressurization process. The simulation involved the following inputs: pipe length, pipe diameter, flow depth, pressure, velocity and friction slope. Flow depth and pressure values were obtained from initial boundary conditions (typical field values) and were measured from the pipe invert. The velocity was obtained from the initial conditions that were assumed. For the programme inputs in Table 1, the outputs are shown in Figures 2 and 3. Table 2 summarizes key model outputs that are analysed subsequently.

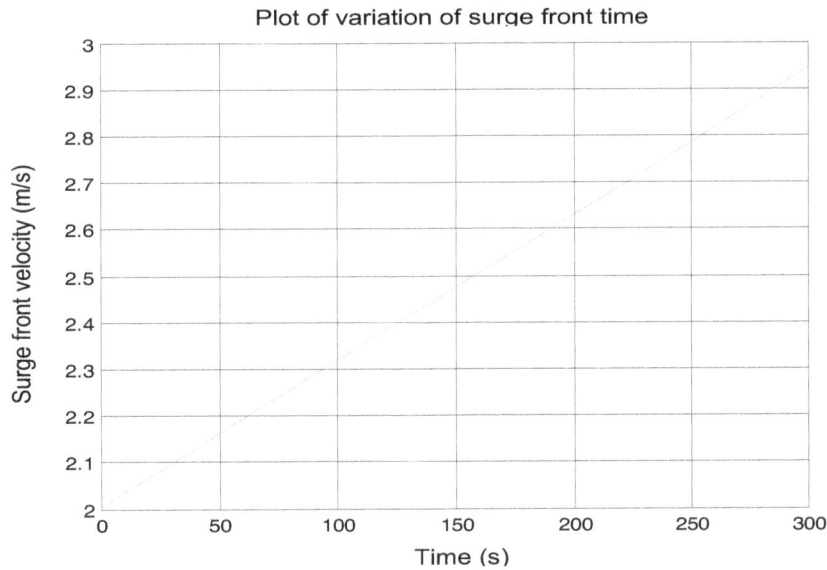

Figure 3. Behaviour of pressure surge velocity.

Table 2. Sample model outputs.

Pressure (m)	Water velocity (m/s)	Interface velocity (m/s)	Time (s)	Location (m)
0.4367	2.0035	2.0056	0	0
0.4358	2.1212	2.1935	60	131.6089
0.4332	2.2389	2.3814	120	285.7622
0.429	2.3566	2.5692	180	462.4551
0.423	2.4742	2.757	240	661.6828
0.4154	2.5919	2.9448	300	883.4408

Surge front characteristics

The plot in Figure 2 reveals the rate of movement of the hydraulic bore (pressure surge or interface) along the pipeline with time. After approximately 45 s, the surge front had travelled 100 m giving a velocity of 2.2 m/s and after 250 s it had travelled 700 m giving a velocity of 2.8 m/s (Figure 3). This is a good indication of the rate of development of pressures along the pipeline and shows that the interface accelerates during pressurization as a result of the net force from the water that overcomes its gravitational and frictional resistance.

Pressure characteristics

The plot in Figure 4 reveals development of pressure along a length of pipe at an instant. It can be observed that pressure builds up gradually along the entire pipeline in accordance with the propagation of the pressure surge or interface. Pressurization occurs in the direction in which the interface moves. As the interface advances forwards along the pipeline, whichever point it touches starts to get pressurized and the pressure at this point continues to increase ahead of the pressure at the subsequent points along the pipeline. It can be interpreted that pressure will increase with time at this point on the pipeline at the same rate as that of the propagation of the interface along the pipeline. It should be noted that the pressure values in Figure 4 and Table 2 are measured from the pipe bottom.

Velocity characteristics

The plot in Figure 5 reveals the variation of water velocity (not interface velocity) as the interface moves along the pipeline during initial pressurization. This shows that

Plot of pressure change with distance along the pipe

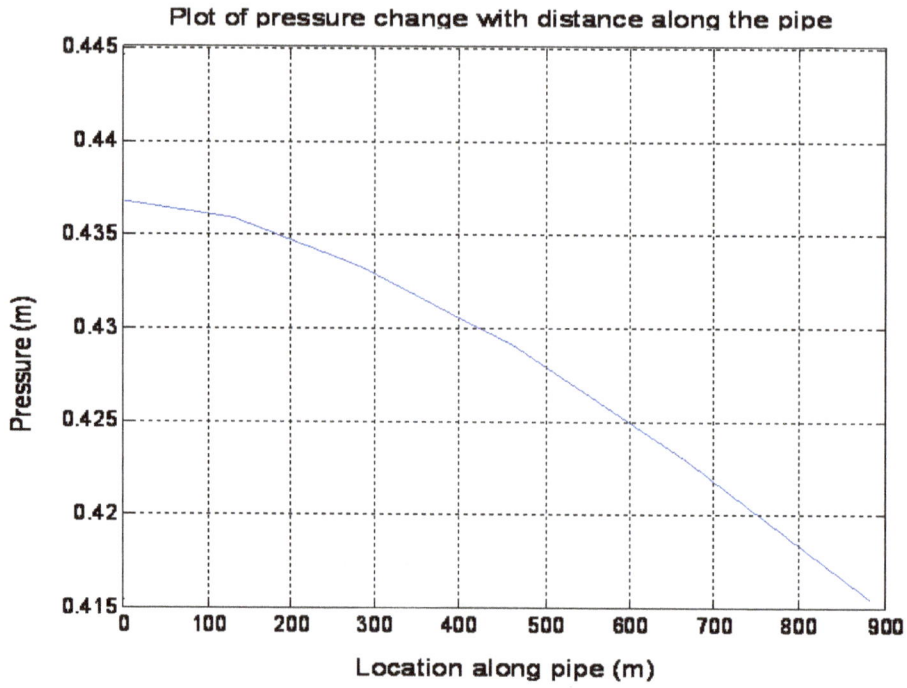

Figure 4. Pressure variation along pipeline.

Plot of velocity change with distance along the pipe

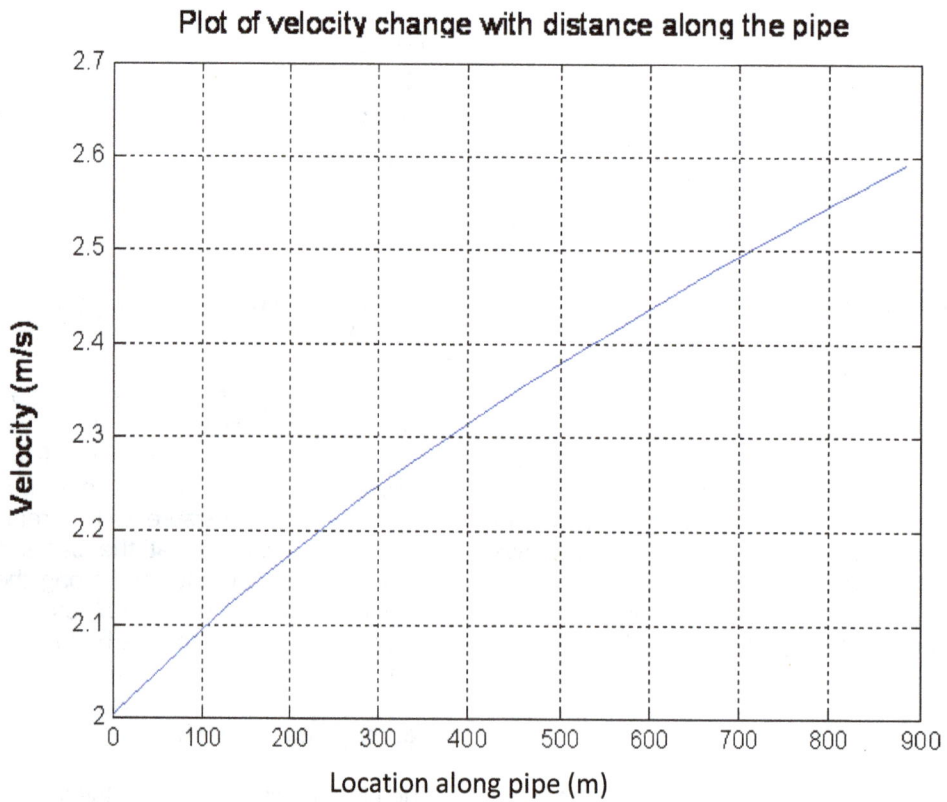

Figure 5. Velocity variation along pipeline.

Table 3. Comparison of surge front and water velocities.

Location (m)	Velocity (m/s)	
	Surge front	**Water**
100	2.2	2.09
700	2.8	2.5

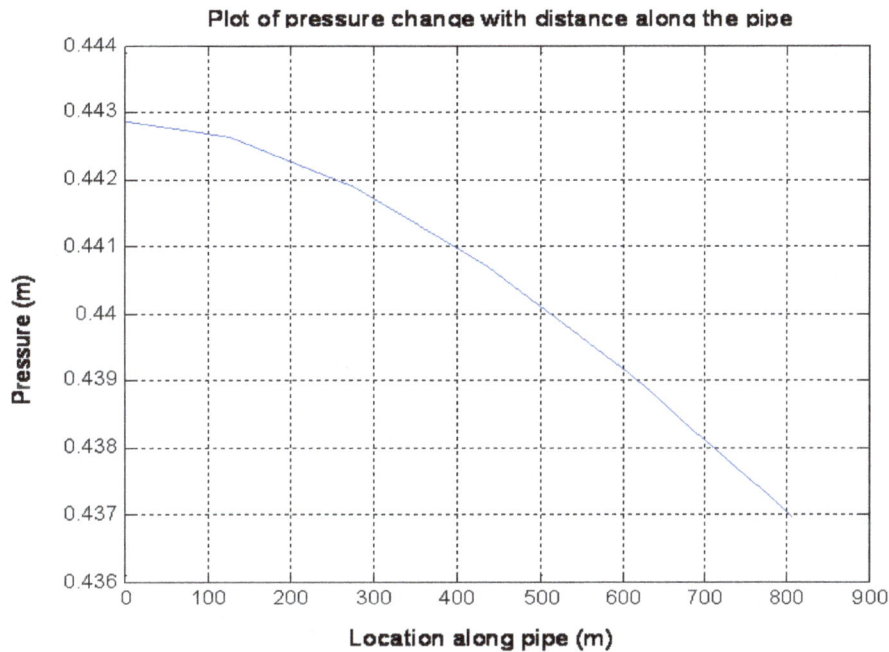

Figure 6. Pressure variation along pipeline.

water accelerates. It can be reasonably explained that the velocity is highest at the points that the interface just touches because it encounters no flow resistance at those points of free surface flow apart from friction. From Figures 3 and 5, it can be observed as would be expected that the surge velocity is higher than the water velocity. A comparison of surge velocity with water velocity is shown in Table 3 for values picked at two locations; 100 and 700 m along the pipeline.

The results obtained highlight the advantages of developing a fully dynamic and transient model. Not only is the surge front location and propagation accurately predicted during the transient flow phase between free surface and pressurized flows but flow conditions as well.

Pipe size and pressurisation

It is shown that the pressure values obtained during the

pressurization of the pipeline increase with pipe size. For example, the pressures at the various points are higher with a 200 mm pipe (Figure 6) than with the 100 mm pipe used in Figure 4. Table 4 illustrates a comparison of pressures along the pipeline in Figures 4 and 6. Figure 7 shows a comparison of both pressure results in one graph. The bigger the pipe size, the lower the frictional head losses arising from interaction of water with pipe walls. This consequently leads to higher pressures and further justifies the use of bigger diameter pipes in water supply.

DISCUSSION

Results show that if we can track the movement of the hydraulic bore in the pipelines, and by obtaining the pressures along the pipeline in the network, we are able to predict what pressures are available at particular

Table 4. Comparison of pressures along pipelines for different pipe diameters.

Pipe diameter (mm)	Location along pipeline (m)				
	0	200	400	600	800
100	437.0	434.5	430.5	425.0	418.5
200	442.8	442.3	441.0	439.2	437.0

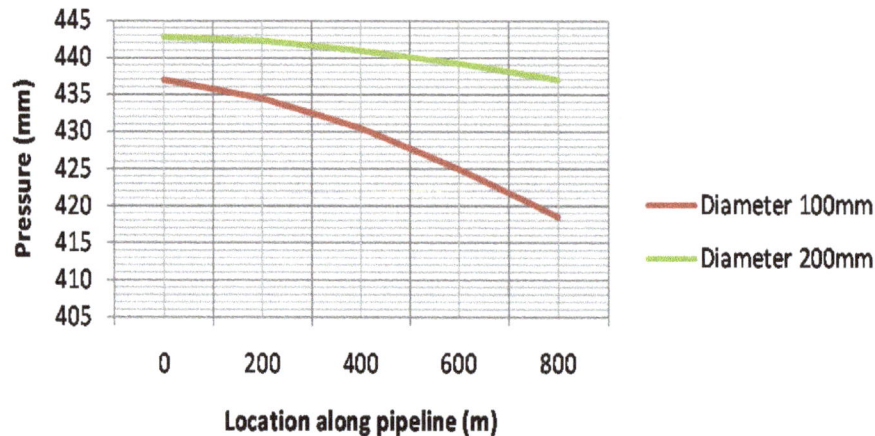

Figure 7. Pressure comparison for different pipe sizes.

locations in the network and the subsequent flows that can be enabled by these pressures. Pragmatic management decisions can then be made to ensure that water is available at different sections of the network. Such decisions can include closure or opening of valves, installation of adequate pipe and reservoir sizes in the network or a logical rationing program for water supply (Nyende-Byakika et al., 2010). In addition, if downstream conditions are adjusted accordingly, for example, by varying the valve aperture, then the model can predict the various pressures and flows that are enabled by the implemented actions while the network undergoes pressurization and depressurization.

Pressurization in a pipe occurs due to a variety of reasons. If the supply head is sufficiently high, or if inflow rate is significantly larger than the outflow rate, the speed with which the pressurization wave moves can be very significant. It should be noted that pressurization does not always occur when the pipeline is full; it can also occur when the pipeline is not completely full as long as the initial head and discharge can allow sufficiently rapid filling. It can be realized that if the initial discharge and head are insufficient, lower water depths cannot produce the transition to pressure flow, but only an increase in water depth. On the other hand, greater water depths may not show a pressure front but almost an

instantaneous transition to the pressure flow for the whole pipeline (Gomez and Achiaga, 2008), a case that is not envisaged under normal conditions because of the large pipeline lengths, diameters and slopes. In all cases, the transition occurs (and was modeled) through a moving interface that advances into the free surface portions of the system.

During rapid filling, a surge moving against the flow may develop a steep front but in cases of gradual filling, gentle slopes or depressurization, the surge may have a very smooth interface. Even if the interface is smooth, the transition between free-surface flow and pressurized flow cannot be continuous because the gravity wave speed (Equation 8) would be infinite at the point of transition where $T = 0$ i.e. $c \rightarrow \infty$ as $T \rightarrow 0$ and this represents an area of abrupt flow change creating a discontinuity which gives the Saint Venant equations their hyperbolic character (Fuamba, 2003). For this reason, it is always necessary to assume a discontinuity at the interface.

Conclusion

In this paper, the pressurization process of pipes was studied and involved tracking the movement of the interface with the aim of determining where and when pressures would start to build up along the pipes. The process was analyzed and modeled in this work with a view of clearly understanding what happens during this phenomenon and consequently aiding engineers in ultimately designing systems and operations that take full advantage of this phenomenon. The motivation to study this flow phenomenon arose from the fact that it is the 'pressurization' stage that leads to the 'pressurized' state that is of profound interest to water supply managers and engineers. As a management tool, this would help inform when particular sections (nodes) of the pipes would build pressures thereby starting to release water and consequently, what actions should be taken for this to happen.

The results obtained highlight the advantages of developing a fully dynamic and transient model in the solution of transient LPOCF conditions in water distribution networks. Not only was the surge front location and propagation accurately predicted during the transient flow phase between free surface and pressurized flows but flow conditions as well.

Further studies should target setting up experiments and field tests to validate the results. They should also consider production of commercial models that tackle the dual character exhibited by several water supply systems, that is, co-existence of pressurized and free-surface flow conditions in the same network.

REFERENCES

Chadwick A, Morfett J, Borthwick M (2004). Hydraulics in Civil and Environmental Engineering. UK: Spon Press, p. 59.

Chou T (2009). The Method of Characteristics. University of California. Los Angeles.

Franz DD, Melching CS (1997). Full Equations model for the solution of the full, dynamic equations of motion for one-dimensional unsteady flow in open channels and through control structures. U.S. Geological Survey Water-Resources Investigations Report.

Fuamba M (2003). Contribution on Transient Flow Modelling in Storm Sewers. J. Hydraul. Res., 40(6): 685-693.

Gomez M, Achiaga V (2008). Mixed Flow Modelling Produced by Pressure Fronts from Upstream and Downstream Extremes. Proc. American Society of Civil Engineers.

Hamam MA, Mccorquodale JA (1982). Transient conditions in the transition from gravity to surcharged sewer flow. Canadian J. Civ. Eng., 9: 189-196.

Ingeduld P, Svitak Z, Pradhan A, Tarai A (2006). Modelling intermittent water supply systems with EPANET. 8th Annual WD Symposium EPA Cincinnati. August.

Leon AS (2007). Improved modelling of unsteady free surface, pressurised and mixed flows in storm sewer systems. PhD Thesis. University of Illinois at Urbana-Champaign.

Li J, Mccorquodale A (1999). Modelling mixed flow in storm sewers. J. Hydraul. Eng., 125(11): 1170-1180.

Nyende-Byakika S (2011). Modelling of Pressurised Water Supply Networks that may exhibit Transient Low Pressure-Open Channel Flow Conditions. PhD Thesis. Vaal University of Technology. Vanderbijlpark. South Africa.

Nyende-Byakika S, Ngirane-Katashaya G, Ndambuki JM (2010). Behaviour of stretched water supply networks. Nile Water. Sci. Eng. J., 3(1).

Politano M, Odgaard J, Klecan W (2005). Numerical Simulation of Hydraulic Transients in Drainage Systems. Mecanica Computacional, p. 24. Argentina.

Rossman AL (2000). EPANET Users' Manual. National Risk Management Laboratory. United States Environmental Protection Agency, Ohio.

Song CCS, Cardle JA, Leung KS (1983). Transient mixed flow models for storm sewers. J. Hydraul. Eng., 109(11): 1487-1504.

Trajkovic B, Ivetic M, Calomino F, Dippolito A (1999). Investigation of transition from free surface to pressurized flow in a circular pipe. Water Sci. Technol., 39(9): 105-112.

Tucciarelli T (2003). A new algorithm for a robust solution of the fully dynamic St. Venant equations. J. Hydraul. Res., 41(3): 239-246.

Vasconcelos JG (2005). Dynamic approach to the description of flow regime transition in storm water systems. PhD Thesis. University of Michigan.

Wiggert DC (1983). Transient flow in free-surface, pressurised systems. J. Hydraulics Division, ASCE, 98, No. HY1, 1972.

3

Structural numerical analysis of a three fingers prosthetic hand prototype

José Alfredo Leal–Naranjo[1], Christopher René Torres-San Miguel[2,3] , Manuel Faraón Carbajal–Romero[1] and Luis Martínez-Sáez[3]

[1]Sección de Estudios de Posgrado e Investigación, Escuela Superior de Ingeniería Mecánica y Eléctrica, Instituto Politécnico Nacional, Unidad Azcapotzalco, Av. de las granjas No. 682 col. Sta. Catarina, delegación Azcapotzalco C.P. 02550, México D.F.
[2]Escuela Superior de Ingeniería Mecánica y Eléctrica, Instituto Politécnico Nacional, Unidad Profesional "Adolfo López Mateos", edificio 5, 2do piso, col. Lindavista, delegación Gustavo A. Madero. C.P. 07738, México D.F.
[3]Instituto Universitario de Investigación del Automóvil, Escuela Técnica Superior de Ingenieros Industriales, Universidad Politécnica de Madrid, Carretera de Valencia, km.7, 28031, Madrid, España.

In spite of the importance of the finite element method analysis in the mechanical design, there are a few research works in the prosthetic hands design area that use this tool. This work uses the finite element method to perform a structural analysis of a prosthetic hand prototype during the cylindrical and tip grasp. Through geometric relations, a set of equations was developed in order to define the position of the prosthesis elements from the position of the actuator. The loads in the numerical analysis were calculated by modeling a static analysis, in which the forces that the fingers of the prosthesis can exert at the tip were obtained. The results are presented in a graphic form showing the load of the fingertip as function of the position of the actuator and the maximum force that can be exerted. It is shown as well the field stress present in the mechanism of the hand simulator for the most critical position and the field stress of the prosthesis prototype. Both parameters were calculated with the use of ANSYS Workbench 2.0 $^{®}$

Key words: Upper limb prosthesis, prehensile force, finite element method, four-bar linkage, structural analysis.

INTRODUCTION

Human hand is able to realize a wide range of sophisticated movements that provides the ability to interact with the environment and communicate with other people (Clement et al., 2011). The loss of the upper limb may cause serious physical and psychological disorders in the life of an amputee (El Kady et al., 2010).

During the last 30 years, several prosthetic hands have been developed (Carrozza et al., 2005). There are mainly three kind of upper limb prosthesis available for people who lacks an extremity due to congenic factors or because they suffered an amputation (Watve et al., 2011). Otto Bock hand® is a popular and robust prosthesis that allows to exert a maximum force of 100 N but only has one degree of freedom (DOF) and weighs 600 g (Carrozza et al., 2004).

The human hand has 22 DOF and a maximum prehensile force of 500 N however most daily life activities require prehensile forces under 70 N (O'Toole, 2007), and an average weight of 500 g (Carozza et al., 2006). Many prosthetic hand prototypes been developed like the one that is shown in Cipriani et al. (2011) which has 16 DOF and is actuated by 4 electric motors, weighs 530 g and is able to lift a briefcase of 10 kg. Different kinds of actuators and transmissions have been used to produce the prosthesis movement like in Kargov et al. (2007) which presents a prosthesis that use an hydraulic pump and valves as actuators (Zhang et al., 2008); a rack-pinion transmission was used for its design.

The advancements in anatomy, material design and computer technology have made significant contributions to the evolution of the design of prosthetic hands during the last decades, and this has allowed a better approach to imitate the movement and mechanical behavior of a healthy hand, as in Pérez et al. (2012) that presents an anthropometric prosthetic hand prototype designed using computer tomography scan obtained from the healthy hand of a man. The use of analytical and experimental models in biomechanics generates important information that permits to understand the behavior of different systems subjected to mechanical loads (Naaji and Gherghel, 2009). Using the finite element method (FEM), it is possible to determine the stress distribution and strains, which could be difficult to obtain through experimental or analytical methods.

Examples of using FEM in biomechanics can be seen in Bougherara et al. (2009) and Omasta et al. (2012) where the stress distribution was analyzed in a knee prosthesis and in a trans-tibial prosthesis, respectively; these analysis were made using Ansys ® Workbench. Geng et al. (2008) shows the use of FEM in dentistry.

In Stanciu and Stanciu (2009), an analysis was presented using the finite element method in a prosthetic hand prototype that uses hydraulic actuators. The analysis is carried out using cosmosworks ® and was performed in two positions, one for the hand wide open and the other for an intermediate position. A 10 N load was considered and applied at the fingertip of the middle finger. All the fingers have the same structure so the analysis was not realized for the other fingers.

The mechanical design and theoretical analysis of a hand prosthesis actuated by muscle wires is shown in O'Toole (2007). The device that holds the wire was analyzed and the optimum wall thickness of the finger was determined through FEM.

The design of an end effector with anthropomorphic shape is presented in Ohol and Kajale (2009). Using ANSYS ® the FEM only for the critical parts was made, positioning the mechanism in extreme positions.

A static analysis in order to determine the prehensile force exerted by a prosthetic hand is presented in Jung and Moon (2008). Kargov et al. (2004) shows a comparison of the prehensile force distribution between a human hand, a non-adaptative hand prosthesis, and an adaptative prosthetic hand.

Despite the importance that the finite element method has in mechanical design, there are just a few research works in the prosthetic hands design area that uses this tool and there is no existence of any work in which all the mechanical components of the prosthesis are numerically analyzed during a grasping task. Taking into account these aspects, in this work the numerical simulation of the stress distribution in the components of a prosthetic hand prototype both in cylindrical and tip grasp was developed.

MATERIALS AND METHODS

Due to the computing capacity needed to perform the analysis, this work was divided in two phases. The first phase consisted the analysis of the finger mechanism operation and for this objective was used a hand simulator. Once the mechanism operation was assimilated; the second phase is aimed to perform the structural analysis of the prototype doing a cylindrical grasp and a tip grasp.

In order to apply the FEM to the prosthesis prototype, the following steps were considered:

1. Define a function that allows to know the geometrical position of the prototype elements knowing the actuator position,
2. Perform a static analysis of the hand simulator in order to determinate a set of equations that predicts the force that can be exerted at the tip as a function of the finger position,
3. Solve for different positions within the operating range of the mechanism the equations obtained in the static analysis,
4. Perform a FEM analysis of the hand simulator in ANSYS Workbench 2.0,
5. Determinate the areas with the highest magnitude of mechanical stress in the hand simulator,
6. Realize a static analysis of the prototype in order to determine the forces that take place during the cylindrical grasp, which is a grasp used to grab objects with a continues form like bars, and the tip grasp,
7. Execute a FEM analysis of the hand prosthesis prototype doing a cylindrical, and a tip grasp.

All this previous steps were followed in the mentioned sequence in order to perform the analysis.

In this work SolidWorks® Premium 2012 x64 Edition was used to handle the CAD design and fix the prototype in the desired position; Matlab R2010a was used to solve the set of equations and ANSYS Workbench 2.0 Framework Release 13.0.0 in order to realize the FEM analysis.

Fingertip force analysis

The prototype proposed has 3 fingers, the thumb, index and middle finger (Figure 1). Index and middle finger has 1 DOF and the thumb has 2 degrees of freedom, one for the open-close movement and the other for the opposition movement.

The hand simulator structure is the four-bar linkage shown in Figure 2. The movement of the finger is given by a crank and slider mechanism, which is driven by a servomotor.

Figure 3 shows the components that form the hand simulator.

The lengths of the components are shown in Table 1. These lengths were used for the calculation of the prostheses elements position and for the estimation of the forces that take place during the grasp.

Figure 4 shows the angles of the hand simulator elements. Within the operating range of the mechanism, the elements do not present any interference, and do not align, so the mechanism does not present a critical position, this is why it is proposed to perform a

Figure 1. Design of the prosthetic hand prototype.

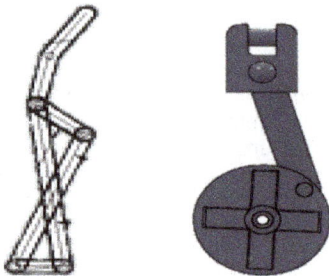

Figure 2. Four-bar linkage and crank-slider mechanism.

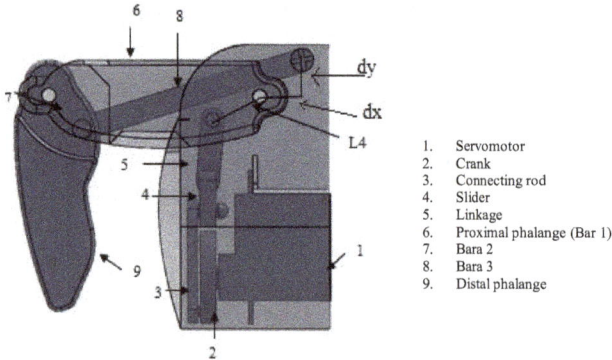

1.	Servomotor
2.	Crank
3.	Connecting rod
4.	Slider
5.	Linkage
6.	Proximal phalange (Bar 1)
7.	Bara 2
8.	Bara 3
9.	Distal phalange

Figure 3. CAD design of the hand simulator.

static analysis instead of a dynamic one.

The first step to make the static analysis of the hand simulator, is obtain the geometric positions for each one of the elements.

The angle θ_2 can be obtained through geometric relations, so θ_2 is given by the following equation:

$$\theta_2 = a\sin\frac{r\cdot\cos\theta_1}{L_1} \tag{1}$$

Elements L2, L3 and L4 can be analyzed as a four-bar linkage, then

Table 1. Lengths of the components.

Component	Length [mm]	Nomenclature
2	8	R
3	21.5	L1
4	6.2	L2
5	13	L3
6	45.03	L5
7	10	L6

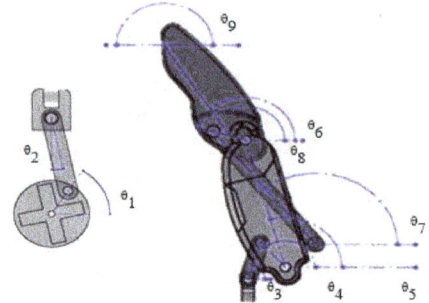

Figure 4. Reference angles.

by using the Freudenstein equation (Cardona and Clos, 2001) we have:

$$\tan\frac{\theta_3}{2} = \frac{-B\pm\sqrt{B^2+A^2-C^2}}{C-A} \tag{2}$$

Where:

$$A = -2\cdot X\cdot L_3 \tag{3}$$

$$B = -2\cdot L_3\cdot(H-h_2) \tag{4}$$

$$C = (H-h_2)^2 + L_3^2 + X^2 - L_4^2 \tag{5}$$

H_2 is the vertical distance between servomotor's shaft and the bottom of element L_3, which can be obtained by the following equation:

$$h_2 = L_2 + R\cdot\sin\theta_1 + L_1\cdot\cos\theta_2 \tag{6}$$

H and X are the vertical and horizontal distance between the servomotor's shaft and the beginning of the distal phalange, respectively, with values H=38.15 mm, X=11.05 mm.

Once θ_3 is obtained, the horizontal and vertical projection of element L_4 are calculated in order to obtain θ_4.

$$h_4 = H - h_2 - L_3\cdot\sin\theta_3 \tag{7}$$

$$x_4 = X - L_3\cdot\cos\theta_3 \tag{8}$$

$$\theta_4 = a\tan\frac{h_4}{x_4} \tag{9}$$

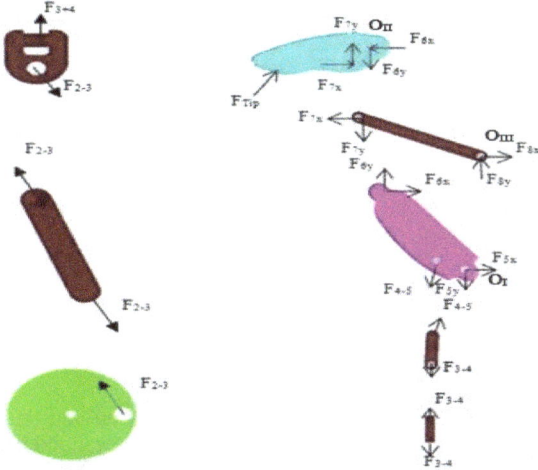

Figure 5. Mechanism free body diagram.

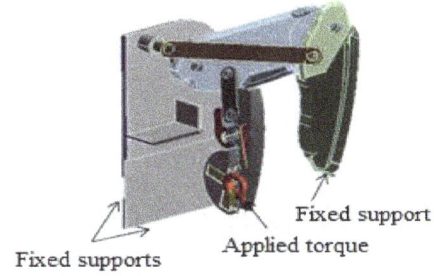

Figure 6. Position of the analysis.

static analysis of the finger was done in order to obtain the fingertip force. This force is the resultant of the pressure exerted by the finger on the manipulated object and this force is used as a simplification in the calculations. Figure 5 shows the forces considered for the analysis.

Assuming a concentrated force at the finger tip (Figure 6), this can be represented by the following non-linear equation:

$$F_{Tip} = f(\theta_1) \tag{20}$$

In the free body diagram it is possible to observe that the force transmitted by the crank to the connecting rod is given by the following equation:

$$F_{2-3} = \frac{M}{R \cdot (\sin\theta_2 \cdot \sin\theta_1 + \cos\theta_2 \cdot \cos\theta_1)} \tag{21}$$

This force is transmitted to the slider. As the horizontal force is the only one that produces work, it is considered just this force for the next equation:

$$F_{3-4} = F_{2-3} \cdot \cos\theta_2 \tag{22}$$

The force transmitted to element 5 is represented by:

$$F_{4-5} = \frac{F_{3-4}}{\sin\theta_3} \tag{23}$$

In order to obtain the fingertip force, it is necessary to analyze simultaneously the elements 6, 7 and 8.
For equilibrium, for the element 6 we obtain:

$$\sum F_x = F_{6x} + F_{5x} + F_{4-5} \cdot \cos\theta_3 = 0 \tag{24}$$

$$\sum F_y = F_{6y} - F_{5y} + F_{4-5} \cdot \sin\theta_3 = 0 \tag{25}$$

$$\sum M_{0_I} = F_{4-5} \cdot \cos\theta_3 \cdot L_4 \cdot \sin\theta_4 - F_{4-5} \cdot \sin\theta_3 \cdot L_4 \cdot \cos\theta_4$$
$$- F_{6x} \cdot L_5 \cdot \sin\theta_5 + F_{6y} \cdot L_5 \cdot \cos\theta_5 = 0 \tag{26}$$

Analogously, for the element 7 we get:

$$\sum F_x = F_{7x} - F_{6x} + F_{Tip} \cdot \cos\theta_9 = 0 \tag{27}$$

$$\sum F_y = F_{7y} - F_{6y} + F_{Tip} \cdot \sin\theta_9 = 0 \tag{28}$$

Knowing that the angle between θ_4 and θ_5 is constant, with a value of 26.5651°, θ_5 is obtained as follows:

$$\theta_5 = 180 - (26.561 - \theta_5) \tag{10}$$

The elements L5, L6 and L7 are analyzed as a four-bar linkage, and using the Freudestein equation we have:

$$\tan\frac{\theta_7}{2} = \frac{B \pm \sqrt{B^2 + A^2 - C^2}}{C + A} \tag{11}$$

Where:

$$A = 2 \cdot dx \cdot L_7 - 2 \cdot L_7 \cdot L_5 \cos\theta_5 \tag{12}$$

$$B = 2 \cdot dy \cdot L_7 - 2 \cdot L_7 \cdot L_5 \sin\theta_5 \tag{13}$$

$$C = dx^2 + dy^2 + L_7^2 + L_5^2 - L_6^2 - 2 \cdot dx \cdot L_5 \cdot \cos\theta_5$$
$$- 2 \cdot dy \cdot L_5 \cdot \sin\theta_5 \tag{14}$$

Calculating the horizontal and vertical projections of element L_6 in order to get θ_6:

$$x_6 = dx + L_7 \cos\theta_7 - L_5 \cos\theta_5 \tag{15}$$

$$y_6 = dy + L_7 \sin\theta_7 - L_5 \sin\theta_5 \tag{16}$$

$$\theta_6 = a\tan\frac{y_6}{x_6} \tag{17}$$

Angles θ_8 and θ_6 have a constant angle of 40.82°, so:

$$\theta_8 = \theta_6 + 40.82 \tag{18}$$

Finally we can get θ_9 using the next equation:

$$\theta_9 = 56.31 + (\theta_8 - 180) \tag{19}$$

We can proceed to describe the position of all the components; a

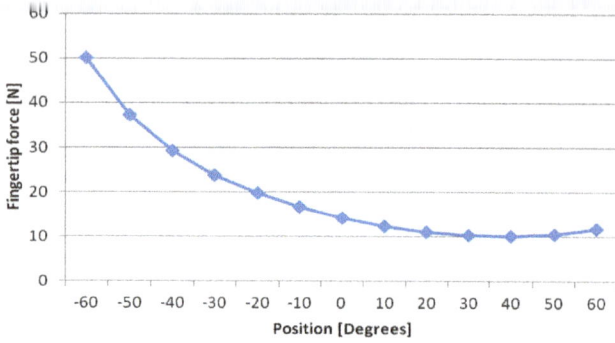

Graph 1. Fingertip force VS position.

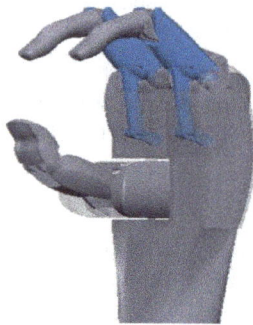

Figure 7. Index and middle finger transmission

$$\sum M_{0_{II}} = -F_{Tip} \cdot \cos\theta_9 \cdot L_9 \cdot \sin\theta_8 + F_{Tip} \cdot \sin\theta_9 \cdot L_9 \cdot \cos\theta_8$$

$$- F_{Tip} \cdot \cos\theta_9 \cdot \frac{L_6}{2}\sin\theta_6 + F_{Tip} \cdot \sin\theta_9 \cdot \frac{L_6}{2} \cdot \cos\theta_6$$

$$- F_{7x} \cdot L_6 \cdot \sin\theta_6 + F_{7y} \cdot L_6 \cdot \cos\theta_6 = 0 \tag{29}$$

The corresponding equations for element 8 are:

$$\sum F_x = F_{8x} - F_{7x} = 0 \tag{30}$$

$$\sum F_y = F_{8y} - F_{7y} = 0 \tag{31}$$

$$\sum M_{0_{III}} = F_{8x} \cdot L_7 \cdot \sin\theta_7 - F_{8y} \cdot L_7 \cos\theta_7 = 0 \tag{32}$$

The previous equation 9 form a 9×9 linear system of equations that can be represented in a matrix form as follows:

$$A = -\cos\theta_9 (L_8 \cdot \sin\theta_8 + \frac{L_6}{2} \cdot \sin\theta_6)$$

$$+ \sin\theta_9 (L_8 \cdot \cos\theta_8 + \frac{L_6}{2} \cdot \cos\theta_6) \tag{33}$$

Where:

$$B = F_{4-5} \cdot \cos\theta_3 \cdot L_4 \cdot \sin\theta_4 + F_{4-5} \cdot \sin\theta_3 \cdot L_4 \cdot \cos\theta_4 \tag{34}$$

A program was made using MATLAB® to solve Equation 20. In an iterative way, a torque of 1 N-m was chosen to represent the torque the motor develops.

Within the range of -60° to 60°, the fingertip force was calculated for every 10° (Graph 1), and in the position where the mechanism transmits the highest force, a structural static analysis was performed.

FEM analysis of the hand simulator

After the critical position of the mechanism was calculated, a FEM analysis was performed in order to assure that the hand simulator will be able to withstand the loads produced during its operation.

In order to carry out a FEM simulation, the next steps should be followed:

1. Build a geometric model,
2. Create a finite element mesh for the prototype,
3. Define loading conditions,
4. Apply boundary conditions,
5. Solve the analysis,
6. Generate the results analysis.

During the discretization process, tetrahedral elements were used due to the complexity of the prototype surface, and a mesh refinement was used in the regions on high stress, resulting in 67872 elements.

The material employed in the simulation, because of its low density and high strength, was a titanium alloy with the following properties:

1. Density 4620 [Kg/m^3],
2. Young modulus 96 [GPa],
3. Poisson ratio 0.36,
4. Yield stress 930 [MPa].

The simulation was performed for the position $\theta_1 = -60°$ as in this position the force transmit is the greatest, with a torque of 1 N-m applied on the crank.

Grasping force analysis

Once this simulator analysis has been fully understood, a second analysis for the hand prosthesis prototype was made in order to determine the stress field during the grasping task.

Cylindrical grasp

Movement of the index and middle finger of the prosthesis through a four-bar linkage is shown in Figure 7, while the thumb is moved through a crank-slider mechanism.

For the cylindrical grasp, it was considered the handling of a cylindrical body with a diameter of 57 mm which represents a 600 ml water bottle like the one used in Kargov et al. (2004).

For the analysis is assumed that each finger touches the body in just one point and that the forces are perpendicular to the surface on the contact point. Figure 8 shows the forces involved in the cylindrical grasp, neglecting gravity.

Since the entire system is at static equilibrium, each individual part showed is in equilibrium. The analysis was performed for each

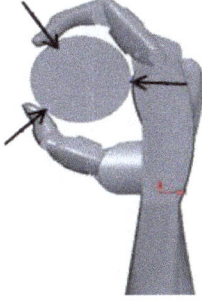

Figure 8. Forces involved in the cylindrical grasp.

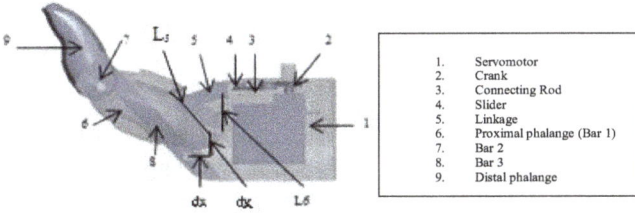

Figure 10. Reference angles.

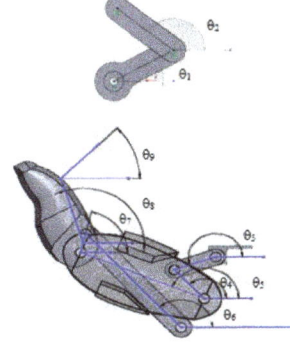

Figure 9. Thumb elements.

Table 2. Lengths and nomenclature of the elements.

Component	Lengths [mm]	Nomenclature
2	14	L1
3	14.6	L2
4	7	L3
5	12	L4
6	37	L8
7	6	L10
8	39.5	L9
9	21	L11
L5	12.1	L5
L6	13	L6
Dx	6.3	Dx
Dy	9	Dy

finger in the same way as in the hand simulator. Figure 9 shows the components of the thumb. The lengths of the components are given in Table 2.

Figure 10 shows the reference angles of the thumb elements, these angles were calculated using the same methodology employed for the hand simulator.

From the free body diagram shown in Figure 11, we can derive the next equation:

$$F_1 = \frac{M_1}{L_1 \cdot (\cos\theta_2 \cdot \sin\theta_1 - \sin\theta_2 \cdot \cos\theta_1)} \tag{35}$$

This force is transmitted to the slider. As the vertical force does not produce any work, it is neglected and the horizontal force is the only one considered, this one is calculated by the equation:

$$F_2 = F_1 \cdot \sin\theta_2 \tag{36}$$

The force over the element 5 is defined by:

$$F_3 = \frac{F_2}{-\cos\theta_2} \tag{37}$$

Following the same steps as with the hand simulator, a set of equations is obtained in order to calculate the fingertip force of the thumb:

$$
\begin{bmatrix}
0 & 1 & 0 & -1 & 0 & 0 & 0 & 0 & 0 \\
0 & 0 & -1 & 0 & -1 & 0 & 0 & 0 & 0 \\
0 & 0 & 0 & L_8 \cdot \sin\theta_8 & L_8 \cdot \cos\theta_8 & 0 & 0 & 0 & 0 \\
-\cos\theta_9 & 0 & 0 & 1 & 0 & -1 & 0 & 0 & 0 \\
-\sin\theta_9 & 0 & 0 & 0 & 1 & 0 & -1 & 0 & 0 \\
A & 0 & 0 & 0 & 0 & L_{10}\cdot\sin\theta_7 & -L_{10}\cdot\cos\theta_7 & 0 & 0 \\
0 & 0 & 0 & 0 & 0 & 1 & 0 & -1 & 0 \\
0 & 0 & 0 & 0 & 0 & 0 & 1 & 0 & -1 \\
0 & 0 & 0 & 0 & 0 & 0 & 0 & -L_9\cdot\sin\theta_6 & L_9\cdot\cos\theta_6
\end{bmatrix}
\begin{bmatrix}
F_{Tip} \\ F_{4x} \\ F_{5y} \\ F_{6x} \\ F_{6y} \\ F_{7x} \\ F_{7y} \\ F_{8x} \\ F_{8y}
\end{bmatrix}
=
\begin{bmatrix}
F_3\cos\theta_3 \\ F_3\sin\theta_3 \\ B \\ 0 \\ 0 \\ 0 \\ 0 \\ 0 \\ 0
\end{bmatrix}
$$

Where:

$$A = -L_{11}\,(\cos\theta_9 \cdot \sin\theta_8 + \cos\theta_8 \cdot \sin\theta_9)$$
$$-\frac{L_{10}}{2}(\sin\theta_9 \cdot \cos\theta_7 - \sin\theta_7\cos\theta_9) \tag{38}$$

$$B = F_3 \cdot L_5(-\cos\theta_3 \cdot \sin\theta_4 + \sin\theta_3 \cdot \cos\theta_4) \tag{39}$$

In a vector form, the force applied by the thumb is given by the equation:

$$\overline{F}_{Thumb} = F_{Thumb} \cdot \cos\theta_9\, i + F_{Thumb} \cdot \sin\theta_9\, j \tag{40}$$

We now proceed to calculate the forces of the index and middle finger. These fingers have the same structure (Figure 12), and in order to know the position of its elements a set of reference angles were defined (Figure 13).

Elements l_1, l_2 and l_3 are modeled as a four-bar linkage using the Freudestein equation (Table 3), θ_3 can be obtained by the next equation:

$$\tan\frac{\theta_3}{2} = \frac{-B \pm \sqrt{B^2 + A^2 - C^2}}{C - A} \tag{41}$$

Where:

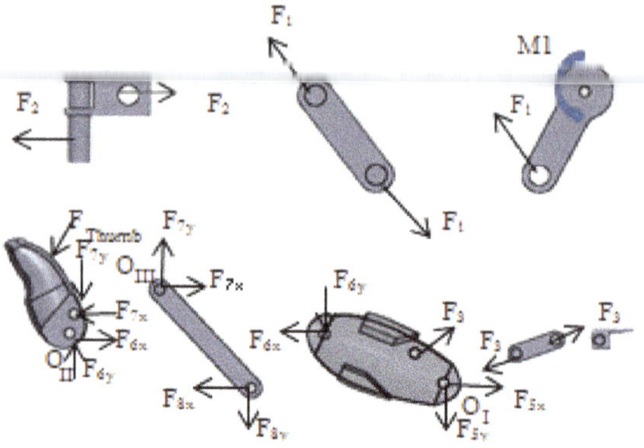

Figure 11. Free body diagram for the thumb elements.

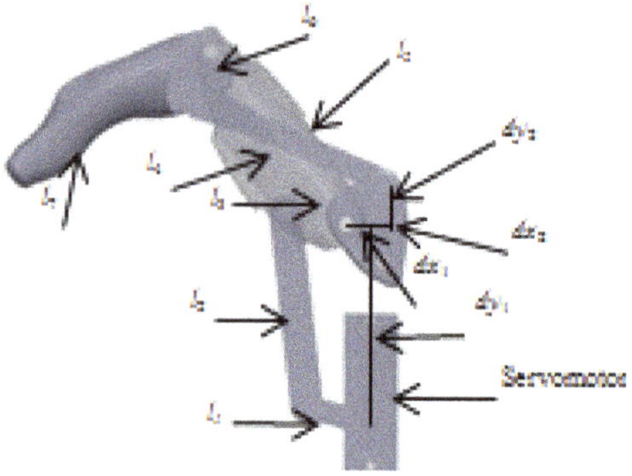

Figure 12. Middle and index finger

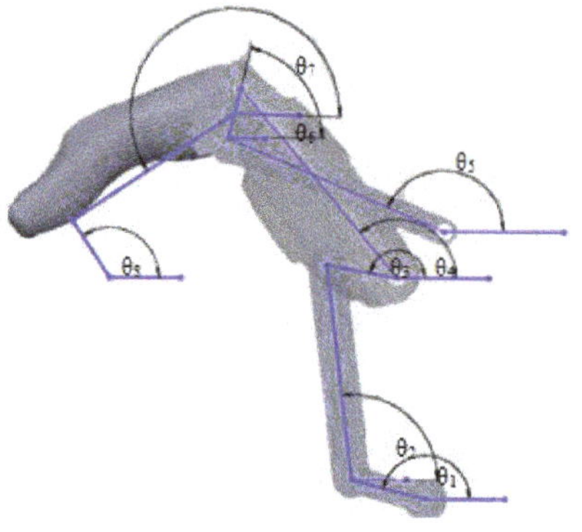

Figure 13. References angles.

$$A = -2 \cdot dx \cdot l_3 - 2 \cdot l_1 \cdot l_3 \cos\theta_1 \tag{42}$$

$$B = 2 \cdot dy \cdot l_3 - 2 \cdot l_3 \cdot l_1 \sin\theta_1 \tag{43}$$

$$C = dx^2 + dy^2 + l_1^2 + l_3^2 - l_2^2 + 2 \cdot dx \cdot l_1 \cdot \cos\theta_1 - 2 \cdot dy \cdot l_1 \cdot \sin\theta_1 \tag{44}$$

θ_2 is obtained from the following equation:

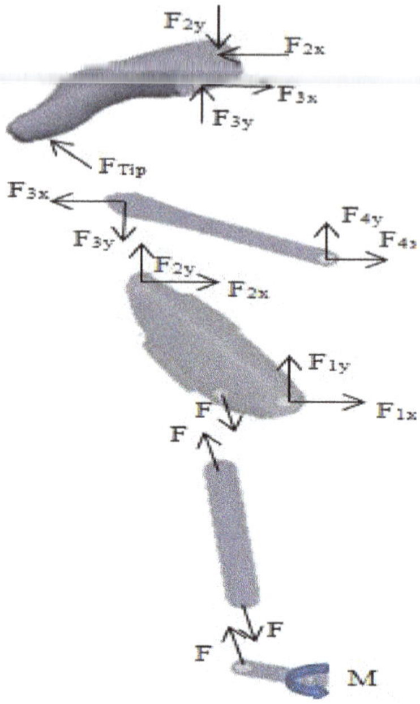

Figure 14. Free body diagram of the index and middle finger.

$$l_1 \cos\theta_1 + l_2 \cos\theta_2 = l_3 \cos\theta_3 - dx \tag{45}$$

And solving for θ_2:

$$\theta_2 = a\cos(\frac{l_3 \cos\theta_3 - dx - l_1 \cos\theta_1}{l_2}) \tag{46}$$

The remaining reference angles are obtained the same way as with the hand simulator and the force exerted by the finger is calculated. Figure 14 shows the forces acting on the finger elements.

From the crank free body diagram the next equation can be derived:

$$F = \frac{M_2}{L_1 \cdot (\cos\theta_2 \cdot \sin\theta_1 - \sin\theta_2 \cdot \cos\theta_1)} \tag{47}$$

This force is transmitted to the proximal phalange and a set of equations can be obtained in order to calculate the force at index and middle fingertips.

$$\begin{bmatrix} 0 & 1 & 0 & 1 & 0 & 0 & 0 & 0 & 0 \\ 0 & 0 & 1 & 0 & 1 & 0 & 0 & 0 & 0 \\ 0 & 0 & 0 & -L_4 \cdot \sin\theta_4 & L_4 \cdot \cos\theta_4 & 0 & 0 & 0 & 0 \\ \cos\theta_8 & 0 & 0 & -1 & 0 & 1 & 0 & 0 & 0 \\ \sin\theta_9 & 0 & 0 & 0 & -1 & 0 & 1 & 0 & 0 \\ A & 0 & 0 & 0 & 0 & L_6 \cdot \sin\theta_6 & -L_6 \cdot \cos\theta_6 & 0 & 0 \\ 0 & 0 & 0 & 0 & 0 & -1 & 0 & 1 & 0 \\ 0 & 0 & 0 & 0 & 0 & 0 & -1 & 0 & 1 \\ 0 & 0 & 0 & 0 & 0 & 0 & 0 & L_5 \cdot \sin\theta_5 & -L_5 \cdot \cos\theta_5 \end{bmatrix} \begin{bmatrix} F_{Tip} \\ F_{1x} \\ F_{1y} \\ F_{2x} \\ F_{2y} \\ F_{3x} \\ F_{3y} \\ F_{4x} \\ F_{4y} \end{bmatrix} = \begin{bmatrix} F\cos\theta_2 \\ F\sin\theta_2 \\ B \\ 0 \\ 0 \\ 0 \\ 0 \\ 0 \\ 0 \end{bmatrix}$$

Where:

$$A = L_7 (-\cos\theta_8 \cdot \sin\theta_7 + \cos\theta_7 \cdot \sin\theta_8) + \frac{L_6}{2}(\sin\theta_6 \cdot \cos\theta_8 - \sin\theta_8 \cos\theta_6) \tag{48}$$

$$B = F \cdot L_3 (\cos\theta_2 \cdot \sin\theta3 - \sin\theta_2 \cdot \cos\theta_3) \tag{49}$$

Figure 15. Position for the tip grasp.

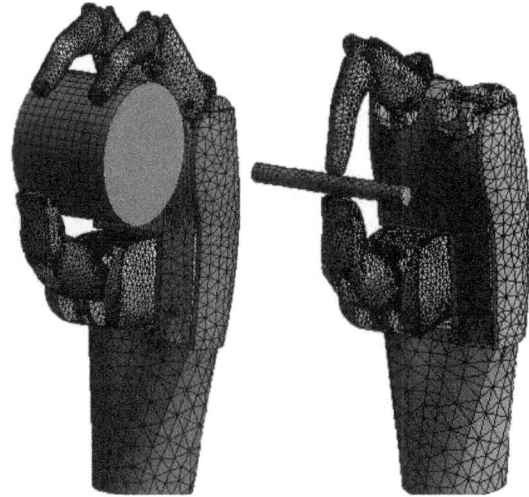

Figure 16. Mesh of the prototype.

In a vector form, the force exerted by the finger is given by:

$$\bar{F}_{Index} = -F_{Index} \cdot \cos\theta_8 \, i - F_{Index} \cdot \sin\theta_8 \, j \tag{50}$$

To determine the torques used in the simulation, it was considered that the 3 fingers exert a 10 N prehensile force on the palm. So it was assumed that the force applied by the index and middle finger has a magnitude of 2.5 N in the X direction. Using Equation 50 and knowing that for this position $\theta_1 = 164.9°$ and $\theta_8 = 123.08°$, we can obtain the index finger force $F_{Index} = 4.58$ N. The linear system equation that represents the index finger force is solved in an iterative way in order to obtain the torque that has to be applied, its magnitude is $M_{Index} = 0.398$ N-m.

With this applied torque, the index and middle finger exert a force of 3.84 N in the vertical direction which should be compensated by the thumb in order to achieve static equilibrium. From Equation 40, the vertical force exerted by the thumb is $F = F_{Thumb} \cdot \sin\theta_9$, so the force applied is

$$F_{Thumb} = \frac{2*3.84}{\sin\theta_9} \tag{51}$$

For this position $\theta_1 = 27.08°$ and $\theta_9 = 45.93°$. Solving for Equation 51 $F_{Thumb} = 10.68$ N. Using the set of equations that models the thumb force, the torque was calculated being its magnitude $M_{Thumb} = 1.88$ N-m. With this torque, the horizontal force exerted by the thumb is 7.41 N.

Using the calculated values of the torques, the total force in the palm direction has a value of 12.41 N; this value is greater than the 10 N that was assumed. Recalculating the torques but taking into account the ratio $F_{Index}/F_{Thumb} = 0.337$, it was obtained that $M_{Thumb} = 1.51$ N-m and $M_{Index} = 0.32$ N-m, so the force in the palm has a magnitude of 9.96 N. These are the values used in the simulation.

Tip grasp

For the tip grasp analysis (Figure 15) it was considered a cylindrical body of 7 mm diameter which represents a pen. For this grasp it is considered that only acts the index finger and the thumb so the only set of different parameters are the positions of the cranks and the point where the force is applied, this makes it possible to use the same equations developed previously.

In order to determine the torques needed for the simulation, a 10 N grasping force was considered. The vertical force exerted by the thumb and the index was established to be of 10 N, and the force in the horizontal direction was neglected.

Knowing that for this position $\theta_1 = 176.92°$ y $\theta8 = 69.80°$ and solving Equation (50) the force of the index is obtained and has a magnitude of $F_{Index} = 10.65$ N. From the set of equations for the index finger force the torque has a magnitude of $M_{Index} = 0.691$ N-m. Considering that for the thumb $\theta_1 = 156.45°$ and $\theta_9 = 96.32°$, and solving Equation (40) $F_{Thumb} = 10.06$ N. Using the set of equations for the thumb finger force the torque has a magnitude of $M_{Thumb} = 0.668$ N-m.

FEM analysis of the prosthetic hand prototype

Once the torques have been calculated, the FEM analysis was performed with the goal of assure that the prototype would be able to support the loads that are involved both in the cylindrical and tip grasp.

Based on the geometry model, a FE mesh was carried out using the ANSYS automeshing techniques. Hexahedral elements were used in the primitive geometries and tetrahedral elements in complex regions. A mesh refinement was used in the regions of nonlinear contact in order to obtain convergence in the simulation. With this parameters established, the total finite elements in the model were 171,284 for the cylindrical grasp and 83,467 for the tip grasp (Figure 16).

The material used in the simulations was a titanium alloy, the same as the one used in the simulation of the hand simulator. The base of the prosthesis was fixed restraining the movement of the bottom; also the movement along the symmetry axis of the grasped objects was constrained in order to prevent sliding (Figure 17).

The regions where exist contact with bolts, like the one shown in Figure 18, were defined as a frictionless contact, resulting in a total of 28 nonlinear contacts for the cylindrical grasp simulation and 19 for the tip grasp analysis. In the tip grasp analysis, in order to decrease the computing capacity required, the middle finger was suppressed because is not under any load. Due to the high non linearity of the simulation the load was applied in sub-steps

Figure 17. Boundary conditions applied.

Table 3. Lengths of the elements of the index and middle finger.

Component	Length (mm)
L_1	14
L_2	39.5
L_3	13.12
L_4	45
L_5	44.9
L_6	10
L_7	35
Dx_1	5.2
Dy_1	40.6
Dx_2	8.5
Dy_2	8.5

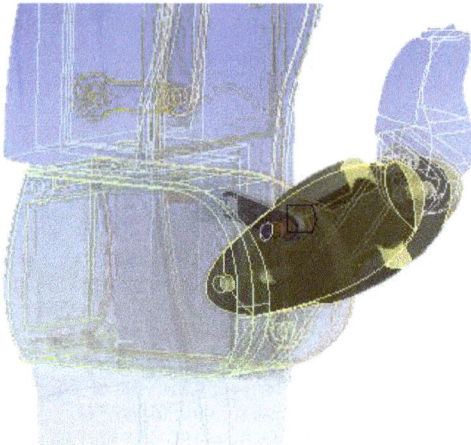

Figure 18. Contact region example.

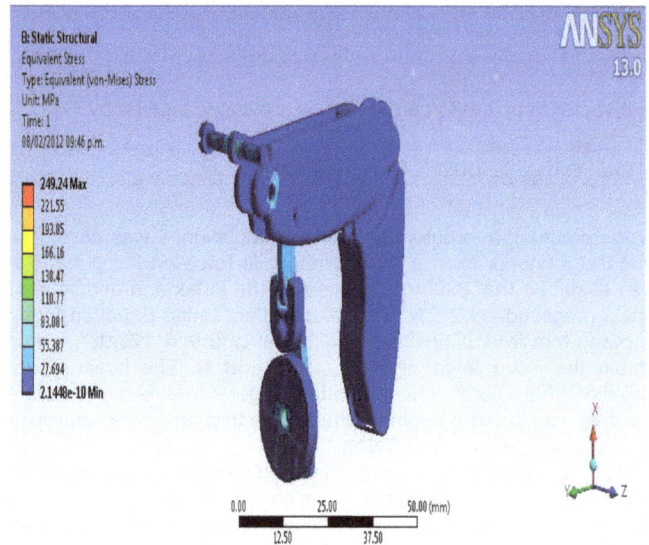

Figure 19. Hand simulator stress distribution.

RESULTS

Graph 1 shows the fingertip force applied by the hand simulator. The maximum force occurs for the position $\theta_1 = -60°$ and has a magnitude of 49.95 N, and the minimum force within the operating range is 10 N. This last value could be used as in Jung and Moon (2008), De Laurentis and Mavroidis (2002), Yang et al. (2009), Huang et al. (2006), Kawasaki et al. (2002), and Carozza et al. (2004) through static analyses it is shown that these designs can exert 10 N or less at the fingertip.

As a result of the hand simulator analysis it was obtained the stress distribution of the mechanism (Figure 19). The force reaction obtained at the fingertip was 54 N, which is a value close to the one calculated analytically.

Observing the field stress for the analyzed position, the greatest value obtained, based on the von Mises yield criterion, is located in the element showed in Figure 20, which is used to link the servomotor to the crank, and has a magnitude of 249.46 MPa.

As the greatest stress in the simulator is lower than the yield stress, the design of the elements will not fail. The minimum safety factor obtained is:

$$\text{Safety Factor} = \frac{\text{Yield Stress}}{\text{Maximum Stress}} = \frac{930}{249.46} = 3.72$$

The area where it is located the highest stress is a zone that has an abrupt change of direction, acting this as a stress concentrator, so adding a fillet would reduce this value.

All the remaining elements present lower stress values so its mechanical integrity is assured. A set of modifications were done in the component at zones where there were located the maximum stresses, adding radius in the corners in order to eliminate the stress concentrators and the hand simulator was analyzed again. With these modifications, the maximum stress was located in the same component but with a magnitude of 227.01 MPa. With this maximum stress, the minimum safety factor obtained is 4.09.

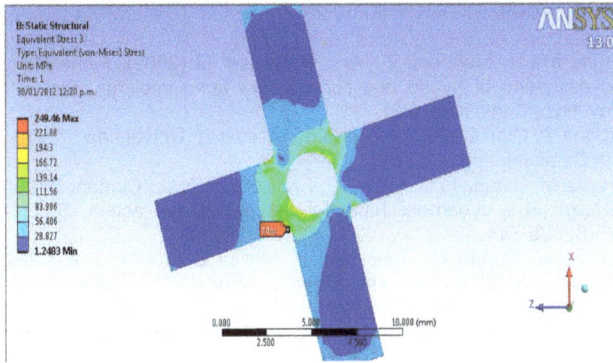

Figure 20. Element with the greatest stress.

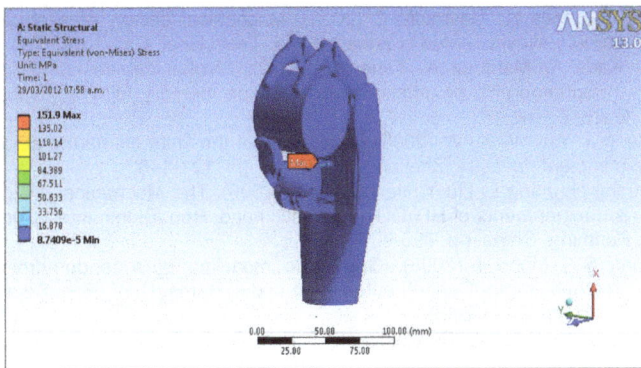

Figure 21. Stress distribution during cylindrical grasp.

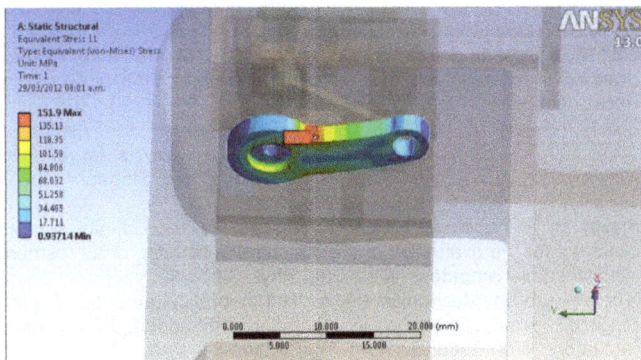

Figure 22. Highest stress element.

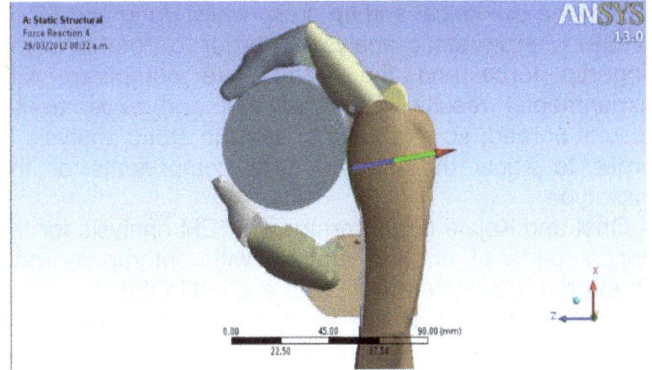

Figure 23. Applied prehensile force.

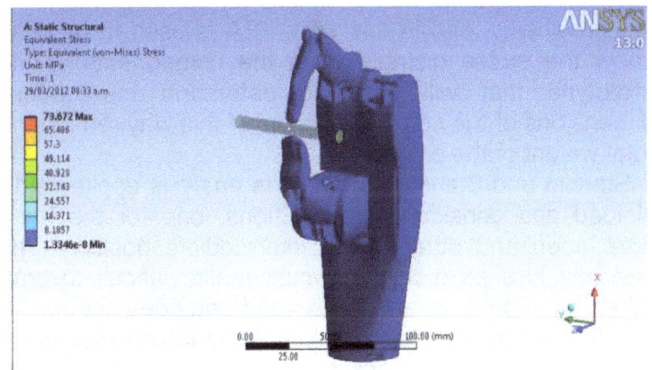

Figure 24. Tip grasp stress distribution.

With the cylindrical grasp simulation, a maximum stress with a value of 151.9 MPa was obtained (Figure 21). This stress was located in the crank of the thumb mechanism (Figure 22), which is an expected result as for this element, it applied the largest torque considered in the simulation. A safety factor of 6.12 is obtained with this maximum stress.

The applied force in the horizontal direction by the thumb has a magnitude of 6.05 N and the index and middle finger exert a force of 1.85 N in the simulation, while analytically values of 5.97 N and 2.01 N was calculated respectively, hand the prehensile force has a magnitude of 9.75 in the horizontal direction (Figure 23). For the tip grasp simulation, it was obtained a maximum stress of 73.7 MPa, and the minimum safety factor was 12.6 (Figure 24).

For this case, the maximum stress was located at the crank of the index finger mechanism, while the crank of the thumb presented a maximum stress of 72.11 MPa.

The most important result is that the highest stress at the exterior of the prosthetic hand prototype has a magnitude of 28.19 MPa, and this makes it possible to use another material instead of titanium and this would benefit the cost of the prosthesis as the cost of this material is high.

DISCUSSION

In this work, static analysis was carried out in order to predict the loads in the components of a prosthetic hand prototype, and this load were used in a FEM analysis in order to determine the mechanical stress in the prototype during the cylindrical and tip grasp whilst Jung and Moon (2008) realizes static analysis in order to determine the fingertip force and the results are compared with experimental results. The analytical and experimental results agreed, so it is valid to use the static analysis in order to know the loads in the components of the prototype.

Ohol and Kajale (2009) realizes a FEM analysis for the critical parts of an end effector with anthropomorphic shape but it is unclear the loads applied in the simulation on the other hand, in this work, the FEM analysis was realized for the entire prototype and the loads applied were the maximum loads that are developed on the elements of the prototype during its operation.

Through the use of FEM analysis, O'Toole (2007) determines the material for a hand prosthesis, it also determined optimal wall thickness for some elements of the prosthesis, and in this work FEA is used in order to know the stress distribution in the components of the prototype that will allow to determine the optimal dimensions of the components and in this way reduce the total weight of the prosthesis.

Stanciu and Stanciu (2009) in its analysis applies a 10 N load and considers two positions, one for the hand wide open and other for an intermediate position. The first one is a position that would make difficult to grab things with the hand and the second one does not assure that in this position the maximum loads will develop while in this work it is considered only positions where maximum loads take place.

Conclusion

It has been shown that, in order to have a grasping force of 10 N, it is necessary to apply a torque with a magnitude of 1.51 N-m. Also, it is notorious as well that the friction at the unions was neglected in the calculations, so the real torque in order to achieve the force with the required magnitude would be higher.

The actual actuators used in this prototype have a torque with a lower magnitude than the needed. Taking into account this, it is proposed to study the option of using a different servomotor with a higher torque in order to transmit the movement to the fingers. It must be noted that the fabrication of this prototype is based in a stereolithography technic and it is necessary to generate the development through the use of advanced manufacturing techniques.

ACKNOWLEDGEMENTS

We kindly acknowledge the financial support from the Instituto Politécnico Nacional through grants SIP; from CONACyT; and also from COFAA-IPN.

REFERENCES

Bougherara H, Mahboob Z, Miric M.Youssef M (2009). Finite element investigation of hybrid and conventional knee implants. Fini. Elem. Inv. Hyb. Conv. Kne. 3(3).

Cardona S, Clos C (2001). Teoría de máquinas. Universidad Politénica de Catalunya. pp. 139-158.

Carozza M, Cappiello G, Micera S, Erdin B, Beccai L, Cipriani C (2006.) Design of a cybernetic hand for perception and action. Biol. Cyb. 95(6):629-644.

Carozza M, Suppo C, Sebastiani F, Massa B (2004). The SPRING hand: development of a self-adaptive prosthesis for restoring natural Grasping. Auto. Rob. 16(2):125-141.

Carrozza M, Capiello G, Stellin G, Zaccone F, Vecchi F (2005). On the development of a novel adaptive prosthetic hand with compliant joints: Exp. Plat. EMG Con. pp. 1271-1276.

Cipriani C, Controzzi M, Carozza M (2011). The Smart hand transradial prosthesis. J. Neur. Reh. 8:29.

Clement R, Bugler K, Oliver C (2011). Bionic prosthetic hands: A review of present technology and future aspirations. Surgeon 9:336-340.

De Laurentis K, Mavroidis C (2002). Mechanical design of a shape memory alloy actuated prosthetic hand. Tech. Heal. Car. 10:91-106.

El Kady A, Mahfouz A, Taher M (2010). Mechanical design of an antrophomorphic prosthetic hand for shape memory Alloy actuation. Cairo, Egypt.

Geng J, Yan W, Xu W (2008). Application of the finite element method in implant dentistry. Zhejiang Univ Press. pp. 121-148.

Huang H, Jiang L, Liu Y, Hou L, Cai H (2006). The Mechanical design and experiments of HIT/DLR prosthetic hand. Rob. Biomi. Inter. Conf. Kunming, China. pp. 896-901.

Jung SY, Moon I (2008). Grip force modeling of a tendon-driven prosthetic hand. Control, automation and systems. Conf. Pub. Seoul, Korea. pp. 2006-2009.

Kargov A, Pylatiuk C, Martin J, Schulz S, Doderlein L (2004). A comparison of the grip force distribution in natural hands and in prosthetic hands. Disab. Reh. 26(12):705-711.

Kargov A, Pylatiuk C, Oberle R, Klosek H, Wender T, Roessler W, Schulz S (2007). Development of a multifunctional cosmetic prosthetic hand. Conf, Pub. 550-553.

Kawasaki H, Komatsu T, Uchiyama K (2002). Dexterous anthropomorphic robot hand with distributed tactile sensor: Gifu hand II. IEEE Trans. Mech. 7(3):296-303.

Naaji A, Gherghel D (2009). The Application of the finite element method in the biomechanics of the human upper limb and of some prosthetic components. WSEAS Trans. Comp. 8(8):1296-1305.

Oho R, Kajale S (2009). Biomimetic approach for design of multifingered robotic gripper (MRG) and Its analysis for effective dexterous grasping. Inter. Conf. Mach. Lear. Comp. Singapore 3:213-221.

Omasta M, Palousek D, Návrat T, Rosicky J (2012). Finite element analysis for the evaluation of the structural behaviour, of a prosthesis for trans-tibial amputees. Med. Eng. Phys. 34:38-45.

O'Toole K (2007). Mechanical design and theoretical analysis of a four fingered prosthetic hand incorporating embedded SMA bundle actuators. Wor. Acad. Sci. Eng. Tech. Issue 31.

Pérez M, Velázquez A, Torres C, Martínez L, Huerta P, Urriolagoitia G (2012). Sub- actuadedanthropometricroboticprototypehand. Rev. Fac. Ing. Uni. Antio. 65:46-59.

Stanciu L, Stanciu A (2009). Designing and implementing a human hand prosthesis. Inter. Conf. Advan. Med. Heal. Car. Throu. Tech. Springer Berlin Heidelberg. pp. 399-404.

Watve S, Dodd G, MacDonal R, Stoppard E (2011). Upper limb prosthetic rehabilitation. Orth. Trau. 25(2):135-142.

Yang D, Zhao J, Gu Y, Wang X, Li N, Jiang L, Liu H, Huang H, Zhao D (2009). An anthropomorphic robot hand developed based on underactuated mechanism and controlled by EMG signals. J. Bionic. Eng. 6(3):255-263.

Zhang W, Qiu M, Ma X (2008). Super under-actuated humanoid robot hand with gear-rack mechanism. Intell. Rob. Appl. Springer Berlin / Heidelberg. pp. 597-606.

4

Modeling the machining parameters of AISI D2 tool steel material with multi wall carbon nano tube in electrical discharge machining process using response surface methodology

S. Prabhu[1]* and B. K. Vinayagam[2]

[1]School of Mechanical Engineering, S.R.M. University, Chennai 603 203, Tamil Nadu, India.
[2]Department of Mechatronics, S.R.M. University, Chennai 603 203, Tamil Nadu, India.

This work investigates the machining characteristics of American Iron and Steel Institute (AISI) D2 tool steel with copper as a tool electrode during electrical discharge machining (EDM) process. The multi wall carbon nano tube (MWCNT) is mixed with dielectric fluids in EDM process to analyze the surface characteristics of surface roughness. Regression model were developed to predict the surface roughness (SR) in EDM process. In the development of predictive models, machining parameters of pulse current, pulse duration and pulse voltage were considered as model variables. The collection of experimental data adopted Box-Behnken central composite design (CCD). Analysis of variance (ANOVA) and F-test were used to check the validity of regression model and to determine the significant parameter affecting the surface roughness. Later, the AISI D2 tool steel was analyzed and the parameters are optimized using MINITAB software, and regression equation are compared with and without MWCNT used in EDM process. The average 34% of surface finish was improved by using carbon nano tube (CNT) mixed as dielectric fluid. The maximum test errors for regression model using copper electrode are 8.18% for without CNTs and 5.44% for with CNTs. The R^2 value of developed empirical model for SR with MWCNTs is 69.45% compared without CNT is 55.4%. The high R^2 value indicates that the better the model fits the data.

Key words: Multiwall carbon nanotube, electrical discharge machining (EDM), roughness, Box-Behnken central composite design, analysis of variance (ANOVA).

INTRODUCTION

Electrical discharge machining (EDM) is one of the most successful and widely accepted processes for production of complicated shapes and tiny apertures with high accu-racy. This method is commonly used for profile truing of metal bond diamond wheel, micro nozzle fabrication, drilling of composites and manufacturing of moulds, and dies in hardened steels. These hard and brittle materials fabricated by conventional machining operation cause excessive tool wear and expense. The mechanical properties of tool steels have been studied extensively for many years. During EDM, the tool and the work piece are separated by a small gap, and submerged in dielectric fluid. The discharge energy produces very high tempera-tures on the surface of the work piece at the point of the spark. The specimen is subject to a temperature rise of up to 40,000 K causing a minute part of the work piece to be melted and vaporized. The top surface of work piece subsequently resolidifies and cools at very high rate.

EDM technology is increasingly being used in tool, die and mould making industries, for machining of heat treated tool steels and advanced materials (super alloys, ceramics and metal matrix composites) requiring high precision, complex shapes and high surface finish. Traditional machining technique is often based on the material removal

*Corresponding author. E-mail: prabhume@yahoo.co.in.

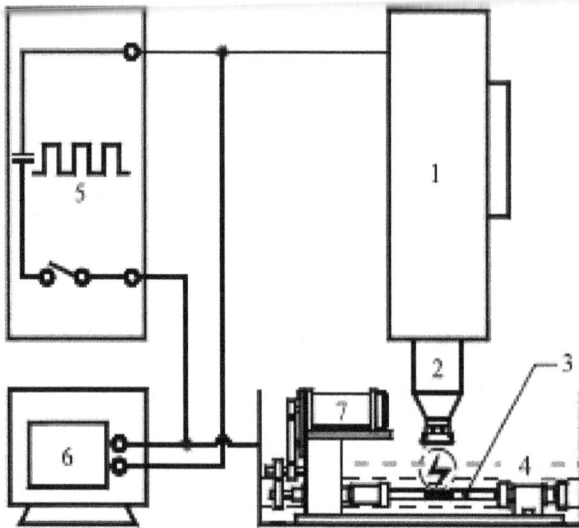

Figure 1. Schematic drawing of EDM. 1, Servo-control; 2, electrode; 3, work piece; 4, dielectric fluid; 5, pulse generator; 6, oscilloscope; 7, DC motor.

using tool material harder than the work material and is unable to machine them economically. Heat treated tool steels have proved to be extremely difficult-to-machine using traditional processes due to rapid tool wear, low machining rates, inability to generate complex shapes and imparting better surface finish.

Guu et al. (2003) proposed and investigated the EDM of American Iron and Steel Institute (AISI) D2 tool steel. The surface characteristics and machining damage caused by EDM were studied in terms of machining parameters. Based on the experimental data, an empirical model of the tool steel was also proposed. Surface roughness (SR) was determined with a surface profilometer. Guu (2001) proposed the surface morphology, SR and micro-crack of AISI D2 tool steel machined by EDM process were analyzed by means of the atomic force microscopy (AFM) technique. Pecas and Henriques (2008) presented on EDM technology with powder mixed dielectric and to compare its performance to the conventional EDM when dealing with the generation of high-quality surfaces. Kansal et al. (2005) study has been made to optimize the process parameters of powder mixed electrical discharge machining (PMEDM). Response surface methodology has been used to plan and analyze the experiments. Izquierdo et al. (2009) pro-posed and presented a new contribution to the simulation and modeling of the EDM process. Prabhu and Vinayagam (2008a) analyzed the surface characteristics of tool steel material using multiwall carbon nanotube (MWCNT) to improve the surface finish of material to nanolevel. Prabhu and Vinayagam (2008b) proposed nanosurface generation in grinding process using MWCNT with lubricant mixture to improve the surface finish of grinding process to nanolevel due to good thermal conductivity of

carbon nanotubes (CNTs). Ozlem and Cengiz (2008) proposed roughness values obtained from the experiments that have been modeled by using the genetic expression programming (GEP) method and a mathematical relationship has been suggested between the GEP model and SR and parameters affecting it. Moreover, EDM has been used by applying copper, copper–tungsten (W–Cu) and graphite electrodes to the same material with experimental parameters designed in accordance with the Taguchi method. Yan-Cherng et al. (2009) developed the force assisted standard EDM machine. The effects of magnetic force on EDM machining characteristics were explored. Moreover, this work adopted an L18 orthogonal array based on Taguchi method to conduct a series of experiments and statistically evaluated the experimental data by analysis of variance (ANOVA). Ko-Ta et al. (2007) proposed a methodology for modeling and analysis of the rapidly resolidified layer of spheroidal graphite (SG) cast iron in the EDM process using the response surface methodology. The results of ANOVA indicate that the proposed mathematical model obtained can adequately describes the performance within the limits of the factors being studied.

Mustafa and Ali (2011) proposed dependent and independent variables which were also modeled by regression analysis. The results showed that cutting force, surface roughness, cylindricity and vibration were minimised in machining process and production quality was improved. Yang et al. (2009) proposed an optimization methodology for the selection of best process parameters in EDM. Regular cutting experiments are carried out on die-sinking machine under different conditions of process parameters. The system model is created using counter-propagation neural network using experimental data. Jegaraj and Babu (2007) attempted to make use of Taguchi's approach and ANOVA using minimum number of experiments for studying the influence of parameters on cutting performance in abrasive water jet (AWJ) machining considering the orifice and focusing tube bore variations to develop empirical models. Chattopadhyay et al. (2009) investigated the machining characteristics of EN8 steel with copper as a tool electrode during rotary EDM process. Three independent input parameters of the model viz: peak current, pulse on time and rotational speed of tool electrode are chosen as variables for evaluating the output parameters such as metal removal rate (MRR), electrode wear ratio (EWR) and SR.

EDM shown in Figure 1 is an important non-traditional manufacturing method to produce plastics moldings, die castings and forging dies etc. New developments in the field of material science have led to new engineering metallic materials, composite materials and high tech ceramics, having good mechanical properties and thermal characteristics as well as sufficient electrical conductivity so that they can readily be machined by spark erosion. The die sinking EDM SD35-5030 model is used to do the machining. AISI D2 tool steel is one of the

Table 1. Chemical composition of the AISI D2 tool steel (wt. %).

Element	C	Si	Mn	Mo	Cr	Ni	V	Co	Fe
Wt.%	1.5	0.3	0.3	1.0	12.0	0.3	0.8	1.0	Balance

Table 2. Mechanical properties of the AISI D2 tool steel at room temperature.

0.2% offset yield strength	1532 MPa
Tensile strength	1736 MPa
Hardness (HRC)	57

Table 3. Specification of MWCNTs.

Outer diameter	10 to 20 nm
Length	10 to 30 μm
Purity	> 95 wt%
Ash	< 1.5 wt%
Specific surface area	> 233 m^2/g
Electrical conductivity	> 10^{-2} S/cm

carbon steels alloyed with Mo, Cr, and V and is widely used for various dies and cutters for its high strength, and wear resistance due to formation of chrome carbider in heat treatment. Table 1 shows the chemical composition (wt.%) of the material, while Table 2 shows the mechanical properties of the AISI D2 tool steel.

CNTs are related to graphite. The molecular structure of graphite resembles stacked, one-atom-thick sheets of chicken wire, a planar network of interconnected hexagonal rings of carbon atoms. In conventional graphite, the sheets of carbon are stacked on top of one another, allowing them to easily slide over each other. That is why graphite is not hard, but it feels greasy and can be used as a lubricant. When graphene sheets are rolled into a cylinder and their edges joined, they form MWCNT (Table 3). Only the tangents of the graphitic planes come into contact with each other, and hence their properties are more like those of a molecule as mentioned in the above passage clearly. Furthermore, the high-frequency carbon-carbon bond vibrations provide an intrinsic thermal conductivity higher than even diamond. The TEM images of multi wall carbon nano tubes are shown in Figure 2 was received from Cheap tubes Inc., USA.

CNT nanofluids, is of special interests to researchers because of the novel properties of CNTs -extraordinary strength, unique electrical properties and efficient conductors of heat. CNTs are fullerene-related structures that consist of either a grapheme cylinder or a number of concentric cylinders (Wen and Ding, 2004). Choi et al. (2001) measured the effective thermal conductivity of MWCNTs dispersed in synthetic (poly-α-olefin) oil and reported the enhancement up to a 150% in conductivity at approximately 1 vol% CNT, which is by far the highest thermal conductivity enhancement ever achieved in a liquid (Lockwood and Zhang, 2005). Solid lubricants are useful for conditions when conventional liquid lubricants are inadequate such as high temperature and extreme contact pressures. Their lubricating properties are attributed to a layered structure on the molecular level with weak bonding between layers. Such layers are able to slide relative to each other with minimal applied force, thus giving them their low friction properties. CNT is having high strength to weight ratio used in aero space industry. Young's modulus of CNT is over 1 TPa versus 70 GPa for aluminium, steel 200 Gpa and 700 GPa for C-fibre. The strength to weight ratio is 500 times greater than aluminium. Maximum strain will be 10% much higher than any material. Thermal conductivity of 3,000 W/mK in the axial direction is with small values in the radial direction. Conductivity of CNTs is 109 A/cm^2 and copper is 106 A/cm^2. CNT's having very high current carrying capacity, excellent field emitter and high aspect ratio. Model Hommel Tester TR500 SR tester is a multi-application measuring instrument for component surface quality evaluation. It is capable of checking the work piece SR on plane, cylinder, groove and bearing raceway.

In this paper, CNT mixed dielectric fluids are used in the EDM process to analyze the surface characteristics of AISI D2 tool steel material. Till now, no work has been carried out by using CNT mixed dielectric machining. CNT based nano fluid is used to improve the surface finish from micro level to nano level which improves the accuracy of the work piece. The collection of experimental data adopted Box-Behnken Central composite Design (CCD) using Table 5 coded level of three machining parameters. ANOVA and *F*-test were used to check the validity of regression model and to determine the significant parameter affecting the surface roughness. Later the AISI D2 tool steel was analyzed and the parameters are optimized using MINITAB software and regression equation are compared with and without multiwall carbon nanotubes used in EDM process.

EXPERIMENTAL WORK

The specimen was made of the AISI D2 tool steel, which is widely used in the mold industry. The electrode material used is copper. The raw materials were machined as using conventional methods such as turning, parting and grinding. The specimens were made to a size of diameter 20mm and length 20.5 mm and the electrode were made to a size of 24 mm diameter and length 50 mm. The machined material was heated to 1030°C at a heating rate of 20°C/min in muffle furnace. It was kept at 1030°C for 1 h and then quenched. After quenching, the specimens were tempered at 520°C for 2 h and then air cool. The hardness obtained for the specimen is 58HRC Table 2 shows the chemical composition (wt. %) of the material. The EDM specimens were sparked on a die-sinking EDM machine model type SD35 – 5030. The experiment was carried out in kerosene dielectric covering the work piece by 40 mm. A cylindrical copper rod machined was used as the electrode for sparking the work piece. The copper electrode was the negative polarity and the specimen was the positive polarity during the EDM process. During EDM, the primary parameters are pulsed current, pulse-on duration, and pulse-off duration. Table 4 shows the EDM conditions.

During the EDM process, the varying pulse-off duration setting from 1 to 10 µs could effectively control the flushing of the debris from the gap, giving machining stability. Hence, the effect of the pulse-off duration on the machined characteristics was not considered in the present work. After each experiment, the machined surface of the EDM specimen was studied by means of a scanning electron microscope. The dielectric fluid was mixed in a proportion of 2 g of MWCNT for 0.5 litre of kerosene. The sparking was carried out in this setting (Figure 3). After experiment, the machined surface of the EDM specimen was studied by means of a scanning electron microscope. Different samples were examined.

A separate tank was made to hold the dielectric fluid containing MWCNT in which the specimen was placed. After experiment, the machined surface of the EDM specimen was studied by means of scanning electron microscope. Statistical technique was used for investigating and modeling the relationship between variables. Models of functional relationships are usually approximations. Uses of regression: data description, parameter estimation, prediction and estimation, control. Iterative procedure from data to selection of model, model fitted to data, model adequacy checking and modification of the model or fit. The SR was measured using SR tester Hommel Tester TR500 for given working condition of EDM for with and without CNT (Table 6). The collection of experimental data adopted Box-Behnken CCD. It is an independent quadratic design in that it does not contain an embedded factorial or fractional factorial design. In this design the treatment combinations are at the midpoints of edges of the process space and at the center. These designs are rotatable (or near rotatable) and require 3 levels of each factor. These designs require fewer treatment combinations than a CCD in cases involving 3 or 4 factors. This property prevents a potential loss of data in those cases. Each factor with low (-1) and

Table 4. Electrical discharge machining conditions.

Work material	AISI D2 tool steel
Dielectric	Kerosene
Electrode material	Copper
Pulsed current	4.5, 5, 5.5A
Pulse-on duration	6 to 25 µs
Pulse voltage	40- 80 v

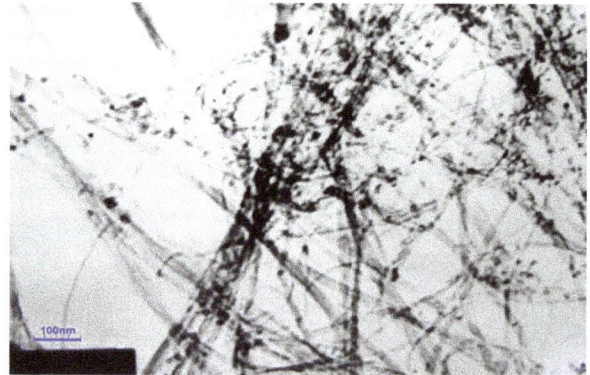

Figure 2. A TEM image of our MWCNTs 95wt% <8nm OD.

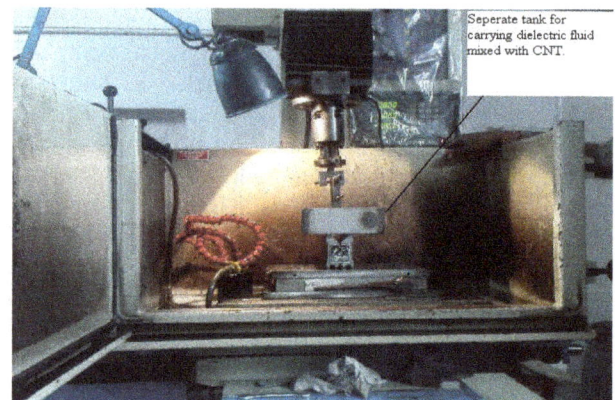

Figure 3. EDM set up using MWCNT.

high (+1) value of variables also center full factorial value (0). To determine the number of measurement points needed to take a measurements are based on no of factorial runs $\alpha = 2^k$, where k is no of parameters are used. Here, 3 parameters are used so 8 corner points, 6 star points are used and 1 central point is used. So totally, 15 readings were taken for Box-Behnken CCD. The input parameters are chosen within the processing guides of material and correlated the machine.

The coded values are obtained by using the following transformation

Pulse current $X1 = I - I_0 / \Delta I$ (1)

Table 5. The coded level of three machining parameters and their range.

S/N	Parameter	Unit	Coded value		
			-1(low)	0	+1(High)
1	Pulse current (I)	Amp	4.5	5	5.5
2	Pulse duration (τ)	Second	6	12	25
3	Pulse voltage (V)	Voltage	40	60	80

Table 6. Experimental results along with design matrix.

S/N	Coded value			Actual value			With CNTs surface roughness (Ra)	Without CNTs surface roughness (Ra)
	X1	X2	X3	X1	X2	X3		
1	-1	-1	0	4.5	6	60	1.30	4.02
2	+1	-1	0	5.5	6	60	2.03	4.60
3	-1	+1	0	4.5	25	60	1.90	4.87
4	+1	+1	0	5.5	25	60	3.04	4.83
5	-1	0	-1	4.5	12	40	2.98	3.08
6	+1	0	-1	5.5	12	40	2.34	3.38
7	-1	0	+1	4.5	12	80	2.98	3.71
8	+1	0	+1	5.5	12	80	2.60	3.28
9	0	-1	-1	5	6	40	3.00	4.30
10	0	+1	-1	5	25	40	3.95	5.89
11	0	-1	+1	5	6	80	2.34	4.75
12	0	+1	+1	5	25	80	2.74	3.10
13	0	0	0	5	12	60	3.04	4.58
14	0	0	0	5	12	60	3.48	4.72
15	0	0	0	5	12	60	4.25	5.74
	Mean (μ)						**2.79**	**4.32**

Pulse duration $X2 = \tau - \tau_0 / \Delta\tau$ (2)

Pulse voltage $X3 = V - V_0 / \Delta V$ (3)

Where $X1, X2, X3$ are coded value of I, τ and V respectively. The I_0, τ_0, $V0$ are cutting speed, feed and depth of cut at zero level. ΔI, $\Delta\tau$, ΔV are the units of variation in I, Γ and V. The experimental matrixes with coded and actual values are shown in Table 6.

RESPONSE SURFACE MODELING

Regression model is determining the relationship between independent variable with dependent variables. Here, pulse current, pulse duration and pulse voltage were used as independent variable and SR was used as dependent variable. Empirical expressions have been developed to evaluate the relationship between input and output parameters. The average output values of SR have been used to construct the empirical expressions. The empirical model was developed based on relationship between SR with pulse current, pulse on duration and pulse voltage in EDM process.

The empirical model was

$$Y = A \left(X_1\right)^a \left(X_2\right)^b \left(X_3\right)^c \qquad (4)$$

Y, Surface roughness (μm); A, coefficient; X_1, pulse current (A); X_2, Pulse on Duration (μs); X_3, pulse voltage (V).

The non- linear Equation 4 is converted to linear form by:

$$\log Y = \log(A) + a\log(x1) + b\log(x2) + c\log(x3) \qquad (5)$$

Now, Equation (2) can be written as:

$$\bar{Y} = \beta_0 + \beta_1 x_1 + \beta_2 x_2 + \beta_3 x_3 \qquad (6)$$

where $\bar{Y}$ is a true value of dependent machining output on a logarithmic scale, x_1, x_2, x_3 are the logarithmic transformation of the different input parameters $\beta_0, \beta_1, \beta_2, \beta_3$ are corresponding parameters to be estimated. MINITAB 15 software was used to estimate the parameters of the above first order model using the data shown in Tables 6a, b and c are coefficients determined by regression analysis.

RESULTS AND DISCUSSION

Surface roughness

To determine the effect of the EDM process on the SR of the tool steel, the surface profiles of the EDM specimens

Were measured by SR tester (Hommel Tester TR500). Figure 4 shows the surface roughness (Ra) value and its graph results for the specimen which was sparked using MWCNT. The parameters used are pulse current 5 amps, pulse duration 12 µs and pulse off duration 60 µs. The roughness value obtained for the specimen using this parameter is 4.25 µm.

The developed empirical model for surface roughness without carbon nanotube

$$R_a = AI^a \tau^b V^c \qquad (7)$$

Where A, 4.21; a, 0.100; b, 0.0207; c, - 0.0113; R^2, 6.9%. Here R^2 is defined as a measure of the amount of reduction on the variability of machining output. The regression analysis of the experimental data yields the semi empirical model

$$R_a without CNT = 4.21 I^{0.100} \tau^{0.0207} V^{-0.0113} \qquad (8)$$

The developed empirical model for surface roughness with carbon nanotube:

$$R_a = AI^a \tau^b V^c \qquad (9)$$

Where A, 1.97; a, 0.213; b, 0.0271; c, - 0.0101; R^2, 11.8%.

The regression analysis of the experimental data yields the semi empirical model

$$R_a with CNT = 1.97 I^{0.213} \tau^{0.0271} V^{-0.0103} \qquad (10)$$

Results of regression analysis are compared with experiments in Table 6 for 15 check sets. The comparison results are shown in Table 7. The maximum test errors for regression model using copper electrode are 8.18% for without carbon nanotubes and 5.44% for with CNTs. This method is suitable for estimating SR in an acceptable error ranges. The model generation of regression model took just a couple of seconds. From the results, errors of measurements occurs in SR with CNT is less than without CNTs.

Figure 5 represents error showing actual SR of measurement results with predicted SR through empirical model with and without using multi wall carbon nanotube used in EDM machining process. ANOVA test produce total variability of experimental results into components of variance and then their significant. F-test is utilized for comparing variances for this purpose.

Non-significant effects of the unusual observations were removed from the model because it will affect the

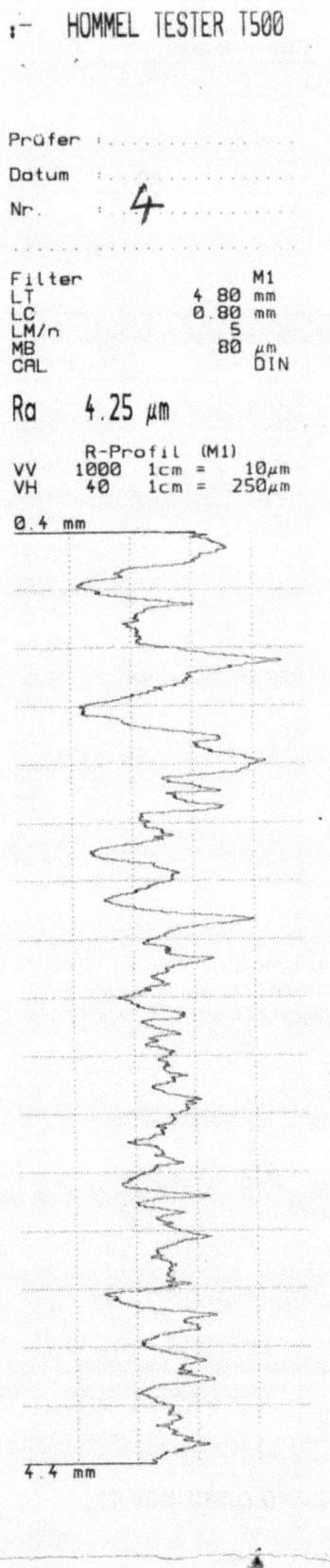

Figure 4. Surface roughness (Ra) value for 5 amps with MWCNT.

Modeling the machining parameters of AISI D2 tool steel material with multi wall carbon nano tube in electrical...

37

Table 7. Comparison of regression model with experiment measurements for copper electrode.

Experiment number	With CNTs surface roughness (µm)		Without CNTs surface roughness (µm)	
	Experimental measurement	Regression model	Experimental measurements	Regression model
1	1.30	2.485	4.02	4.1062
2	2.03	2.698	4.60	4.2062
3	1.90	3.00	4.87	4.4995
4	3.04	3.213	4.83	4.5995
5	2.98	2.8497	3.08	4.4564
6	2.34	3.0627	3.37	4.5564
7	2.98	2.4457	3.71	4.0044
8	2.60	2.6587	3.28	4.1044
9	3.00	2.7936	4.30	4.3822
10	3.95	3.3085	5.89	4.7755
11	2.34	2.3896	4.75	3.9302
12	2.74	2.9045	3.10	4.3235
13	3.04	2.7542	4.58	4.2804
14	3.48	2.7542	4.72	4.2804
15	4.25	2.7542	5.74	4.2804

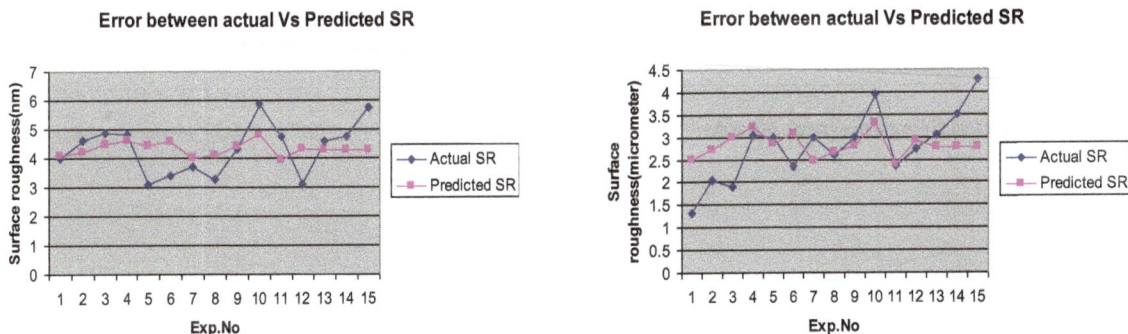

Figure 5. Error showing actual Vs predicted surface roughness without and with carbon nanotubes.

output parameters. The R^2 value of developed empirical model for SR with multi wall carbon nanotubes is 69.45% compared without CNT is 55.4%. The high R^2 value indicates that the better the model fits your data. The more variance 69.45% that is accounted for by the regression model the closer the data points will fall to the fitted regression line using MWCNTs.

In determining the appropriateness of rejecting the null hypothesis in a hypothesis test, the smaller the p-value, the smaller the probability that rejecting the null hypothesis is a mistake. A commonly used value is 0.05. Here p-value 0.04 of cutting speed is less than the alpha value which means fewer mistakes occurs than without CNT is 0.211 p-values. The results predicted by regression model are compared with experimental measurements results with multi wall carbon nanotubes and without CNT used in EDM process.

The main output from an ANOVA study arranged in a table, shows the sources of variation, their degrees of freedom, the total sum of squares and the mean squares. The ANOVA table also includes the F-statistics and p-values. Use these to determine whether the predictors or factors are significantly related to the response. The CNT used in machining process pulse current and pulse durations are significant parameters which influence the surface finish when compared without CNT no parameters are significant. In this, error occurs during CNT 30.54% is less than the without using CNT 44.59%. Also the percentage contribution of each parameter which influences the surface finish is analyzed. Tables 8 and 9 show the results of ANOVA for the SR of CNT and without CNT. Larger FA_o value indicates that the variation of the process parameter makes a big change on the SR and P denotes its percent contribution on surface roughness.

Table 8. The results of ANOVA for the surface roughness with MWCNT copper electrode.

Machining parameter	Degree of freedom (f)	Sum of squares (SSa)	Variance (Va)	FA$_o$	P	Contribution (%) with CNT
I	2	2.8756	1.5600	4.93	0.040*	34.7
T	2	2.5506	1.2649	4.00	0.063*	30.78
V	2	0.3283	0.1642	0.52	0.614	3.98
Error	8	2.5310	0.3164			30.54
Total	14	8.2855				100

*, Significant; S, 0.562471; R^2, 69.45%; R^2 (adj), 46.54%.

Table 9. The results of ANOVA for the surface roughness without MWCNT copper electrode.

Machining parameter	Degree of freedom (f)	Sum of squares (SSa)	Variance (Va)	FA$_o$	P	Contribution (%) without CNT
I	2	2.1521	1.1671	1.90	0.211	19.55
T	2	0.7997	0.3021	0.49	0.629	7.28
V	2	3.1464	1.5732	2.56	0.138	28.58
Error	8	4.9090	0.6136			44.59
Total	14	11.0073				100

S, 0.783344; R^2, 55.40%; R^2 (adj), 21.95%.

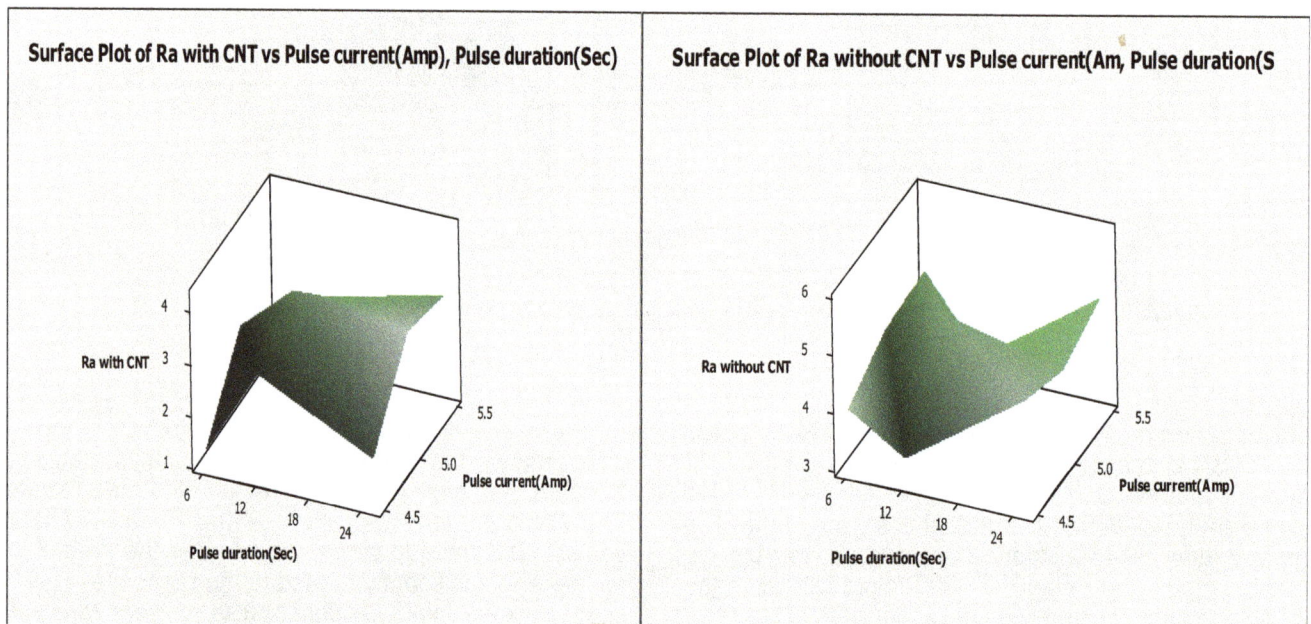

Figure 6. Response surface of Surface roughness (Ra with and without CNT) to pulse current (A) and pulse duration (T).

Effect of parameters on surface roughness using carbon nano tubes

The effect of pulse current and pulse on duration with and without CNT is shown in Figure 6. These Figure 6 indicate that SR increase with increase in pulse current.

CONCLUSION

In this investigation, experimental data adopted Box-Behnken CCD. The first order empirical models were developed for prediction of various output process parameters during EDM. The proposed models were successfully applied to estimate the values of SR under various machining conditions. Through this investigation of EDM of AISI D2 tool steel with copper electrode, the following conclusions are summarized that the main influencing factors for SR, in order of importance, include pulse current and pulse on duration, respectively. The developed empirical formulae can be used to evaluate SR produced by EDM with CNT, with low prediction error. The proposed empirical models were validated to conclude and as well fit for predictions of machining output such as SR with low prediction error.

REFERENCES

Choi SUS, Zhang ZG, Yu W, Lockwood FE, Grulke EA (2001). Anomalous thermal conductivity enhancement in nanotube suspensions. Appl. Phy. Let., 79: 2252-2254.

Guu YH (2001). AFM surface imaging of AISI D2 tool steel machined by the EDM process. Appl. Sur. Sci., 242: 245–250.

Guu YH, Hocheng H, Chou CY, Deng CS (2003). Effect of electrical discharge machining on surface characteristics and machining damage of AISI D2 tool steel. Mat. Sci. Eng., 358: 37-43.

Izquierdo B, Sancheza JA, Plazaa S, Pomboa I, Ortegaaet N (2009). A numerical model of the EDM process considering the effect of multiple discharges. Int. J. Mach. Tools Manuf., 49(3-4): 220-229.

Kansal HK, Singh S, Kumar P (2005). Parametric optimization of powder mixed electrical discharge machining by response surface methodology. J. Mat. Proc. Tech., 169: 427–436.

Lockwood FE, Zhang ZG (2005). The current development on nanofluid research. SAE 2005 World Congress and Exhibition, Detroit, MI, USA, Session: Military Vehicle Fuels, Lubes and Water, p. 7.

Ozlem S, Cengiz KM (2008). Evolutionary programming method for modeling the EDM parameters for roughnesses. J. Mat. Proc. Tech., 200: 347–355.

Prabhu S, Vinayagam BK (2008a). Analysis of Surface Characteristics of AISI D2 Tool Steel Material using Carbon Nano Tube. Int. J. Nano. Appl., 2(2-3): 107-122.

Prabhu S, Vinayagam BK (2008b). Nano surface generation in grinding process using carbon nano tube with lubricant mixture. Int. J. Nano. appl., 2(2-3): 149-160.

Pecas P, Henriques E (2008). Electrical discharge machining using simple and powder-mixed dielectric: The effect of the electrode area in the surface roughness and topography. J. Mat. Proc. Tech., 200: 250–258.

Wen D, Ding Y (2004). Effective thermal conductivity of aqueous suspensions of carbon nanotubes. J. Thermo. Heat Trans., 18(4): 481-485.

Yan-Cherng L, Yuan-Feng C, Der-An W, Ho-Shiun L (2009). Optimization of machining parameters in magnetic force assisted EDM based on Taguchi method. J. Mat. Proc. Tech., 209: 3374–3383.

Ko-Ta C, Fu-Ping C, De-Chang T (2007). Modeling and analysis of the rapidly resolidified layer of SG cast iron in the EDM process through the response surface methodology. J. Mat. Proc. Tech., 182: 525–533.

Mustafa AY, Ali T (2011). Determination and optimization of the effect of cutting Parameters and work piece length on the geometric tolerances and surface roughness in turning operation. Int. J. Phy. Sci., 6(5): 1074-1084.

Chattopadhyay KD, Verma S, Satsangi PS, Sharma PC (2009). Development of empirical model for different process parameters during rotary electrical discharge machining of copper–steel (EN-8) system. J. Mat. Proc. Tech., 209: 1454–1465.

5

Wavelet based dynamic Mel Frequency Cepstral Coefficients (MFCC) and block truncation techniques for efficient speaker identification under narrowband noise conditions

S. Selva Nidhyananthan, R. Shantha Selva Kumara and D. S. Roland

Department of ECE, Mepco Schlenk Engineering College, Sivakasi-626005, India.

Speaker identification strategies are well convincing in their performance when clean speeches are scrutinized. But the performance degrades when speech samples are corrupted by narrowband noise. Block truncation of the cepstral coefficients ensures that not all the features are affected by narrowband noise but it cannot reduce the extent of degradation. This work is focused towards improving the performance of speaker identification systems by block truncating the features which are subjected to wavelet processing. Wavelet decomposition divides the entire energy spectrum of the speech signal into bands corresponding to the number of levels of decomposition performed in the wavelet transformation thereby segregating the noise affected bands from other bands. In addition to that, wavelet filters provide the smoothening of the noisy speech signals which enhances the identification of the correct speaker. Dynamic Mel filtering of these wavelet coefficients followed by block truncation provides better identification, taking advantage of the fact that some filter bank coefficients remain unaffected by narrowband noise. The features are modeled by Gaussian mixture model - Universal background model (GMM-UBM) that serves as a generic one timed trained model. Speaker identification efficiency of 97.23% is achieved through this wavelet based dynamic MFCC technique which exhibits 7.58% improvement in speaker identification accuracy when compared with non wavelet based block truncation method.

Key words: Wavelet decomposition, block truncation, Dynamic Mel Filtering Cepstral Coefficients (DMFCC), Gaussian mixture model - Universal background model (GMM-UBM), speaker identification.

INTRODUCTION

Speaker identification is a biometric process (ZoranCirovi et al., 2010) of identifying a person by comparing the features extracted from the person to be identified, with the features extracted from the speakers enrolled in the database. The success of speaker identification depends upon the feature extracted and its modeling method. The feature extracted during testing phase will match the feature extracted during enrollment phase perfectly, if the conditions under which it is tested are ideal. But under practical circumstances the testing conditions will include noise disturbances such as car noise, train noise, narrowband noise, etc. The car and train noises can be thought of disturbances that degrade the performance effectively depending upon the signal to noise ratio.

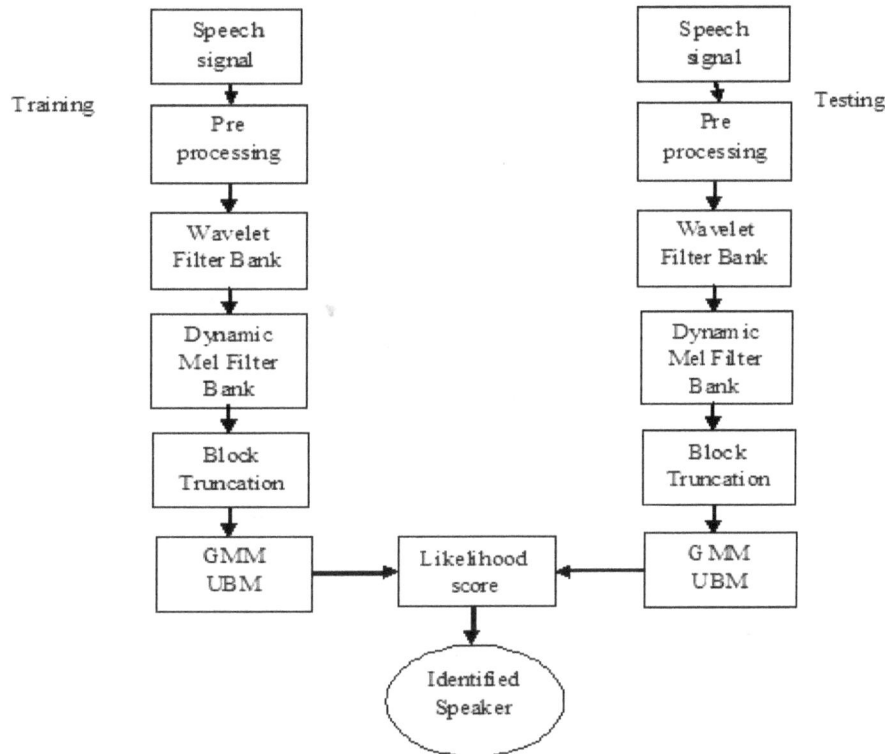

Figure 1. Overall block diagram of the proposed speaker identification system.

Moreover, the testing environments with these conditions are usually avoided in the practical speaker identification circumstances. But the narrowband noises are short time disturbances that are present in almost all the testing environments which lead to performance degradation of the speaker identification system. This problem can be overcome by a strategy called block based transformation, which is detailed in Sahidullah and Saha (2011).

The most common feature used for speaker identification is Mel Frequency Cepstral Coefficients (MFCC). As the human auditory system is most sensitive to the pitch frequency of the speaker, a feature that is based on the pitch frequency will model the speaker much more efficiently than the features that are not based on the pitch frequency of the speaker. In this work, Dynamic Mel Frequency Cepstral Coefficients (DMFCC) are used as features, which are formed by imparting the pitch frequency into the MFCC thereby producing dynamic features. The feature extracted is then used to model the speaker by using Universal Background Models (UBM). But the problem in modeling is that whether the model adapts itself for all the speakers enrolled in the database or not. This is called the bias/variance dilemma problem (Utpal and Kshirod, 2012).

The Discrete Cosine Transform (DCT) is popularly used for MFCC and DMFCC computation, because the correlation matrix of Mel Frequency Log Energy data is similar to the correlation matrix of first order Markov process and DCT provides better energy compaction (Kekre and Vaishali, 2011) than any other linear transform. In this work, DCT is carried out in blocks to mitigate the effects of narrowband noise since the effect of a noise corrupted Mel Filter Log Energies (MFLE) will not pronounce in the MFCC obtained through other DCT blocks. Therefore by combining the narrowband noise overcoming strategies (Qi and Yan, 2011) of Block Truncated DCT and sub band processing, together with the added advantage of eliminating the bias / variance dilemma by sub band concept, enhanced speaker identification can be obtained. TIMIT database has been used in this work. The speeches in TIMIT database (John et al., 2013) was recorded at TI, transcribed at MIT and produced by the National Institute of Standards and Technology (NIST). The TIMIT database consists of 630 speakers.

This paper focuses on segregating the noise affected portions of speech, by making blocks of speech frames to confine the noise spread using wavelet transform, which reduces the effects of noise degradation when higher levels of decomposition is performed.

PROPOSED SYSTEM

The overall block diagram representing the speaker identification system is shown in Figure 1. The speech signal is first pre-processed and the energy spectrum of the speech wave is

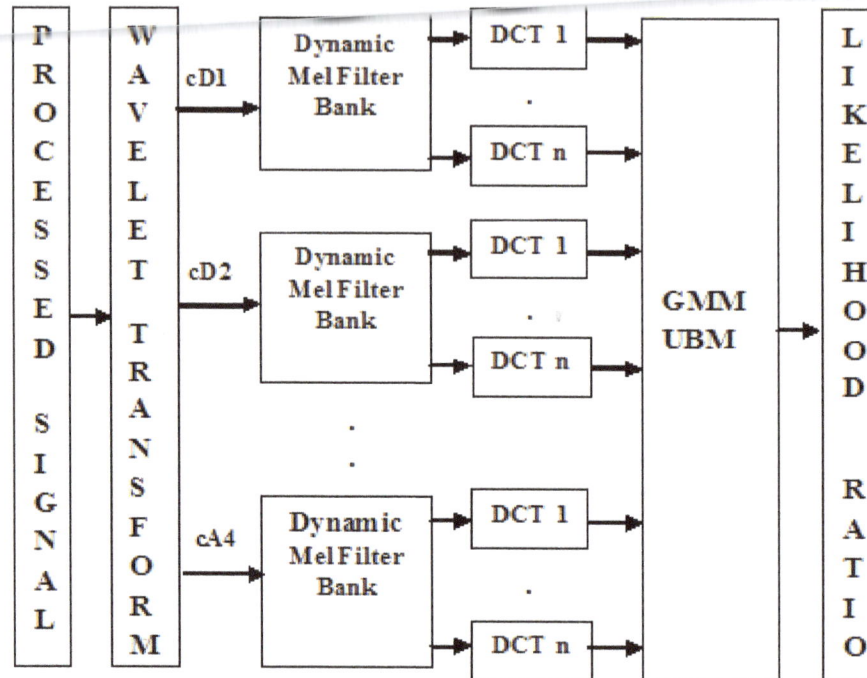

Figure 2. Block diagram of wavelet processed DMFCC.

computed using multilevel wavelet decomposition. The wavelet transformation divides the spectrum into appropriate number of bands corresponding to the number of levels of decomposition performed in the transformation. The wavelet filter outputs are robust to noise degradation due to the smoothening effect of the low pas filters while performing the discrete wavelet transformation on the speech signal. Moreover, the wavelet filter bank acts as dividing the entire bandwidth of the speech signal into bands corresponding to the number of levels used in the wavelet transformation. Moreover, the noise affected band is confined to one or few bands rather than being available at the whole spectrum. Hence it is easy to process the noise affected sub bands uniquely using Block truncation strategy. The methodology of the proposed work is enhancing the speaker identification under narrowband noise condition which makes the speech signal at 10 dB signal to noise ratio (SNR), by the features extracted through wavelet processing. Wavelet filters provide smoothening of the noise affected speech signal which provides reduction of the degradation caused by noise when higher levels of decomposition is performed on wavelet transformation. Further, the block truncation of the cepstral coefficients helps in restricting the effect of narrowband noise in affecting all the energy coefficients. Hence when the features are modeled using GMM-UBM, improvement in speaker identification accuracy can be achieved. The central schematic of this work lies in the sub band and block truncation techniques which is depicted in Figure 2. Each of these bands of wavelet coefficients is then provided as inputs to individual Dynamic Mel Filter Bank to obtain the Wavelet Dynamic Mel Frequency Cepstral Coefficients (WDMFCC). These features are more robust because the noise affected speech signal is smoothened by the low pass filters of the wavelet transformation process.

The features are then extracted by decorrelating the Dynamic Mel Filter Log Energies using block truncated DCT. Here, instead of taking DCT for the entire log energies of the sub band, DCT is performed in a block truncated manner because when DCT is performed on individual blocks narrowband noise in a block will not

spread to the other blocks in the sub band. The features are then used to model the speaker by means of Gaussian Mixture Models-Universal Background Model (GMM-UBM).

Speaker identification steps

Speaker identification process is a kind of pattern classification. In pattern classification problem, the first step is evaluating representation of input pattern. In speaker identification, this step is evaluation of power spectrum. These acoustic representations are extracted within successive analysis windows of 20-30 ms overlapped by 10 ms size. As vocal tract is a slowly varying system, speech signal is nearly stationary over this analysis window. Other pre processing stages are briefly outlined here for the sake of completeness.

Pre-emphasis

Pre-emphasis (Tomi and Haizhou, 2010) is performed to boost the higher frequencies of the signal. It is performed with a pre-emphasis factor of 0.97 according to the equation given by:

$$y(n) = x(n) - \alpha x(n-1) \qquad (1)$$

Pre-emphasis offsets the negative spectral slope of 20 dB per decade that is naturally present in the speech signals.

Windowing

The pre-emphasized signal is then segmented into smaller frames for the stationary property to be satisfied in taking DFT. Hamming window (Rabiner and Biing-Hwang, 2007) is used in this work. It is given by:

$$w(n) = 0.54 - 0.46\cos(\frac{2\pi n}{N-1}) \quad ,0 \le n \le N-1 \tag{2}$$

The windowed signal is given by:

$$S_w(n) = y(n) * w(n) \tag{3}$$

where $y(n)$ is the pre-emphasized signal and $w(n)$ is the window used.

Energy spectrum

The energy spectrum of the windowed frames is computed by taking wavelet transform. Wavelet transforms have advantages over traditional Fourier transforms for representing functions that have discontinuities and sharp peaks, and for perfect deconstructing and reconstructing finite, non-periodic and/or non-stationary signals. The transformation is given by:

$$X(k) = \left| DWT(S_w(n)) \right|^2 \tag{4}$$

Wavelet bands

Successful speaker identification is critically dependent on obtaining good speaker models from training data. Data modeling is subjected to the bias/variance dilemma (Vibha and Jyoti, 2011). According to this, models with too many adjustable parameters will tend to overfit the data, exhibiting high variance and hence the model will generalise poorly. On the other hand, few parameters will make the model biased. Wavelet processing (Pawar and Badave, 2011) helps to solve this problem by dividing the entire energy spectrum of the speech signal into bands corresponding to the number of levels performed in the wavelet transformation. Also, the low pass filters of the wavelet transformation acts as smoothening filters thereby reducing the degradation of signal by noise. The wavelet energy coefficients are then fed to the Mel scale filters for extracting speaker specific features.

DMFCC feature extraction

Features are the representatives of the speech signal in speaker identification task. Feature extraction is the estimation of variables, called a feature vector, from another set of variables called speech samples. The feature extraction will transform the speech signal into feature vectors which present the specific properties of each speaker. Raw speech signals cannot be used as such for speaker identification because of two reasons: (i) direct comparison and identification are complex and unreliable and (ii) requires large storage capability.

The human auditory system can sensitively perceive the changes in pitch. The pitch frequency is calculated by taking the autocorrelation of the signal and then taking maximum value for the autocorrelation function. Therefore by incorporating the pitch information into the MFCC feature, dynamic mel frequency cepstral coefficients can be extracted, which proves to provide strong robustness to background noise compared other features (Wang, et al., 2009) thus increasing the identification rate.

$$Mel(f_{i_p}) = 2595\log\left(1 + f_{i_p}/700\right) \tag{5}$$

where f_{i_p} is the pitch frequency of ith frame. The Mel frequency

energy spectrum is then passed through the Gaussian Mel filter bank (Sandipan and Goutam, 2009) followed by cosine transformation to obtain DMFCC features.

Block truncation transformation

Discrete cosine transform is performed on Dynamic Mel Filter Log Energies (DMFLE) to decorrelate them and so to make the extracted feature suitable for modeling. When such DCT is applied to a narrowband noise affected speech signal's log energies, all the features will be affected by the noise and hence will make it unsuitable for speaker identification. To alleviate this problem, block based transformation is performed. The filter log energies are divided into blocks and DCT is performed on them. This will ensure that a narrowband noise affected block will not pronounce its effect in the feature extracted from the other DMFLE blocks of the speech signal and hence will facilitate correct identification of the speaker. Narrow band noise (Ming et al., 2007) is synthetically generated by adding four frequency components (that is, sinusoidal tones) of 2000, 2100, 2200 and 2300Hz. The amplitudes of the sinusoids are chosen randomly.

The filter bank log energies are decomposed into several blocks unlike standard full band based DCT technique. In this work, the whole signal is divided into non-overlapping blocks (Jingdong et al., 2000) and individual blocks are processed independently. Therefore, the presence of narrowband noise in one block will not affect the other blocks because of truncation. The transformation matrix can be given as:

$$L = \Phi_1 \oplus \Phi_2 \oplus \cdots \oplus \Phi_N = \begin{bmatrix} \Phi_1 & 0 & 0 & \cdot\cdot & 0 \\ 0 & \Phi_2 & 0 & \cdot\cdot & 0 \\ \cdot & & \cdot & \cdot\cdot\cdot \\ \cdot & & \cdot & \cdot\cdot\cdot \\ 0 & 0 & 0 & \cdot\cdot & \Phi_N \end{bmatrix} \tag{6}$$

Where $\Phi_1, \Phi_2, \Phi_3 .... \Phi_N$ are orthogonal discrete cosine transformation matrices applied to individual blocks of Dynamic Mel Filter Log Energies (DMFLE).

Suppose two blocks of same sizes q are considered then the DCT matrix of size $q \times (q-1)$ is given by

$$\phi_1 = \sqrt{\frac{2}{q}} \cos\left[\frac{\pi i(2j+1)}{2q}\right] \quad \text{and} \quad \phi_2 = \sqrt{\frac{2}{p-q}} \cos\left[\frac{\pi(i-q)(2j+1)}{2(p-q)}\right].$$

GMM-UBM

A Universal Background Model (UBM) is a model used in a biometric identification system to represent general, person independent feature characteristics to be compared against a model of person-specific feature characteristics. The likelihood ratio statistic is given by:

$$LR(X) = \frac{p(X/\lambda_p)}{p(X/\lambda_{\bar{p}})} \tag{7}$$

where $p(X/\lambda_p)$ is the probability that the feature models the speaker correctly, $p(X/\lambda_{\bar{p}})$ is the probability that the feature belongs to the alternate hypothesis.

$\lambda = \{w_i, \mu_i, \sigma_i\}$, i=1,2,...M

M is the number of Gaussian components, w_i is the mixture weights, μ_i is the means and σ_i is the variance.

The alternate hypothesis that gives the probability of speaker belonging to the false category is modeled by means of UBM. It is a speaker independent one time trained model. Since UBM is a large GMM (Reynolds, 1995; Reynolds and Rose, 1995) trained to represent the speaker independent features, its idea is to capture the general characteristics of a population and then adapting it to the individual speaker by means of EM algorithm. With training vectors from the hypothesized speaker, $X=\{x_1, x_2...x_T\}$ and for mixture 'i', the probability distribution is given by:

$$\Pr(i/x_i) = \frac{w_i p_i(x_t)}{\displaystyle\sum_{j=1}^{M} w_j p_j(x_t)} \tag{8}$$

Then with the distribution known, the statistics for the weight, mean, and variance parameters are initialized as given as follows:

$$n_i = \sum_{i=1}^{T} \Pr(i/x_t)$$

$$E_i(x) = \frac{1}{n_i} \sum_{t=1}^{T} \Pr(i/x_t)x_t \tag{9}$$

$$E_i(x^2) = \frac{1}{n_i} \sum_{t=1}^{T} \Pr(i/x_t)x_t^2$$

The updated coefficients are given by:

$$w_i^{new} = [\alpha n_i / T + (1-\alpha)w_i]\gamma$$

$$\mu_i^{new} = \alpha E_i(x^2) + (1-\alpha)\mu_i \tag{10}$$

$$\sigma_i^{2new} = \alpha E_i(x^2) + (1-\alpha(\sigma_i^2 + \mu_i^2)) - \mu_i^{2new}$$

where $\alpha = \dfrac{n_i}{n_i + r}$ with relevance factor r = 16

RESULTS AND DISCUSSION

TIMIT database is used for the analysis in this speaker identification work. Each speaker record contains ten speech signals, from which six signals are used for training and the remaining four signals are used for testing. The database consists of clean speech recorded at 16 kHz sampling frequency. The maximum frequency content in the speech waveform is 8 kHz. The TIMIT database speech signals are subjected to Narrow band noise generated by adding four frequency components of 2000, 2100, 2200 and 2300 Hz. The narrow band noise affected speech signal is first pre-processed and the energy spectrum of the speech wave is computed using four level decomposed wavelet transformation. Thus

energy spectrum is obtained into bands corresponding to the number of decomposition levels. Now the noise affected band is confined to one or few bands rather than being available at the whole spectrum. Hence, it is easy to process the noise affected sub bands uniquely using Block truncation strategy. Since the band processing is done well ahead of decorrelarion step, the narrowband noise affected portion can be segregated from the rest of the bands that are not affected by the noise and better decorrelation is achieved at the Discrete Cosine Transform stage of Dynamic Mel Frequency Cepstral Coefficient feature extraction.

To improve the speaker identification efficiency, Dynamic Mel scale filter bank is constructed using Gaussian shaped filters in contrary to the triangular filters used in conventional systems. A triangular filter provides crisp partitions in an energy spectrum by providing non-zero weights to the portion covered by it while giving zero weight outside it. This phenomena cause loss of correlations between a sub band output and the adjacent spectral components that are present in the other sub band; whereas Gaussian shaped filters can provide much smoother transition from one sub band to other preserving most of the correlation between them. The DMFCC feature thus extracted is modeled using GMM-BUM modeling with 2048 mixture components.

During testing, the percentage of correct identification is calculated by using the formula:

$$Percentage\ of\ correct\ identification = \frac{No.\ of\ utterances\ correctly\ identified}{Total\ no.\ of\ utterances\ under\ test} * 100$$

The DMFCC features results have been observed for frame sizes of 1024, 512, and 256 without wavelet decomposition. The identification performance is tabulated in Table 1. The performance deteriorates in full band DCT since the effect of narrowband noise will be spread out to all the filter bank coefficients. But under the pro block based DCT system, the identification rate enhances to 89.65% for a frame size of 256. This proves the advantage of block based transformation under narrowband noise conditions. This accuracy can be further enhanced by the proposed wavelet based DMFCC feature. The WDMFCC features results have been observed for frame sizes of 1024, 512, and 256 with two level wavelet decomposition. The identification performance is tabulated in Table 2.

The percentage of correct identification improves to 94.13% for the frame size of 256. The wavelet filters provides coefficients that represent the smoothened version of the noise affected speech signal which when dynamic mel filtered followed by block truncation, reduces the noise degradation and provides improved identification percentage. Now the WDMFCC features results have been observed for frame sizes of 1024, 512, and 256 with three level wavelet decomposition. The identification performance is tabulated in Table 3. It is inferred from Table 3 that the performance of speaker

Table 1. Identification performance for DMFCC with and without block truncation

S/N	Total number of speakers	Frame size	GMM-UBM	
			Identification accuracy in presence of narrowband noise (%)	
			Without block truncation	With block truncation
1		1024	61.57	73.05
2	630	512	75.25	81.97
3		256	83.73	89.65

Table 2. Identification performance for WDMFCC with and without block truncation for two-level wavelet decomposition

S/N	Total number of speakers	Frame size	GMM-UBM	
			Identification accuracy in presence of narrowband noise (%)	
			Without block truncation	With block truncation
1		1024	71.13	82.88
2	630	512	80.50	87.25
3		256	87.21	94.13

Table 3. Identification performance for WDMFCC with and without block truncation for three level wavelet decomposition

S/N	Total number of speakers	Frame size	GMM-UBM	
			Identification accuracy in presence of narrowband noise (%)	
			Without block truncation	With block truncation
1		1024	72.75	83.5
2	630	512	81.13	88.38
3		256	88.25	96.37

identification system is improved to 96.37% corresponding to frame size of 256. The improvement corresponds to the higher level decomposition of the wavelet transformation which exhibit better noise smoothening compared to the results obtained through two level wavelet processed DMFCC results.

The identification performance for WDMFCC features with four level wavelet decomposition for frame sizes of 1024, 512, and 256 is tabulated in Table 4. From Table 4, it is observed that the identification accuracy is improved to 97.23% for the Dynamic Mel features obtained through four-level wavelet decomposition. The accuracy is enhanced by 7.58%. The results are compared with (Ramaligeswararao et al., 2011) the work on text-independent speaker identification model is developed by integrating MFCC's with Independent component analysis (ICA) for obtaining feature independency and to achieve low dimensionality in feature vector extraction. The work by Ramaligeswararao et al. (2011) evaluated the speaker identification performance for a database of 50 speakers under 0dB, 10dB and 20 dB SNR conditions and obtained a maximum identification performance of 72.34% for 10 dB SNR and 88.45% for 20 dB SNR. But our proposed work achieves 97.23% for 630 speakers even at 10 dB SNR.

Conclusion

In this paper, the speaker identification rate under

Table 4. Identification performance for WDMFCC with and without block truncation for four level wavelet decomposition

S/N	Total number of speakers	Frame size	GMM-UBM	
			Identification accuracy in presence of narrowband noise (%)	
			Without block truncation	With block truncation
1		1024	73.62	84.69
2	630	512	81.63	88.87
3		256	88.50	97.23

narrowband noise conditions is found to be enhanced by the features extracted through wavelet processing. Wavelet filters provide smoothening of the noise affected speech signal which provides reduction of the degradation caused by noise when higher levels of decomposition is performed on wavelet transformation. The block truncation of the cepstral coefficients helps in restricting the effect of narrowband noise in affecting all the energy coefficients. The identification performance stands at 97.23% for the four level wavelet decomposed WDMFCC for a frame size of 256. Further developments such as fusion of several other features with adaptive weights can improve the narrowband noise performance to significant levels of successful identification accuracies.

REFERENCES

Kekre HB, Vaishali K (2011). Speaker Identification using Row Mean of DCT and Walsh Hadamard Transform. Int. J. Comput. Sci. Eng. 3(3):1295–1301.

Jingdong C, Paliwal K, Nakamura S (2000). A block cosine transform and its application in speech identification. In: Proc Int. Conf. Spoken Language Processing (INTERSPEECH 2000 – ICSLP) IV:117–120.

John HL, Hansen L, Jun-Won S, Matthew RL (2013). In-set/out-of-set speaker recognition in sustained acoustic scenarios using sparse data. Speech Commun. 55(6):769–781.

Ming J, Timothy JH, James RG, Senior M, Douglas AR, Senior M (2007). Robust speaker identification in noisy conditions. IEEE Trans. Audio Speech Lang. Process. 15(5):1711–1723.

Ramaligeswararao NM, Sailaja V, Srinivasa R (2011). Text Independent Speaker Identification using Integrated Independent Component Analysis with Generalized Gaussian Mixture Model. Int. J. Adv. Comput. Sci. Appl. 2:12.

Pawar MD, Badave SM (2011). Speaker Identification System Using Wavelet Transformation on Neural Network. Int. J. Comput. Appl. Eng. Sci. I Special Issue on Cns ,July 2011.

Rabiner LR, Biing-Hwang J (2007). Fundamentals of speech Identification. Pearson Education Book.

Reynolds D, Rose R (1995). Robust text-independent speaker Identification using Gaussian mixture speaker models." Speech Audio Process. IEEE Trans. 3(1):72–83.

Reynolds DA (1995). Speaker identification and verification using Gaussian mixture speaker models. Speech Communication.

Sahidullah Md, Saha G (2011). Design, analysis and experimental evaluation of block based transformation in MFCC computation for speaker identification, IEEE.

Sandipan C, Goutam S (2009). Improved Text-Independent Speaker Identification using Fused MFCC & IMFCC Feature Sets based on Gaussian Filter. Int. J. Info. Commun. Eng. 5(1).

Tomi K, Haizhou L (2010). An overview of text-independent speaker recognition: From features to supervectors. Speech Commun. 52:12–40.

Wang Y, Li B, Jiang X, Liu F, Wang L (2009). IEEE April 2009 Speaker Identification based on Dynamic MFCC parameters, pp. 406-409.

Qi L, Yan H (2011). An Auditory-Based Feature Extraction Algorithm for Robust Speaker Identification Under Mismatched Conditions. IEEE Trans. Audio, Speech Lang. Process. 19(6).

Utpal B, Kshirod S (2012). GMM-UBM Based Speaker Verification in Multilingual Environments. Int. J. Comput. Sci. 9(6):2.

Vibha T, Jyoti S (2011). Wavelet Based Noise Robust Features for Speaker Recognition. Sig. Process. An Int. J. (SPIJ) 5(2).

ZoranCirovi C, Milan M, Zoran B (2010). Multimodal Speaker Verification Based on Electroglottograph Signal and Glottal Activity Detection. EURASIP J. Adv. Sig. Process. Article ID 930376, p. 8.

6

Almost quaternionic integral submanifolds and totally umbilic integral submanifolds

Fatma Özdemir

Department of Mathematics, Faculty of Science and Letters, Istanbul Technical University, 34469 Maslak-Istanbul, Turkey.

In this work, we consider a Riemannian manifold which admits an integrable distribution P with an almost quaternionic structure V. We show that the torsion tensor of an almost quaternionic structure V is independent of the choice of bases. We prove that the integral submanifolds are totally umbilic under the condition of the conjugate shape operator $\widetilde{CL}_z$ is skew-adjoint. We give another condition for the shape operator $\widetilde{L}_z$ of P to be quaternionic linear.

Key words: Subbundle, almost complex structure, almost quaternionic structure, shape operator.

MSC 2010. 53C55, 53C15.

INTRODUCTION

In literature (Kobayashi and Nomizu, 1963, 1969; Yano and Ako, 1972; Ishihara, 1974; Özdemir, 2006; Alagöz et al., 2012), almost complex and almost quaternionic structures have been investigated widely. These structures are special structures on the tangent bundle of a manifold. A detailed review can be found in Kirichenko and Arseneva (1997). Let us recall some basic facts and definitions from literature. In this work, by a manifold, we mean a smooth manifold, that is, a Hausdorff space with a fixed complete atlas compatible with the pseudo group of transformations of class C^∞ of R^n. Also, vector fields and plane fields are all supposed to be smooth. We denote by $\chi(M)$ the algebra of vector fields on a manifold M.

An almost complex structure on a manifold M is a tensor field $J: \ TM \to TM$ satisfying the relation $J^2 = -I$, where I denotes the identity transformation of $T_x(M)$. An almost hypercomplex structure on a $4m$-dimensional manifold M is a bundle $S = (F, G, H)$ of almost complex structures F, G and H satisfying the conditions

$$F^2 = G^2 = H^2 = -I, \quad H = FG, \quad FG + GF = FH + HF = GH + HG = 0. \tag{1}$$

We now consider a $4m$-dimensional Riemannian manifold M admitting a three dimensional subbundle V

of the bundle of $(1,1)$ tensors such that on a neighborhood U of each $x \in M$, V has a local base $\{F, G, H\}$. If on each such neighborhood the tensors F, G and H satisfy the conditions 1, then the bundle V is called an almost quaternionic structure on M (Ishihara, 1974). The Nijenhuis bracket or torsion tensor of tensor fields A and B of type $(1,1)$ is a tensor field of type $(1,2)$ and defined by:

$$
\begin{aligned}
[A,B](X,Y) &= [AX,BY] - A[BX,Y] - B[X,AY] \\
&+ [BX,AY] - B[AX,Y] - A[X,BY] + (AB+BA)[X,Y].
\end{aligned} \tag{2}
$$

The famous Newlander-Nirenberg Theorem states that an almost complex structure is a complex structure if and only if it is integrable; that is, it has no torsion (Yano and Ako, 1972). Thus, if the tensor fields F, G and H are integrable, then the Nijhenuis bracket of any two of them vanish (Yano and Ako, 1972); that is,

$$
[F,F] = [G,G] = [H,H] = 0 \quad \text{and}
$$
$$
[F,G] = [H,F] = [G,H] = 0. \tag{3}
$$

If any two of six Nijenhuis tensors are zero, then the others vanish too. It is shown that if any of Nijhenuis tensors vanish, then there exits a torsion-free connection ∇ such that F, G and H are covariantly constant with respect to the connection, that is, $\nabla F = \nabla G = \nabla H = 0$, which means that V is a trivial bundle (Yano and Ako, 1972).

In this work, we consider an almost quaternionic manifold M having an integrable distribution P with an almost quaternionic structure V, and we prove that the torsion tensor of an almost quaternionic structure V is well-defined. By defining the shape operator $\widetilde{L}_Z$ of P, we state and prove that the conjugate shape operators of the tensors F, G, and H are skew-adjoint if and only if they are quaternionic linear. Furthermore, we prove that if the integral submanifolds are totally umbilic then conjugate shape operators are skew-adjoint.

ALMOST QUATERNIONIC SUBSTRUCTURES

We quote the definition of the torsion of bundle of V from (Özdemir, 2006) as:

$$
[V,V] = [F,F] + [G,G] + [H,H] \tag{4}
$$

where $[\,,\,]$ denotes the Nijenhuis bracket.

We now state the following theorem about the torsion tensor of V.

Theorem 1

The torsion tensor of the bundle V is well defined. That is, $[V,V]$ is independent of the choice of bases.

Proof

Let $\{F', G', H'\}$ and $\{F, G, H\}$ be local bases for V defined on neighborhoods U and U', respectively, and assume that $U \cap U' \neq 0$. Since U and U' are not disjoint, then in $U \cap U'$ we have

$$
F' = a_{11}F + a_{12}G + a_{13}H, \tag{5}
$$
$$
G' = a_{21}F + a_{22}G + a_{23}H, \tag{6}
$$
$$
H' = a_{31}F + a_{32}G + a_{33}H, \tag{7}
$$

where a_{ij} are functions defined on $U \cap U'$, and at any point $x \in U \cap U'$, $a_{ij} \in SO(3)$ (Yano and Ako, 1972; Ishihara, 1974).

By using the definition of torsion for primed coordinates, Equation 4 can be written as:

$$
[V,V] = [F',F'] + [G',G'] + [H',H']. \tag{8}
$$

We now compute torsion tensor term by term as follows:

$$
\begin{aligned}
[F',F'](X,Y) &= a_{11}^2[F,F](X,Y) + a_{11}a_{12}[F,G](X,Y) + a_{11}a_{13}[F,H](X,Y) \\
&+ a_{11}a_{12}[G,F](X,Y) + a_{12}^2[G,G](X,Y) + a_{12}a_{13}[G,H](X,Y) \\
&+ a_{11}a_{13}[H,F](X,Y) + a_{12}a_{13}[H,G](X,Y) + a_{13}^2[H,H](X,Y),
\end{aligned} \tag{9}
$$

$$
\begin{aligned}
[G',G'](X,Y) &= a_{21}^2[F,F](X,Y) + a_{21}a_{22}[F,G](X,Y) + a_{21}a_{23}[F,H](X,Y) \\
&+ a_{21}a_{22}[G,F](X,Y) + a_{22}^2[G,G](X,Y) + a_{22}a_{23}[G,H](X,Y) \\
&+ a_{21}a_{23}[H,F](X,Y) + a_{22}a_{23}[H,G](X,Y) + a_{23}^2[H,H](X,Y),
\end{aligned} \tag{10}
$$

and

$$
\begin{aligned}
[H',H'](X,Y) &= a_{31}^2[F,F](X,Y) + a_{31}a_{32}[F,G](X,Y) + a_{31}a_{33}[F,H](X,Y) \\
&+ a_{31}a_{32}[G,F](X,Y) + a_{32}^2[G,G](X,Y) + a_{32}a_{33}[G,H](X,Y) \\
&+ a_{31}a_{33}[H,F](X,Y) + a_{32}a_{33}[H,G](X,Y) + a_{33}^2[H,H](X,Y).
\end{aligned} \tag{11}
$$

By taking into account coordinate transformations of the local bases we see that the torsions $[F,G]$, $[G,F]$, $[F,H]$, $[H,F]$, $[G,H]$, and $[H,G]$ are needed to be calculated. The torsion tensor $[F,G]$ for any vector fields (X,Y) on M is:

$$
\begin{aligned}
[F,G](X,Y) &= [FX,GY] + [GX,FY] + FG[X,Y] + GF[X,Y] \\
&- F[X,GY] - F[GX,Y] - G[X,FY] - G[FX,Y],
\end{aligned} \tag{12}
$$

and similarly, the tensors $[G,F]$, $[F,H]$, $[H,F]$, $[G,H]$ and $[H,G]$ are obtained as follows

$$
\begin{aligned}
[G,F](X,Y) &= [GX,FY]+[FX,GY]+GF[X,Y]+FG[X,Y] \\
&- G[X,FY]-G[FX,Y]-F[X,GY]-F[GX,Y],
\end{aligned} \tag{13}
$$

$$
\begin{aligned}
[F,H](X,Y) &= [FX,HY]+[HX,HY]+FH[X,Y]+HF[X,Y] \\
&- F[X,HY]-F[HX,Y]-H[X,FY]-H[FX,Y],
\end{aligned} \tag{14}
$$

$$
\begin{aligned}
[H,F](X,Y) &= [HX,FY]+[FX,HY]+HF[X,Y]+FH[X,Y] \\
&- H[X,FY]-H[FX,Y]-F[X,HY]-F[HX,Y],
\end{aligned} \tag{15}
$$

$$
\begin{aligned}
[G,H](X,Y) &= [GX,HY]+[HX,GY]+GH[X,Y]+HG[X,Y] \\
&- G[X,HY]-G[HX,Y]-H[X,GY]-H[GX,Y],
\end{aligned} \tag{16}
$$

$$
\begin{aligned}
[H,G](X,Y) &= [HX,GY]+[GX,HY]+HG[X,Y]+GH[X,Y] \\
&- H[X,GY]-H[GX,Y]-G[X,HY]-G[HX,Y].
\end{aligned} \tag{17}
$$

Then, substituting Equations 9-11 in Equation 8, we find

$$
\begin{aligned}
[F',F']+[G',G']+[H',H'] &= (a_{11}^2+a_{21}^2+a_{31}^2)[F,F] \\
&+ (a_{11}a_{12}+a_{21}a_{22}+a_{31}a_{32})[F,G]+(a_{11}a_{13}+a_{21}a_{23}+a_{31}a_{33})[F,H] \\
&+ (a_{11}a_{12}+a_{21}a_{22}+a_{31}a_{32})[G,F]+(a_{12}^2+a_{22}^2+a_{32}^2)[G,G] \\
&+ (a_{12}a_{13}+a_{22}a_{23}+a_{32}a_{33})[G,H]+(a_{11}a_{13}+a_{21}a_{23}+a_{31}a_{33})[H,F] \\
&+ (a_{22}a_{23}+a_{32}a_{33}+a_{12}a_{13})[H,G]+(a_{13}^2+a_{23}^2+a_{33}^2)[H,H].
\end{aligned} \tag{18}
$$

Since $a_{ij} \in SO(3)$, then $a_{11}^2+a_{21}^2+a_{31}^2 = 1$, $a_{12}^2+a_{22}^2+a_{32}^2 = 1$, $a_{13}^2+a_{23}^2+a_{33}^2 = 1$, and the other components are $a_{11}a_{12}+a_{12}a_{22}+a_{31}a_{32} = 0$, $a_{11}a_{13}+a_{21}a_{23}+a_{31}a_{33} = 0$, $a_{12}a_{13}+a_{22}a_{23}+a_{32}a_{33} = 0$.

So, we find

$$
[F',F']+[G',G']+[H',H'] = [F,F]+[G,G]+[H,H] \tag{19}
$$

which shows that V is independent of the choice of bases.

We now consider an almost quaternionic manifold M having an integrable distribution P with an almost quaternionic structure V and we let H denote the division algebra of quaternions by the isomorphism $R^{4m} \to H^m$:

$$
(x_1,\cdots,x_{4m}) \mapsto (q_1,\cdots q_m); \quad q_m = x_a+x_{m+a}i+x_{2m+a}j+x_{3m+a}k; \quad 1 \le a \le m.
$$

At $x \in M$, for $q = a1+bi+cj+dk \in H$, and

$X \in P(x)$, defining Xq as $Xq = aX+bF(X)+cG(X)+dH(X)$, we see that $P(x)$ can be made into a quaternionic right vector space. If $\{X_1,X_2,\cdots,X_m\}$ is a basis of $P(x)$ as a quaternionic vector space, then $\{X_1,\cdots,X_m,FX_1, \cdots,FX_m,GX_1,\cdots,GX_m,HX_1, \cdots,HX_m\}$ is a basis of $P(x)$ regarded as a real vector space. Thus, if m is the quaternionic dimension of $P(x)$, then the real dimension is $4m$, hence rank of V is a multiple of 4. This shows that V defines an almost quaternionic structure on the plane field P, and it is called an almost quaternionic substructure on M (Stong, 1977; Doğanaksoy, 1992; Doğanaksoy, 1993).

Let $\{F,G,H\}$ be a basis for V in some neighborhood U of $4m+n$ dimensional manifold M. It is proved by Stong (1977) that each of F,G and H has a constant rank on U, and from the conditions $F = GH$, $G = HF$, $H = FG$, it is observed that their ranks are all equal (Stong, 1977). Also, the rank of V is defined to be the rank of a basis element on some neighborhood U. By choosing $q = 1+F^2 = I-p$, where I denotes the identity operator, it is obtained that

$$
p+q=1, \quad p^2=p, \quad q^2=q \tag{20}
$$

and that

$$
\phi p = p\phi = \phi, \quad \phi q = q\phi = 0, \tag{21}
$$

for any cross-section ϕ of V. This shows that p and q are complementary projection operators. Then, there exist two distributions P and Q corresponding to p and q, respectively. If the rank of V is $4m$, then P is $4m$-dimensional and Q is n-dimensional (Doğanaksoy, 1992).

Let g' be a Riemannian metric of M. Define g to be the tensor field of degree 2 on M by

$$
g(X,Y) = \begin{cases} 0, & X \in P(x), Y \in Q(x) \\ g'(X,Y), & X,Y \subset Q(x) \\ g'(X,Y)+g'(FX,FY) \\ \quad +g'(GX,GY)+g'(HX,HY), & X,Y \in P(x) \end{cases} \tag{22}
$$

where $\{F,G,H\}$ is a canonical local basis of V and $X,Y \in T_x(M)$. Since g' is a Riemannian metric, g satisfies all the conditions for a Riemannian metric (Doğanaksoy, 1992).

We now consider a Riemannian manifold admitting integrable distribution P and introduce the shape operator of P. We give a condition for the shape operator of P to be quaternionic linear.

TOTALLY UMBILIC INTEGRAL SUBMANIFOLDS

Let V be an almost quaternionic substructure of rank $4m$ on a Riemannian manifold M of dimension $4m+n$ with the metric defined in Equation 22. Let P and Q denote the orthogonal plane fields defined by V, and let p, q be projections $T_x(M) \to P$ and $T_x(M) \to Q$, respectively. Furthermore, let V be an almost quaternionic structure on P and $\{F, G, H\}$ be a local basis for V. We define the shape operator $\widetilde{L}_z$ of P by

$$\widetilde{L}_z(X) = -p(\nabla_X Z) \quad \text{for any vector fields } Z \in Q \text{ and}$$

$X \in P$, where ∇ is the covariant derivative operator of the Riemannian connection determined by the metric defined in Equation 22. On the other hand, it is shown that $\widetilde{L}_z(X)$ is self-adjoint in Özdemir et al. (2002). Here, we define quaternionic linearity of $\widetilde{L}_z(X)$ so that: $\widetilde{L}_z$ is quaternionic linear on P if and only if $\widetilde{L}_z F = F\widetilde{L}_z$, $\widetilde{L}_z G = G\widetilde{L}_z$, and $\widetilde{L}_z H = H\widetilde{L}_z$. Also, we define conjugate shape operators $C_1\widetilde{L}_z$, $C_2\widetilde{L}_z$, $C_3\widetilde{L}_z$, satisfying $C_1\widetilde{L}_z = F\widetilde{L}_z$, $C_2\widetilde{L}_z = G\widetilde{L}_z$ and $C_3\widetilde{L}_z = H\widetilde{L}_z$.

Theorem 2

Let P be an integrable distribution with an almost quaternionic structure V. Conjugate shape operators $C_1\widetilde{L}_z$, $C_2\widetilde{L}_z$ and $C_3\widetilde{L}_z$ are skew-adjoint if and only if $F\widetilde{L}_z = \widetilde{L}_z F$, $G\widetilde{L}_z = \widetilde{L}_z G$, and $H\widetilde{L}_z = \widetilde{L}_z H$. That is, they are quaternionic linear.

Proof

Assume $C_1\widetilde{L}_z$, $C_2\widetilde{L}_z$ and $C_3\widetilde{L}_z$ are skew-adjoint. Using Equation 1, we see that F, G and H are skew-adjoint. As $\widetilde{L}_z$ is self-adjoint (Özdemir et al., 2002), we get:

$$(C_1\widetilde{L}_z)^* = (F\widetilde{L}_z)^* = \widetilde{L}_z^* F^* = -\widetilde{L}_z F.$$

Since $C_1\widetilde{L}_z$ is assumed to be skew-adjoint, we have

$$(C_1\widetilde{L}_z)^* = -C_1\widetilde{L}_z = -F\widetilde{L}_z,$$

which implies that $\widetilde{L}_z F = F\widetilde{L}_z$. Similarly, we see that $\widetilde{L}_z G = G\widetilde{L}_z$ and $\widetilde{L}_z H = H\widetilde{L}_z$. Conversely, by the assumption, if $F\widetilde{L}_z = \widetilde{L}_z F$ then we have

$$(C_1\widetilde{L}_z)^* = (F\widetilde{L}_z)^* = \widetilde{L}_z^*(F)^* = -\widetilde{L}_z F \tag{23}$$

$$= -F\widetilde{L}_z = -C_1\widetilde{L}_z. \tag{24}$$

In a similar way, since $G\widetilde{L}_z = \widetilde{L}_z G$, and $H\widetilde{L}_z = \widetilde{L}_z H$, we get $(C_2\widetilde{L}_z)^* = -(C_2\widetilde{L}_z)$ and $(C_3\widetilde{L}_z)^* = -(C_3\widetilde{L}_z)$. That is, $C_1\widetilde{L}_z$, $C_2\widetilde{L}_z$ and $C_3\widetilde{L}_z$ are skew-adjoint.

Theorem 3

Let P be an integrable distribution with an almost quaternionic structure V. If the integral submanifolds are totally umbilic, then the conjugate shape operators $C_1\widetilde{L}_z$, $C_2\widetilde{L}_z$ and $C_3\widetilde{L}_z$ are skew-adjoint.

Proof

Let $\widetilde{L}_z = \lambda I$ where $\lambda \in P$, that is, the integral submanifolds are totally umbilic. By Proposition 1, we have

$$F(\lambda I) = (\lambda I)F, \quad G(\lambda I) = (\lambda I)G, \quad \text{and} \quad H(\lambda I) = (\lambda I)H.$$

Then, we obtain

$$(C_1\widetilde{L}_z)^* = (F\widetilde{L}_z)^* = (F(\lambda I))^* = (\lambda I)(F)^* = -(\lambda I)F = -F(\lambda I) = -F(\widetilde{L}_z) \tag{25}$$

which shows that the conjugate shape operator $C_1\widetilde{L}_z$ is skew-symmetric. Similar results can be obtained for the tensors G and H.

Theorem 4

Let P be an integrable distribution. Integral submanifolds are totally umbilic if and only if $\widetilde{L}_z$ is self-adjoint, $\widetilde{L}_z$ is quaternionic linear, and every quaternionic subbundle of P is $\widetilde{L}_z$ invariant for every vector field $Z \in Q$, where Q denote the orthogonal plane fields defined by V.

Proof

($\Rightarrow$) Assume that integral submanifolds are totally umbilic, that is, $\widetilde{L}_z = \lambda I$, $\lambda \in P$. We have $\langle \widetilde{L}_z X, Y \rangle = \langle \lambda X, Y \rangle = \langle X, \lambda Y \rangle = \langle X, \widetilde{L}_z Y \rangle$, implying that $\widetilde{L}_z = \widetilde{L}_z^*$.

Since integral submanifolds are totally umbilic, that is, $\widetilde{L}_z = \lambda I$, then $(\lambda I)F = F(\lambda I)$, $(\lambda I)G = G(\lambda I)$ and $(\lambda I)H = H(\lambda I)$. This shows that $\widetilde{L}_z$ is quaternionic linear. We now take any quaternionic subbundle R of P, $(R \subset P)$. For any $X \in R$, we get $\widetilde{L}_z(X) = \lambda X$ which concludes that $\widetilde{L}_z(X)$ is in R.

($\Leftarrow$) Conversely, we assume that $\widetilde{L}_z = \widetilde{L}_z^*$, $\widetilde{L}_z F = F \widetilde{L}_z$, $\widetilde{L}_z G = G \widetilde{L}_z$, and $\widetilde{L}_z H = H \widetilde{L}_z$ for any quaternionic subbundle R of P, $X \in R$ implies that $\widetilde{L}_z(X) \in R$.

Assume for a moment that $\widetilde{L}_z$ is not totally umbilic, that is, $\widetilde{L}_z \neq \lambda I$. Since $\widetilde{L}_z$ is self-adjoint, it has eigenvalues. Let λ and μ be two real distinct eigenvalues corresponding eigenvectors X and Y such as $\widetilde{L}_z X = \lambda X$, $\widetilde{L}_z Y = \mu Y$. Let R be the quaternionic subbundle of P spanned by $X + Y$. We observe that $\widetilde{L}_z(X + Y) = (\lambda X + \mu Y) \in R$ if and only if $\lambda = \mu$. Hence, if $\widetilde{L}_z$ is self adjoint, quaternionic linear, and R is $\widetilde{L}_z$ invariant, then $\widetilde{L}_z$ has to be totally umbilic.

Conclusion

In this work, by considering Riemannian manifolds admitting integrable distribution P with an almost quaternionic structure V, we show that the torsion tensor of V is independent of choice of the bases. We introduce the shape operator $\widetilde{L}_z$ and conjugate shape operators $C\widetilde{L}_z$ of P and prove that the conjugate shape operators are quaterninonic linear if and only if the conjugate shape operators are skew-adjoint. Furthermore, we prove that integral submanifolds are totally umbilical if and only if the shape operator $\widetilde{L}_z$ is self-adjoint, quaternionic linear, and every quaternionic subbundle of P is $\widetilde{L}_z$ invariant for every vector field in the orthogonal plane fields defined by V under a suitable metric.

ACKNOWLEDGEMENTS

Author would like to thank the referees for their valuble suggestions.

REFERENCES

Alagöz Y, Oral KH, Yüce S (2012). Split Quaternion Matrices, Miskolc Mathematical Notes. 13:223-232.

Doğanaksoy A (1992). Almost Quaternionic Substructures. Turkish J. Math. 16:109-118.

Doğanaksoy A (1993). On Plane Fields With An Almost Complex Structure. Turk. J. Math. 17:11-17.

Ishihara S (1974). Quaternion Kahlerian Manifolds. J. Diff. Geom. 9:483-500.

Kirichenko VF, Arseneva OE (1997). Differential geometry of generalized almost quaternionic structures. I, dg-ga/9702013.

Kobayashi S, Nomizu K (1963, 1969). Foundations of Differential Geometry, A Wiley-Interscience Publication. 1:2.

Özdemir F, Doğanaksoy A, Crasmareanu M (2002). Almost Complex Integral Submanifolds and Totally Umbilic Integral Submanifolds, Memoirs of the scientific sections of the Romanian Academy. Ser. IV(23):45-49.

Özdemir F (2006). A Global Condition for the Triviality of an almost quaternionic structure on complex manifolds. Int. J. Pure Appl. Math. Sci. 3(1):1-9.

Stong RE (1977). The Rank of an f-Structure. Kodai Math. Sem. Rep. 29:207-209.

Yano K, Ako M (1972). Integrability conditions for almost quaternion structure. Hokaido Math J. 1:63-86.

7

Methodological approach of selecting a vibration indicator in monitoring bearings

O. Djebili[1] **, F. Bolaers**[1]**, A. Laggoun**[2] **and J. P. Dron**[1]

[1]Laboratoire Grespi/MAN UFR, Sciences Exactes and Naturelles Moulin de la Housse, BP 1039
51687 REIMS CEDEX 2, France.
[2]Department de physique, Faculté des Sciences UMBB Boumerdes, Algéria.

A rolling bearing is an important element in a rotating machine. Whatever the operating conditions, it is subject to fatigue which causes spalling. In aiming to obtain the most possible real fatigue curve, the vibration level is shown according to different statistical indicators such as the RMS (Root Mean Square), the kurtosis, the crest value, the crest factor and the peak ratio, then to choose the best of them that is able to show the evolution of the bearing degradation. In this work, through the experimental vibratory follow up of the thrust bearing spall using different statistical indicators, we present an optimization methodology in order to find a most significant indicator that is able to characterize the damage evolution.

Key words: Bearing, vibration, analysis, vibration indicator, spalling.

INTRODUCTION

Bearings are among the most precise components in mechanical assemblies and are manufactured to very tight tolerances. They are normally found in most rotational equipment. The condition and health of bearing play an important role in the functionality and performance of these equipments. The main defect that may know a rolling bearing during normal operation is spalling. Really, this defect begins with subsurface cracks initiating within the raceway materiel. During the last state of service life, cracks might initiate and propagate and eventually reach the surface of the race dislodging a piece of metal from the surface. This results in what is known as spall. The vibration analysis is one of the most used methods to follow a spalling bearing of a rotating machine (Nagi and Mark, 2004). To establish the law of the bearing damage, it uses an efficient and appropriate vibration indicator. The purpose of this work focuses on the choice of the vibration indicator based on criteria that reflect greater significance of a default vibration signal. This generates a curve representative of damage.

Most of the research in this field has been done mainly about fatigue failure progression in ball bearings (Michael et al., 2001; Youngsik and Richard, 2007), diagnostics and residual live prediction of bearings (Nagi and Mark, 2004; Tallian, 1992) but most work only identifies the presence of defect, or at best differentiate based on qualitative measures such as light, medium or heavy damage (Hoeprich, 1992).

In this research, we will try to give an approach for improving the vibration measurement by choosing a vibration indicator to better characterize the evolution of the spalling in the bearing. We chose to study bearings with artificial defects. It was too long to tire bearings in normal use. We therefore simulated a spalling defect of

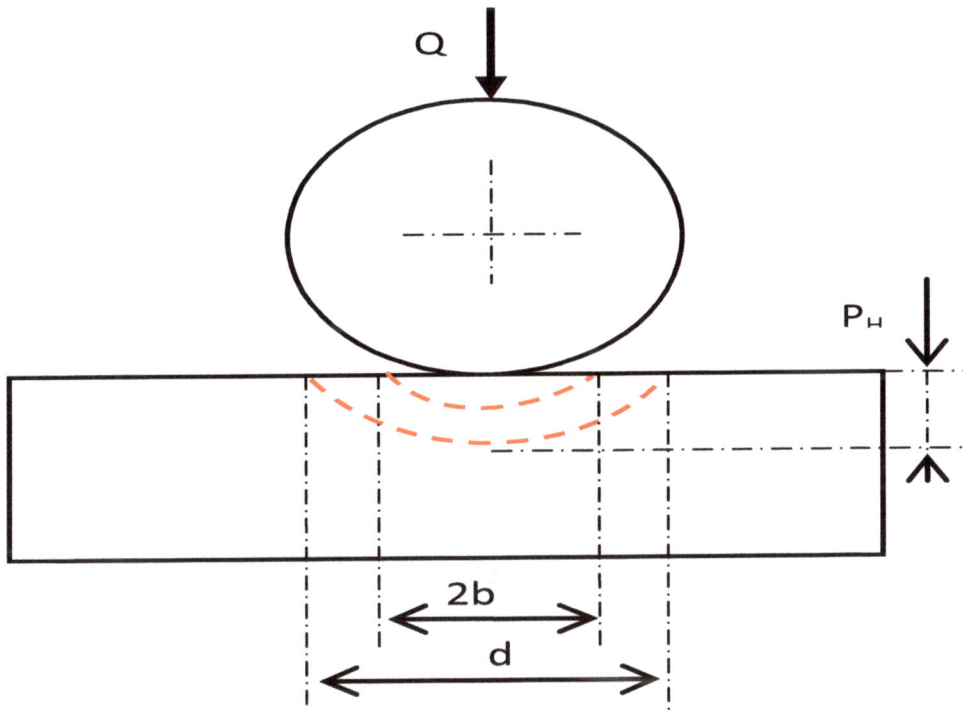

Figure 1. Contact sphere plan.

varying sizes on a series of thrust bearing. These artificial defects are created on the raceway of the thrust bearing ring by electro erosion.

PREPARATION OF EXPERIENCE AND SIMULATION OF FAILURE

We have chosen a thrust bearing type: FAG 51207 CZECH / ATK which has the following characteristics:

(i) Inner diameter D_{i1} = 35 mm,
(ii) Inner diameter D_{i2} = 37 mm,
(iii) Outer diameter D = 62 mm,
(iv) Number of balls N_b = 12.

Before running spalling imprints on the thrust bearing ring, we need to do a calculation for determining the tool dimensions to use in the electro erosion machine. We consider the contact sphere-plan. Knowing the axial load (Q) applied on the bearing and the diameter of the ball (D_b), we determined:

(i) The semi-axis of the contact surface (Michel and René, 2005).

$$b = \sqrt[3]{\frac{3\pi(k_1 + k_2)N}{2(C_1 + C_1' + C_2 + C_2')}} \qquad (1)$$

$$C_1 = \frac{1}{R_1}; C_1' = \frac{1}{R_1'}; C_2 = \frac{1}{R_2}; C_2' = \frac{1}{R_2'} \quad C_1 \text{ and } C_2: \text{ maximum}$$

curvatures, C_1' and C_2' minimum curvatures, R_1 and R_1': ball radius, R_2 and R_2' track radius of a thrust bearing ring, k_1 and k_2: stiffness of the material.

$$k_1 = k_2 = \frac{1 - v^2}{\pi E}$$

in our case, v = 0.3 : Poisson ratio, E = 210000 N/mm^2: elasticity module of the material, N = Q/12 : load on the ball (Figure 1).

(ii) Depth of Hertz (Michel and René, 2005; Daniel et al., 2002).

$$P_H = 0.5b \qquad (2)$$

(iii) Radius of the ball bearing:

$$R = \frac{P_{rH}^2 + \left(\frac{d}{2}\right)^2}{2P_{rH}} \qquad (3)$$

Imprint diameter

After a calculation of experience, we create spalling imprints by sinking using the electro erosion machine. The imprints are obtained according to tools (as ball) of different diameters previously calculated are shown in Figure 2. The hertz depth of the

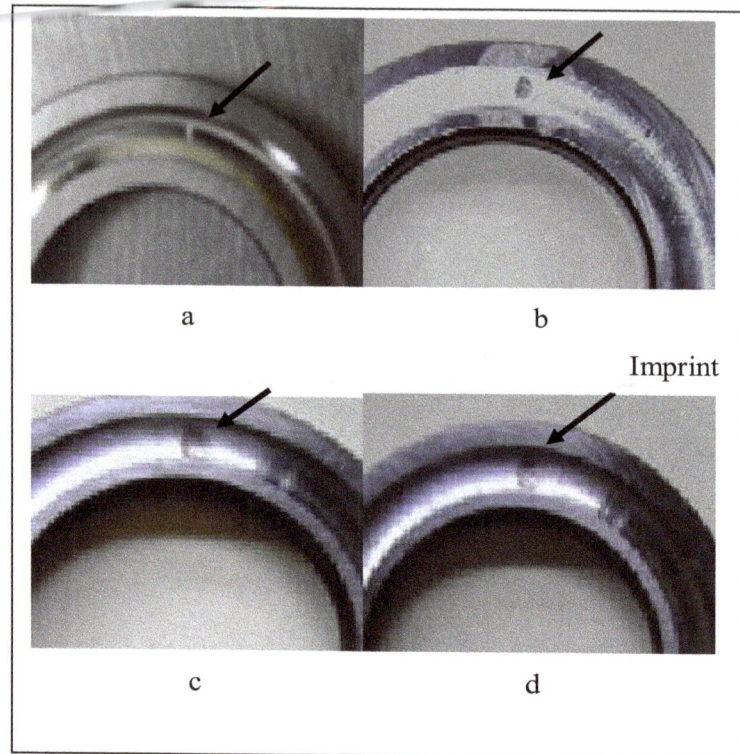

Figure 2. Spalling imprint on a thrust bearing ring (a) imprint Ø2.9; (b) imprint Ø3,9; (c) imprint Ø5,5; (d) imprint Ø6.4.

Table 1. Imprint depth.

Number of bearings	Ball diameter (mm)	Imprint depth (µm)
1	2.9	210
2	3.9	218
3	5.5	228
4	6.4	226

imprints obtained during the experience on four thrust bearings used is given in Table 1.

SPALLING VIBRATION MONITORING OF THE THRUST BEARING RING

The test bench consists of an electric motor, a cooling system, a mandrel to accommodate the thrust bearing and another to transmit the axial load from a hydraulic cylinder (Figure 3).

Vibratory statements of the thrust bearing

We note, at the trial bench, the vibration signals of simulated imprints (Youngsik and Richard, 2007) corresponding to different operating conditions (speed, load and frequency range). The tests are performed for various rotating speeds (600, 1200 and 1800 rpm), with two axial loads (2000 and 3000 daN). Measurements were taken at different frequencies range (1, 5 and 20 kHz).

Signals analysis

Vibration analysis starts with a time-varying, real-world signal from a transducer or sensor. From the input of this signal to a vibration measurement instrument, a variety of options are possible to analyze the signal. Let us take a look at the block diagram for a typical signal path in an instrument, as shown in Figure 4.

Vibration recordings were obtained on a test bench through a chain of acquisition as time signals. These time signals are then processed according to the diagram in Figure 4. To measure a spalling, we used statistical indicators that can be calculated through various stages of signal processing of Figures 5 and 6.

Figure 3. Image of the bearing test bench.

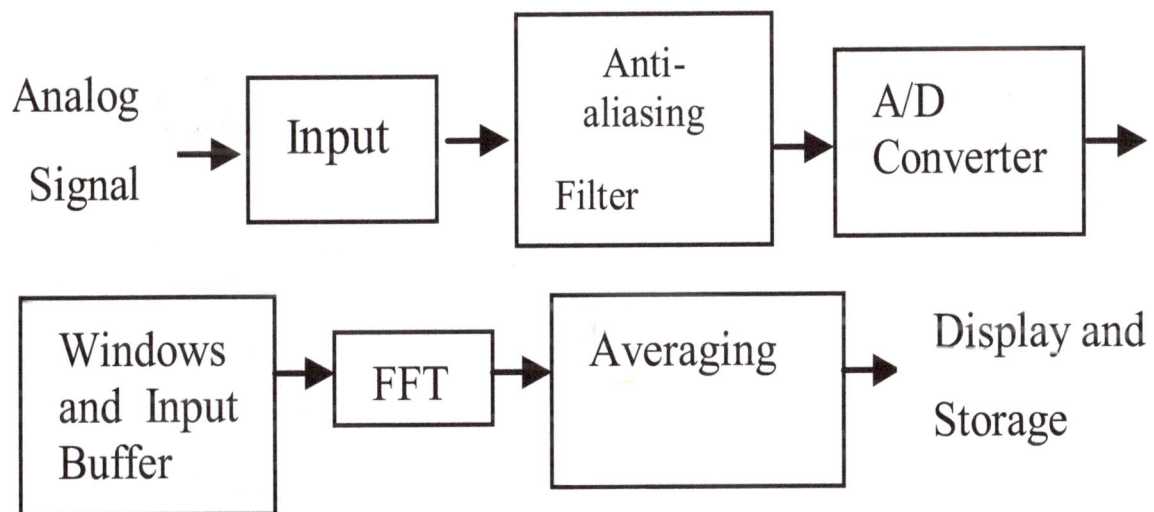

Figure 4. Typical signal path.

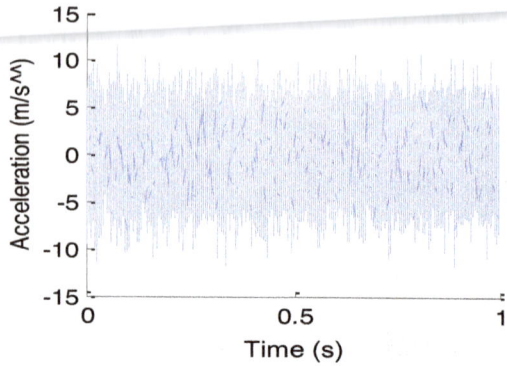

(a)

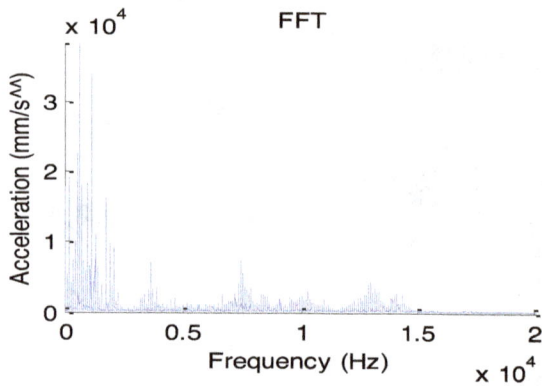

(b)

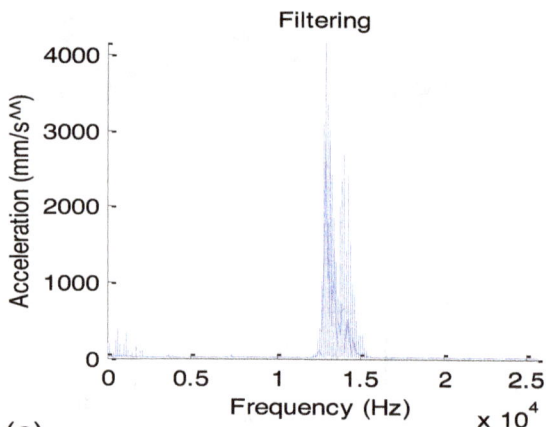

(c)

Figure 5. Signals analysis of undamaged bearing: (a) time domain signal, (b) frequency domain signal, (c) signal filtered at frequency band [13-15 kHz].

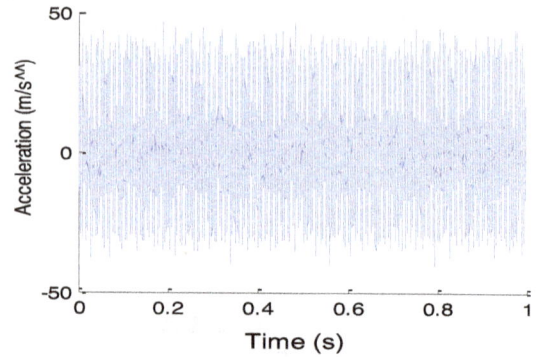

(a)

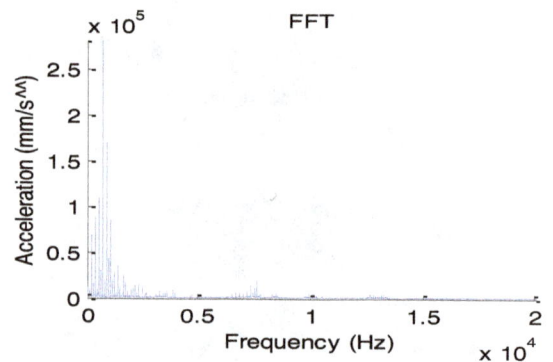

(b)

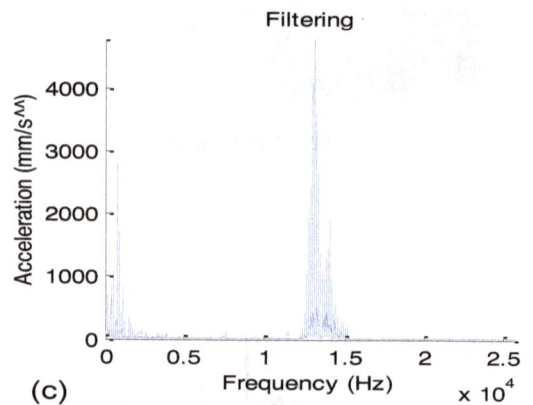

(c)

Figure 6. Signals analysis of failure detection: (a) time domain signal, (b) frequency domain signal, (c) signal filtered at frequency band [13-15 kHz].

Figure 6a shows a vibration signal at the time domain of damage which is the first to get but it is insufficient to quantify the severity of the defect. The time domain signal is converted to a vibration spectrum, which shows the signal in the frequency domain (Figure 6b). The conversion from time domain to frequency domain is done with a fast Fourier transformation (FFT). Then the vibration spectrum is filtered (Figure 6c) to make it more clear and precise. The spalling of the thrust bearing appears with a filtering in a high frequency band (Robert and Jerome, 2011).

Study of the spalling evolution

(1) Using classical vibratory indicators: We use a Matlab program

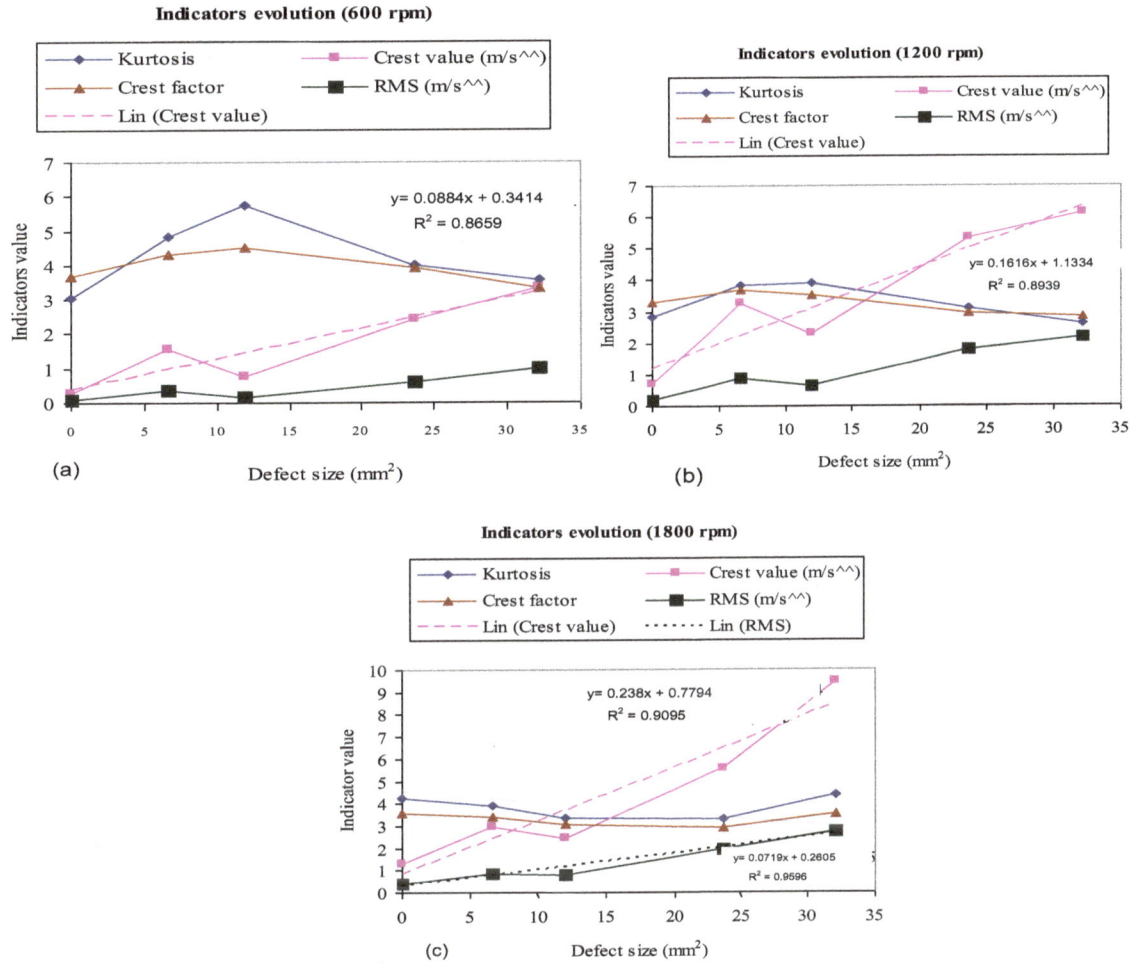

Figure 7. Evolution curve indicators at operating conditions: load 3000 daN. (a) speed 600 rpm; (b) speed 1200 rpm; (c) speed 1800 rpm.

to monitor the defect evolution with different rotation speeds and through several different vibratory indicators (Hoeprich, 1992) known as:

(i) The kurtosis: Is a statistical parameter to analyze the distribution of vibration amplitudes contained in a time signal. It corresponds to the moment of order four and has been shown that for a Gaussian distribution; its value is 3 ± 8%.

$$Kurtosis = \frac{M_4}{M_2^2} = \frac{\frac{1}{N}\sum_{n=1}(x(n) - \bar{x})^4}{\left[\frac{1}{N}(x(n) - \bar{x})^2\right]^2} \qquad (4)$$

where x(n) represents the amplitude of the signal for the sample, $\bar{x}$ the average amplitude, σ^2 variance (moment of order 2) and N the number of samples in the signal.

(ii) The crest value: is the maximum absolute value reached by the representative function of the signal during the time period.

(iii) The crest factor: Is the ratio of the crest value and the signal of root mean square (RMS) value (effective value).

$$Crest\ factor = \frac{Crest\ value}{RMS\ value} = \frac{\sup[x(n)]}{\sqrt{\frac{1}{N}\sum_{n=1}^{N}[x(n)]^2}} \qquad (5)$$

where N is number of samples taken from the signal × (n) the discrete time signal.

(iv) The RMS: The expression of the effective value is given by the following equation.

$$V_{effective} = V_{RMS} = \sqrt{\frac{1}{N}\sum_{n=1}^{N}[x(n)]^2} \qquad (6)$$

The evolution curves of the statistical indicators obtained at different speeds are shown in Figure 7. According to the indicator evolution curves obtained, we note that the crest value and RMS

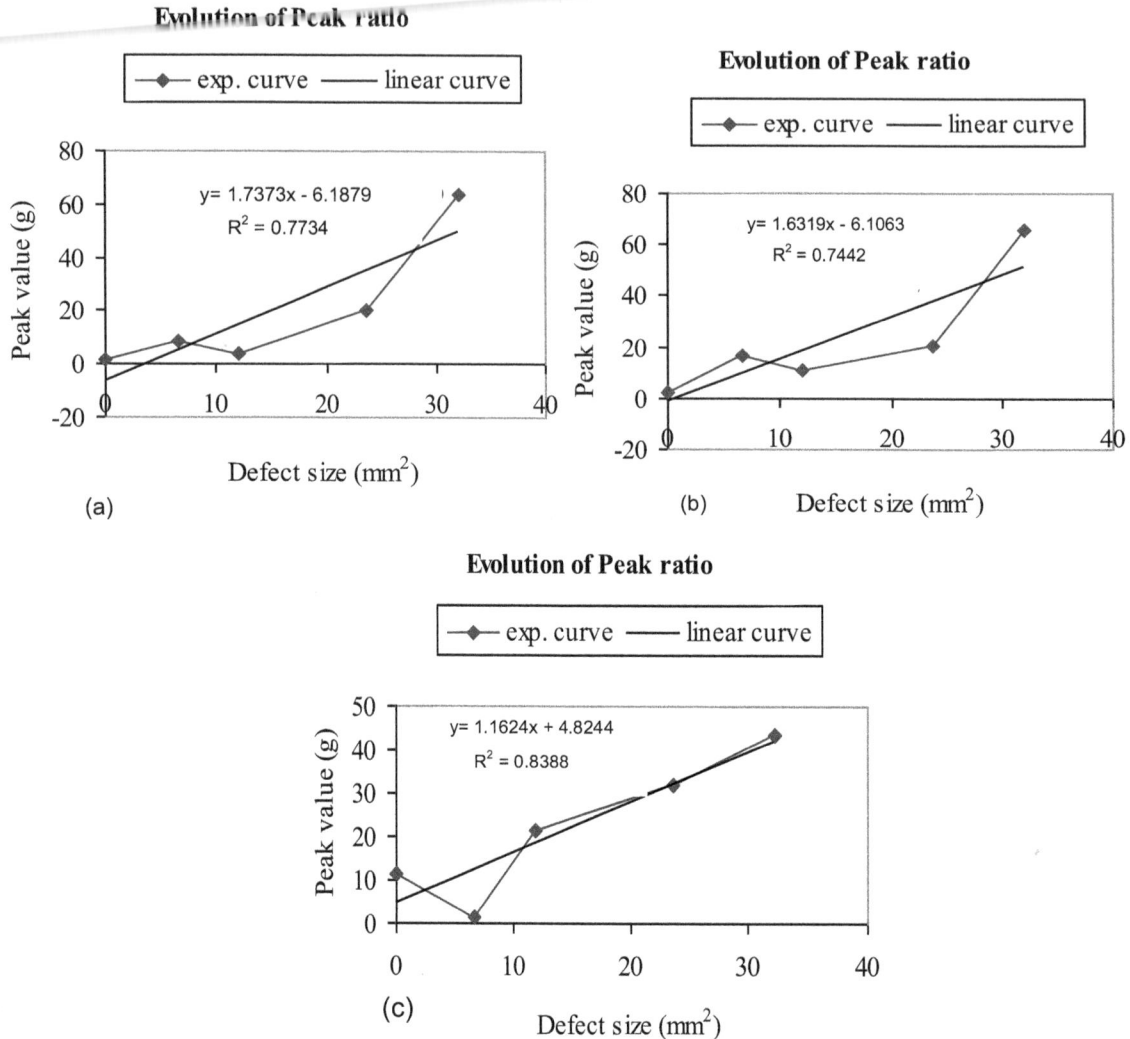

Figure 8. Evolution curve indicator at operating conditions load 3000 daN: (a) speed 600 rpm, (b) speed 1200 rpm, (c) speed 1800 rpm.

indicators are more significant in terms of sensitivity and linearity (see the director coefficient and the correlation coefficient of the linear equation).

(2) Using peak ratio indicator: We use another indicator to monitor the spalling bearing called peak ratio (peak value) (Shiroishi et al., 1997, 1999) of determining the first harmonic amplitude value of the defect frequency for each signal. When it verifies that a defect is present, the peak value is then used to estimate the size (magnitude) of the defect. To calculate the characteristic frequency of the thrust bearing defect, using the following formula (Harris, 1991).

$$F_d = \frac{N_b}{2}(RPS)(1 + \frac{d_b}{D_p}cos\beta)$$

(7)

N_b: Number of balls, RPS: shaft speed, db: ball diameter, Dp:

average diameter of the bearing, β: contact angle, for thrust bearing β=0.

A Matlab algorithm called chibora prepared by the LMA-Grespi laboratory is used to transform the vibration signal collected at the test bench to frequency spectrum. We show the evolution of the defect indicator at different speeds in Figure 8.

DISCUSSION

The diameters of the imprints examined, ranged from 2.9 to 6.4 mm and composed of the remarkable defects on the thrust bearing. This was found through the evolution curves of vibratory indicators such as crest value, RMS and peak value. We have proposed an approach based on two criteria to choose the optimum indicator: The sensitivity and linearity. The indicators sensitive to the default progression are the crest value, the peak value and the RMS that we can justify through linear equations as they were given in Figures 7 and 8 for the rotational speed of 1800 rpm (Li et al., 1999). This case was cited as an example because we find that sensitivity increases with the rotational speed. Otherwise, the RMS indicator moves with more linearity according the defect size by the correlation coefficient (R^2=0.956) in Figure 7c for the same rotational speed. Similarly, the linearity increase with the rotational speed as it is shown in Figure 7 for the crest value.

Conclusion

To make a judicious choice of the vibratory indicator, we must establish a compromise between sensitivity and linearity. Consequently, the evolution of the peak ratio (peak value) indicator is more significant in term of sensitivity. Otherwise the evolution of the RMS indicator is more interesting in term of linearity. Therefore, the evolution of the bearing damage could be relatively expressed by the peak ratio (peak value) indicator or crest value indicator.

REFERENCES

Daniel G, Fabrice V, Roger G, Gilles D (2002). Rolling Contact Fatigue Tests to Investigate Surface Initiated Damage Using Surface Dents. Bearing steel technology, ASTM STA. P. 1419.

Harris TA (1991). Rolling Bearing Analysis. John Wiley & Sons, Inc. New York. 3rd Edition.

Hoeprich MR (1992). Rolling Element Bearing Fatigue Damage Propagation. J. Tribol. P. 114.

Li Y, Billington S, Zhang C (1999). Adaptive prognostics for rolling element bearing condition. Mech. Syst. Signal Process. 13(1):103-113.

Michael N, Tedric A, Harris (2001). Fatigue Failure Progression in Ball Bearings. Transactions of the ASME. J. Tribol. 123(238).

Michel A, René B (2005). Systèmes Mechanical systems. Theory and Dimensioning, DUNOD. 661, ISBN-102100491040.

Nagi G, Mark L (2004). Residual Life Prediction From Vibration-Based Degradation Signals: A Neural Network Approach. IEEE Trans. Ind. Elect. 51(3).

Robert BR, Jerome A (2011). Rolling element bearing Diagnostics. Mech. Syst. Signal Process. pp. 485-520.

Shiroishi J, Li Y, Liang S, Kurfess T, Danyluk S (1997). Bearing Condition Diagnostics Via Vibration and Acoustic Emission Measurements. Mech. Syst. Signal Process. pp. 693-705.

Shiroishi J, Li Y, Liang S, Kurfess T, Danyluk S (1999). Vibration Analysis for Bearing Outer Race Condition Diagnostics. J. Braz. Soc. Mech. Sci. 21(3):484-492.

Tallian TE (1992). Simplified Contact Fatigue Life Prediction Model – Part 1: Review of Published Models. Trans. ASME. J. Tribol. 114:207-213.

Youngsik C, Richard L (2007). Spall progression live model for rolling contact verified by finish hard machined surfaces. Wear 262:24-35.

8

A method for placement of distributed generation (DG) units using particle swarm optimization

Noradin Ghadimi

Department of Electrical Engineering, Ardabil Branch, Islamic Azad University, Ardabil, Iran.

Nowadays, the penetration of distributed generation (DG) in power networks takes special place worldwide and is increasing in developed countries. In order to improve voltage profile, stability, reduction of power losses etc, it is necessary that, this increasing of installation of DGs in distribution system should be done systematically. This paper introduces an optimal placement method in order to sizing and sitting of DG in IEEE 33 bus test system. The algorithm for optimization is particle swarm optimization (PSO). The proposed objective function is the multi objective function (MOF) that considers active and reactive power losses of the system and the voltage profile in nominal load of system. High performance of the proposed algorithm is proved by applying algorithm in 33 bus IEEE system using MATLAB software and in order to illustrate the feasibility of the proposed method optimization in three cases: one DG unit, Two DG units, and Three DG units- will achieved.

Key words: Distributed generation (DG), placement, particle swarm optimization, multi objective function (MOF), optimization.

INTRODUCTION

The anguish about rising environmental population and also the concern about the fossil fuels problems and limitations led to the installation of Distributed Generation (DG) which increases annually. In order to improve voltage profile, stability, reduction of power losses and etc, it is necessary that this increasing of installation of DGs in Distribution system should be systematically (Hedayati et al., 2008). The best choosing size and site of DGs in a distribution system is a complex optimization problem and if this problem contain the Multi Objective Function (MOF), this problem become much complex. Nowadays, meta heuristics optimization methods are being successfully applied to combinatorial optimization problems in distribution systems (Carmen and Djalma, 2006; Thong et al., 2007).

Gandomkar et al. (2005) determined the optimum location of the DG in the distribution network. The work was directed towards studying several factors related to the network and the DG itself such as the overall system efficiency, the system reliability, the voltage profile, the load variation, network losses, and the DG loss adjustment factors.

A Tabu search (TS) search method to find the optimal solution of their problem was explained by Katsigiannis and Georgilakis (2008), but the TS is known to be time consuming algorithm also it is may be trapped in a local minimum. In order to minimize the real power losses of power system in Lalitha et al. (2010), a Particle Swarm Optimization (PSO) algorithm was developed to specify the optimum size and location of a single DG unit. The problem was converted to an optimization program and the real power loss of the system was the only aspect considered in this study in order to determine optimally the location and size of only one DG unit.

El-Khattam et al. (2005), a deferent scenario was investigated to determine the optimum location of DG in order to modify the voltage profile and minimize the investment risk. The placement of one DG unit with specific size was explained by Ochoa et al. (2006). In this paper, MOF such as power line losses, modification of voltage profile, line loading capacity, and short circuit level were considered. P-V curves in Singh and Goswami (2010) have been used for analyzing voltage stability in electric power system to determine the optimum size and location of multiple DG units to minimize the system losses under limits of the voltage at each node of the system.

A genetic algorithm (GA) based fuzzy multi-objective approach for determining the optimum values of fixed and switched shunt capacitors was used to improve the voltage profile and maximize the net savings is proposed in Das (2008).

Particle Swarm Optimization (PSO) is used in this paper in order to find solution to optimization problems (Hashemi et al., 2011), optimal size and site of DG in 33-bus radial system of IEEE test system (Kashem et al., 2000). The aim of this paper is to proffer solution to sitting and sizing problem for optimization of MOF. Objective function of this paper is formed by combining on real power losses, reactive power reduction, voltage profile improving, and short circuit level improving of the mention system.

Problem formulation containing the objective function and constrains is explained in the next section. Section 3 presents the PSO algorithm in order to solve the optimization problem. The test system used to verify the effectiveness of the proposed technique is describe in Section 4 which explores the effectiveness of the proposed technique applied on simulation test system, Section 5 concludes the paper. The simulation test systems were simulated in MATLAB software.

PROBLEM FORMULATION

Objective functions formulation

As mentioned above, this paper introduces MOF optimization. The objective function was procured from the gather of each DG impact by the weighting factor assigned to that impact. This weighting factor is chosen by the planner to reflect the relative importance of each parameter in the decision making of sitting and sizing the DG. The DG location and its corresponding size in the distribution feeders can be optimally determined using the following objective function:

Max f (P_{loss}, Q_{loss}, I_{SC}, V_{level}).

Where:

$$f(P_{loss}, Q_{loss}, Isc, V_{level}) = w_1 F_p + w_2 F_q + w_3 F_i + w_4 F_v \qquad (1)$$

F_p relates to increase of active power loss index in percent of system due to installation of DG which is given by:

$$F_p = \frac{P_{Loss}^{withoutDG} - P_{loss}^{withDG}}{P_{Loss}^{withoutDGl}} \qquad (2)$$

Where, P_{Loss}^{withDG} is the real power loss in study system after installation of DG and $P_{Loss}^{withoutDG}$ is active power losses before installation.

F_q is a factor in order to determine the effect of DG in reactive power losses in mentioned system that given by:

$$F_q = \frac{Q_{Loss}^{withoutDG} - Q_{loss}^{withDG}}{Q_{Loss}^{withoutDGl}} \qquad (3)$$

Where, Q_{Loss}^{withDG} and $Q_{Loss}^{withoutDG}$ are total reactive power losses in study system with installation DGs and without DGs respectively.

One of the avails of optimizes location and size of the DG is the improvement in voltage profile. This index penalizes the size-location pair which gives higher voltage deviations from the nominal value (Vnom). In this way, the closer index to zero, the better is the network performance. The F_v can be defined as:

$$F_v = \max_{i=2}^{n} \left(\frac{|V_{nom}| - |V_i|}{|V_{nom}|} \right) \qquad (4)$$

At last, in order to improve the short circuit level of system, F_I given in Equation 5, is gathered with the objective function:

$$F_I = \frac{I_{sc}^{withoutDG} - I_{sc}^{withDG}}{I_{sc}^{withoutDGl}} \qquad (5)$$

The sum of the absolute values of the weights assigned to all impacts should add up to one as shown in the following equation:

$$|w_1| + |w_2| + |w_3| + |w_4| = 1 \qquad (6)$$

The MOF in this paper in order to achieve the performance calculation of distribution systems for DG size and location is given by:

$$MOF = 0.4F_p + 0.2F_q + 0.15F_I + 0.25F_v \qquad (7)$$

Constrains formulation

The MOF Equation 7 minimized is subjected to various operational constraints to satisfy the electrical requirements for distribution network. These constraints are the following.

Power-conservation limits

The algebraic sum of all incoming and outgoing power including line losses over the whole distribution network and power generated from DG unit should be equal to zero.

$$P_{Gen} + P_{DG} \sum_{i=1}^{n} P_D - P_{total}^{loss} = 0 \tag{8}$$

Distribution line capacity limits

Power flow through any distribution line must not exceed the thermal capacity of the line:

$$S_{ij} < S_{ij}^{max} \tag{9}$$

Voltage limits

The voltage limits depend on the voltage regulation limits should be satisfied:

$$V_i^{min} \leq V_i \leq V_i^{max} \tag{10}$$

This paper employs PSO technique to solve the above optimization problem and search for the optimal or near optimal set of problem. Typical ranges of the optimized parameters are (0.01 to 100) KW for P_{DG} and (0.95-1.05) for voltage of buses.

PARTICLE SWARM OPTIMIZATION ALGORITHM

PSO was formulated by Edward and Kennedy in 1995 (Randy et al., 2004). The thought process behind the algorithm was inspired by the social behavior of animals, such as bird flocking or fish schooling. PSO is one of the most recent developments in the category of combinatorial met heuristic optimizations (Gaing, 2003). In PSO, each individual is referred to as a particle and represents a candidate solution to the optimization problem (Yoshida et al., 2000).

In first, a population of random solutions "particles" in a D-dimension space are composed. Each particle is a solution. The ith particle is represented by $X_i = (x_{i1}, x_{i2}, \ldots, x_{iD})$. Situation of each particle will be change in the next stage. The best situation of each particle will be determined by fitness function. If the fitness functions has a minimum value so far it is called best situation and save in $Pbest_i$. The global version of the PSO keeps track of the overall best value (gbest), and its location, obtained thus far by any particle in the population (Mandal et al., 2008). The particles update their velocities and positions based on the local and global best solutions. According to Equation 11, the velocity of particle i is represented as $V_i = (v_{i1}, v_{i2}, \ldots v_{iD})$. Acceleration is weighed by a random term, with separate random numbers being generated for acceleration toward pbest and gbest. The position of the ith particle is then updated according to Equation 12 (Binghui et al., 2007):

$$v_{id} = w \times v_{id} + c_1 \times rand() \times (P_{id} - x_{id}) + c_2 \times rand() \times (P_{gd} - x_{id}) \tag{11}$$

$$x_{id} = x_{id} + cv_{id} \tag{12}$$

Where, P_{id} and P_{gd} are pbest and gbest, c_1 and c_2 are constant values, ω will be determined by this equation:

ω_{max} and ω_{min} are the maximum and minimum value of ω respectively. At first ω start with large value that in the end of problem the value of the ω will be minimum.

In this optimization problem, the number of particles and the number of iterations are selected 30 and 40, respectively. Dimension of the particles will vary for each condition.

$$\omega = \omega_{max} - \frac{\omega_{max} - \omega_{min}}{iter_{max}} * iter \tag{13}$$

CASE STUDY AND PLACEMENT RESULTS

In this section, we illustrate that, DG placement affects the active power loss, reactive power losses and voltage profile. The placement of only a single DG, two DGs and three DGs are considered. In order to prove the efficiency of the proposed placement algorithm, IEEE 33-bus test system without tie lines that was presented in Kashem et al. (2000) as shown in Figure 1 is considered and the system details are given in Table 1.

In order to demonstrate variable number of DGs effect, we assume that, one-two and three DG unit which its size varying between 25 to 10 MW will be place in the mention network. The optimization results are given in Figure 2. This figure shows the value of MOF value in 40 iteration of PSO. From these results, it was obvious that the amount of MOF of the three DGs placement is least at the 40th iteration. The size and site location of one, two and three DGs are given in Table 2.

As can be seen from Table 2, the active power loss of the network without DG has maximum value and with three DG sitting have minimum amount and with comparing of power loss in four cases that is obvious that, the DG placement can has positive effect in power loss in the whole mention network.

Figure 3 illustrates buses voltage in four cases. With attention to this figure, the voltage profile with DG unit is better than without DG and the increasing number of DG unit's affect the DGs in voltage profile become well.

In the next study, we assume that, three DG units in order to optimal placement are considered. The result of this study is represented by power system. The results of line power loss were presented in Table 2 and in this case this power loss becomes less than other cases and in Figure 4 the voltage profile is showed. The voltage profile in this case is better than the previous cases.

Conclusion

In this paper, a different approach based on PSO in order to multi objective optimization analysis, including one, two and three DG units, for size and site planning of DG in distribution system were presented. In solving this problem, at first problem was written in the form of the optimization problem which its objective function was defined and written in time domain and then the problem has been solved using PSO. The proposed optimization algorithm was applied to the 33-bus test system with tie lines.

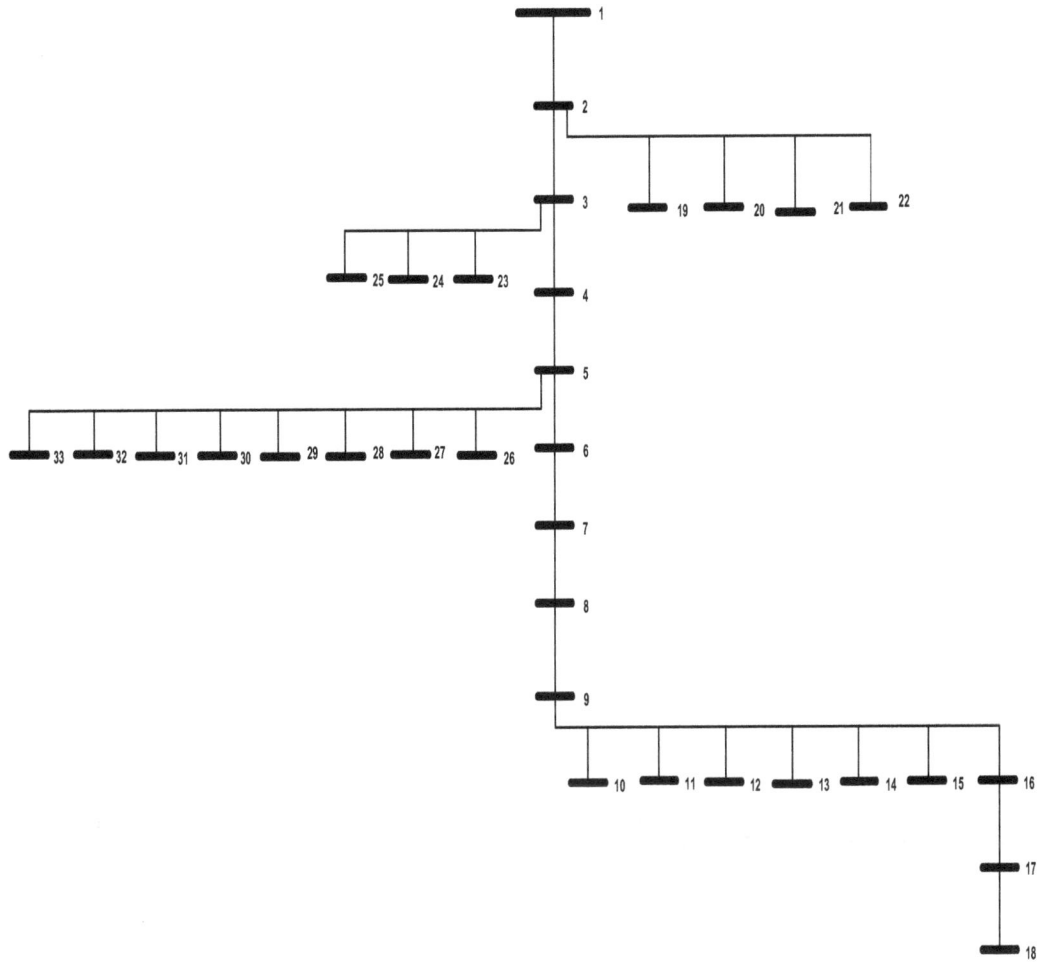

Figure 1. IEEE 33 bus study system with tie lines.

Table 1. Lines, active and reactive power details in study system.

Branch nom	Sen. node	Rec. node	Active power of rec. node KW	Reactive power of rec. node KVAr	Resistance ohms	Reactance ohms
1	1	2	100	60	0.0922	0.0470
2	2	3	90	40	0.4930	0.251 1
3	3	4	120	80	0.3660	0.1 864
4	4	5	60	30	0.3811	0.1941
5	5	6	60	20	0.8190	0.7070
6	6	7	200	100	0.1872	0.6188
7	7	8	200	100	1.7114	1.2351
8	8	9	60	20	1.0300	0.7400
9	9	10	60	20	1.0440	0.7400
10	10	11	45	30	0.1966	0.0650
11	11	12	60	35	0.3744	0.1238
12	12	13	60	35	1.4680	1.1550
13	13	14	120	80	0.5416	0.7129
14	14	15	60	10	0.5910	0.5260
15	15	16	60	20	0.7463	0.5450

Table 1. Contd.

16	16	17	60	20	1.2890	1.7210
17	17	18	90	40	0.7320	0.5740
18	2	19	90	40	0.1640	0.1565
19	19	20	90	40	1.5042	1.3554
20	20	21	90	40	0.4095	0.4784
21	21	22	90	40	0.7089	0.9373
22	3	23	90	50	0.4512	0.3083
23	23	24	420	200	0.8980	0.7091
24	24	25	420	200	0.8960	0.7011
25	5	26	60	25	0.2030	0.1034
26	26	27	60	25	0.2842	0.1447
27	27	28	60	20	1.0590	0.9337
28	28	29	120	70	0.8042	0.7006
29	29	30	200	600	0.5075	0.2585
30	30	31	150	70	0.9744	0.9630
31	31	32	210	100	0.3105	0.3619
32	32	33	60	40	0.3410	0.5302

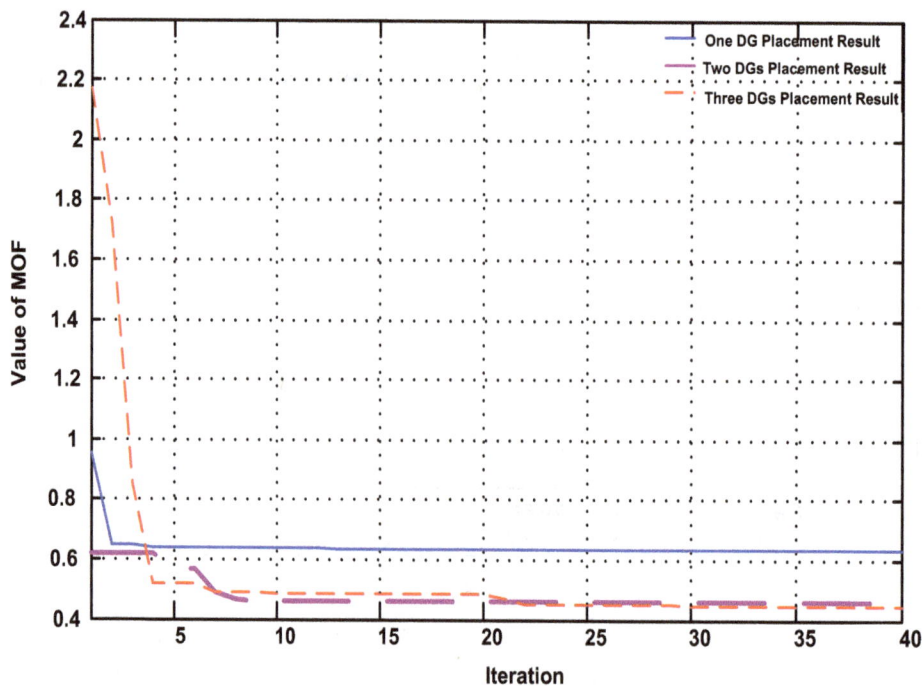

Figure 2. Value of MOF for one, two and three DGs.

Table 2. Optimum results of PSO for location and size of DGs.

Number of DG	DG size			DG site			Network loss
Without DG	-			-			0.8920
One DG	1136.6			11			0.6340
Two DGs	1143.2	1044.2		24	11		0.4583
Three DGs	1700.1	668.7	506.0	3	8	15	0.4436

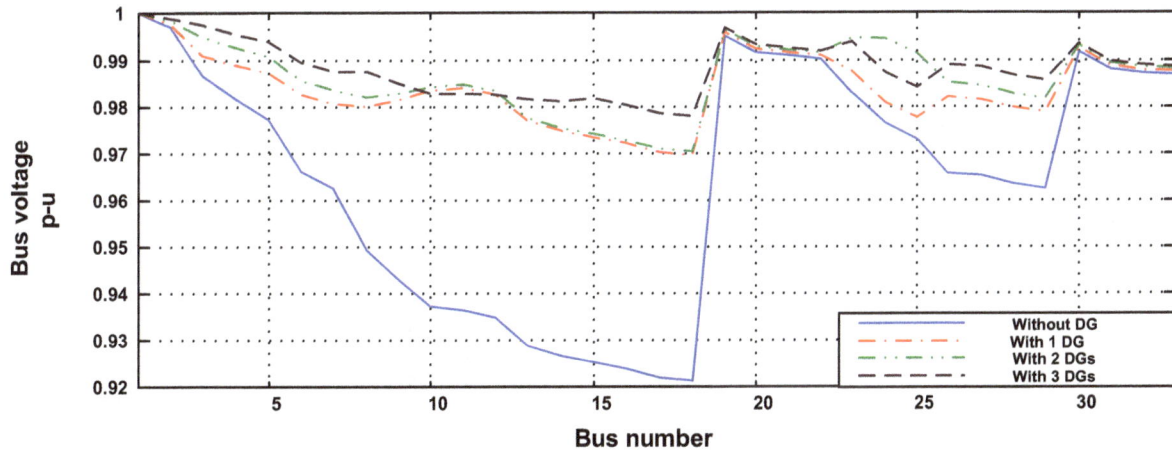

Figure 3. Voltage profile of study system with three DG units, two DG units, single DG unit, and without DG.

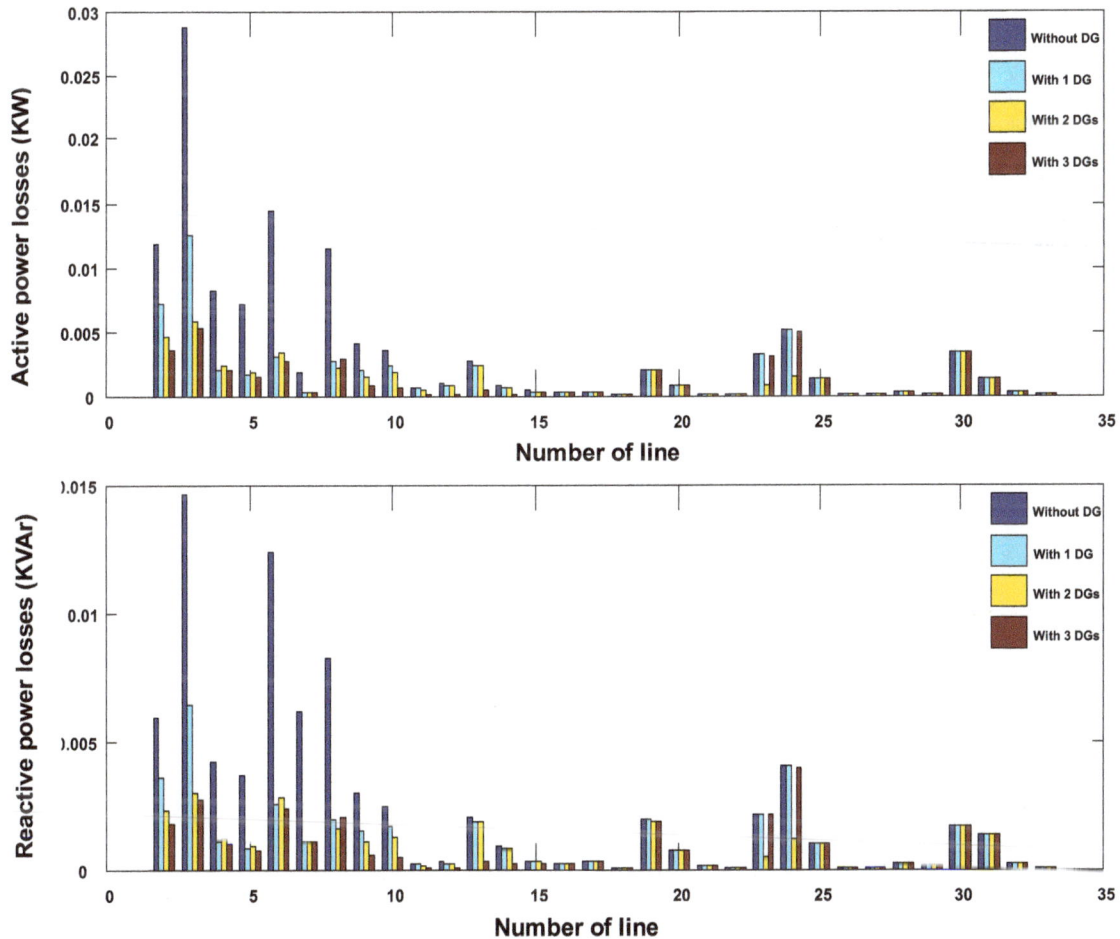

Figure 4. Active and reactive of lines with three DG units, two DG units, single DG unit and without DG.

The results clarified the efficiency of this algorithm for the improvement of voltage profile and reduction of power losses in study system.

REFERENCES

Binghui Y, Xiaohui Y, Jinwen W (2007). Short-term hydro-thermal scheduling using particle swarm optimization method. Energy

Convers. Manage. 48:1902-1908.

Carmen LTB, Djalma MГ (2006). Optimal distributed generation allocation for reliability, losses, and voltage improvement. Elect. Power Energy Syst. 28:413-420.

Das D (2008). Optimal placement of capacitors in radial distribution system using a Fuzzy-GA method'. Elect. Power Energy Syst. 30:361-367.

El-Khattam W, Hegazy YG, Salama MMA (2005). "An integrated distributed generation optimization model for distribution system planning". IEEE Trans. Power Syst. 20(2):1158-1165.

Gaing Z (2003). Particle swarm optimization to solving the economic dispatch considering the generator constraints, IEEE Trans. PWRS. 18(3):1187-1195.

Gandomkar M, Saveh AU, Vakilian M, Ehsan M (2005). Acombination of genetic algorithm and simulated annealing for optimal DG allocation in distribution networks, in: Proc. IEEE Electr. Comput. Eng. Can. Conf. 1-4:645-648.

Hashemi F, Alizade AR, Zebardast SJ, Ghadimi N (2011). "Determining the optimum cross-sectional area of the medium voltage feeder for loss reduction and voltage profile improvement based on particle swarm algorithm" 10th Int. Conf. Environ. Elec. Eng. (EEEIC), 8–11:813-817.

Hedayati HS, Nabaviniaki A, Akbarimajd A (2008). "A method for placement of DG units in distribution networks", IEEE Trans. Power Deliver. 23(3):1620-1628.

Kashem MA, Ganapathy V, Jasmon GB, Buhari MI (2000). A Novel Method for Loss Minimization in Distribution Networks, International Conference on Electric Utility Deregulation and Restructuring and Power Technologies 2000, London, 4-7 April. pp. 251-256.

Katsigiannis YA, Georgilakis PS (2008). "Optimal sizing of small isolated hybrid power systems using tabu search". J. Optoelectron. Adv. Mater. 10(5):1241-1245.

Lalitha MP, Reddy VCV, Usha V (2010). "Optimal DG Placement for Minimum Real Power Loss in Radial Distribution Systems Using PSO". J. Theor. Appl. Inform. Technol. pp. 107-116.

Mandal KK, Basu M, Chakraborty N (2008). Particle swarm optimization technique based short-term hydrothermal scheduling. Appl. Soft Comput. 8:1392-1399.

Ochoa LF, Padilha-Feltrin A, Harrison GP (2006). "Evaluating distributed generation impacts with a multiobjective index". IEEE Trans. Power Deliver. 21(3):1452-1458.

Randy LH, Sue EH (2004). PRACTICAL GENETIC ALGORITHMS, Wiley & Sons, Inc., Hoboken, New Jersey.

Singh RK, Goswami SK (2010). "Optimum allocation of distributed generations based on nodal pricing for profit, loss reduction and voltage improvement including voltage rise issue". Int. J. Elect. Power Energy Syst. 32:637-644.

Thong VV, Driesen J, Belmans R (2007). "Transmission system operation concerns with high penetration level of distributed generation", in Proc. of Inter. Universities Power Engineering Conference, Brighton. pp. 867-871.

Yoshida H, Kawata K, Fukuyama Y, Takayama S, Nakanishi Y (2000). A particle swarm optimization for reactive power and voltage control considering voltage security assessment. IEEE Trans. PWRS. 15(4):1232-1239.

9

Widely tunable micro electromechanical systems (MEMS)-vertical-cavity surface-emitting lasers with single transverse mode operation

Seyed Mahdi Hatamian[1] , Vahid Ahmadi[2] and Elham Darabi[3]

[1]Department of Electrical Engineering, Science and Research Branch, Islamic Azad University, Tehran, Iran.
[2]Department of Electrical Engineering, Tarbiat Modares University, Tehran, Iran.
[3]Plasma Physics Research Center, Science and Research Branch, Islamic Azad University, Tehran, Iran.

In this paper, a micro electromechanical systems vertical-cavity surface-emitting laser (MEMS-VCSEL) with asymmetric double oxide aperture and highly strained GaInNAsSb quantum wells for widely tunable, single mode and high temperature operation has been investigated. The MEMS-VCSEL is based on an integrated two-chip concept which allows us to extend the single wavelength performance to a continuously tunable, selectively wavelength-addressable spectrum of 40 nm. It has a much larger tuning range as compared with previous works. We present a comprehensive model including electrostatic membrane equation coupled with thermal, spatial and temporal rate equations considering Shockley Read Hall (SRH), Auger and carrier diffusion effects. These coupled equations are solved numerically by finite difference method (FDM). Using the simulation results, we design a single mode, high power, high temperature and tunable VCSEL, which is suitable for C-band dense-wavelength-division- multiplexing (DWDM) optical communication systems.

Key words: Micro electromechanical systems (MEMS), tunable vertical-cavity surface-emitting laser (VCSELs), strained GaInNAsSb QWs, single mode operation.

INTRODUCTION

Tunable, single mode and high power vertical-cavity surface-emitting laser (VCSELs) at 1550 nm have a great potential to replace the distributed feedback (DFB) and fabry-perot (FP) edge emitting lasers that are currently used in optical communication (Kogel et al., 2011). Low threshold current together with easily couple light into optical fiber is possible due to their small active volume compare to edge-emitting lasers. Recently many structures of micro electromechanical systems vertical-cavity surface-emitting laser (MEMS-VCSELs) have been reported. These devices can be divided into three categories: cantilever VCSELs (Chang-Hasnain, 2000),

membrane VCSEL devices (Guan et al., 2009), and tunable VCSELs utilizing a half-symmetric cavity (Tayebati et al., 1999).

For tuning in cantilever VCSELs, the technique is based on changing the cavity length using cantilever arm and in half-symmetric cavity VCSELs; the curved top mirror is designed to match the Gaussian curvature of the light oscillating within the optical cavity that creates a single spatial lasing mode. We are primarily concerned with VCSEL tuning utilizing electrostatically actuated membrane with minimum actuation voltages which is comparable to existing MEM tunable VCSEL structures.

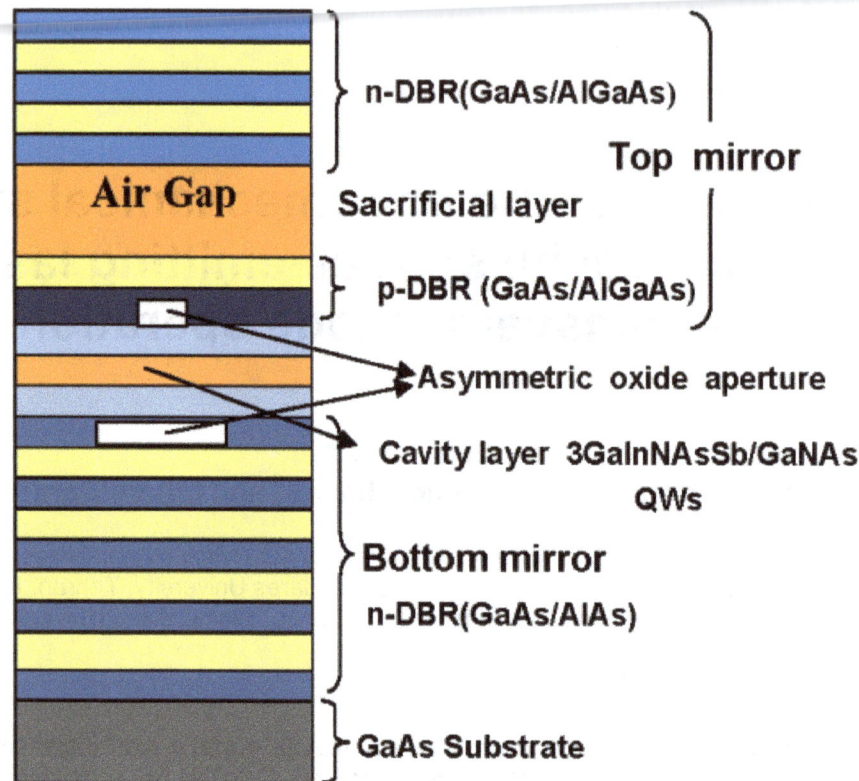

Figure 1. The schematic structure of the proposed MEMS-VCSEL.

In surface micro machined material layers are deposited and patterned one at a time and it is possible to create membrane VCSELs. Many approaches to achieve single mode, high temperature, and high power operations have been reported (Zhou and Mawst, 2002). There are different effects such as pump induced current spreading, spatial hole burning and thermal gradients inside the cavity on the carrier distribution which lead to VCSELs multi mode behaviour (Samal et al., 2005; Nakwaski and Sarzala, 1998). Our aim is to optimize a tunable VCSEL based on the GaAs material system with dilute nitride/antimonide quantum wells for the long wavelength. To resolve the issue of multi-mode behavior, a new structure design using asymmetric double apertures is proposed and theoretically modeled. Here an electrostatic tunable 1.55 µm MEMS-VCSEL based on two-chip concept with a tuning range more than 40 nm is proposed and single mode operation is achieved by engineering the spatial distribution of the injection current profile using asymmetric oxide apertures.

Using finite difference method (FDM), the device performance by solving a comprehensive model has been investigated. It includes electrostatic membrane equation coupled with thermal, spatial and temporal rate equations considering Shockley Read Hall (SRH), Auger and carrier diffusion effects. Base on the simulation

results, we design a widely tunable, single mode, high power and high temperature VCSEL appropriate for dense-wavelength-division-multiplexing (DWDM) optical communication system.

TWO-CHIP MEMS-VCSEL DESIGN

The proposed device consists of a bottom n-DBR, a cavity layer and a top mirror as shown in Figure 1. GaAs substrate based materials are the excellent choice for long-wavelength operation due to better thermal performance (Martin, 2001). The bottom stack consists of 22.5 pairs n-doped GaAs/AlAs DBR which is perfectly lattice matched to GaAs substrate. The active region consists of three highly strained $Ga_{0.59}In_{0.41}N_{0.028}As_{0.942}Sb_{0.03}/GaNAs$ QW for long-wave applications (Gutowski and Sazala, 2008).

GaInNAsSb/GaNAs lasers have excellent high-temperature performance, large T_0, greater efficiency and higher output power. The QWs are sandwiched by two stacks of λ/4 brag mirrors for 1.55 µm. The 100 × 100 µm top mirror, includes three parts: a p-DBR, an air gap, and a top n-DBR, which is freely suspended above the laser cavity and supported via a membrane structure. The p-DBR consists of 4 pairs $GaAs/Al_{0.9}Ga_{0.1}As$. The air gap,

Table 1. Device parameters.

Element	Material	Thickness	Refractive Index	Comment
MEMS DBR	AlGaAs/GaAs	5 µm	3.38/2.99	λ/4
Air-gap	AlGaAs	1.4 µm	1.0	Sacrificial layer
QW and barriers	GaInNAsSb /GaNAs	7/20 (nm)	3.6/3.5	3QWs
Bottom DBR	GaAs/AlAs	5.5 µm	3.53/2.97	λ/4

followed by a section of n- DBR consists of 20.5 pairs GaAs/AlGaAs. Also a double asymmetric aperture VCSEL is proposed when p-aperture is at the third pair in p-DBR and n-aperture at the first pair in bottom n-DBR. The DBR mirrors are doped with Be and Si. Be ($N_{Be} = 5 \times 10^{17}$ cm^{-3} for GaAs, $N_{Be} = 2 \times 10^{18}$ cm^{-3} for AlGaAs) is used as p-dopant, where as Si ($N_{Si} = 5.3 \times 10^{17}$ cm^{-3} for GaAs, $N_{Si} = 3 \times 10^{18}$ cm^{-3} for AlAs) is used as the n-dopant. The topest layer is heavily doped with Be ($N_{Be} = 6 \times 10^{19}$ cm^{-3}) for facilitate current spreading and ohmic contact. The device parameters are summarized in Table 1.

THEORY OF ANALYSIS

Electrostatic equations

Wavelength tuning is accomplished by applying a voltage between the top n-DBR and p-DBR, across the air gap. A reverse bias voltage is used to provide the electro static force, which deflects the membrane downward and shortens the air gap, thus shifts the laser wavelength to shorter wavelengths (blue shift) (Ochoa, 2007). The displacement of the membrane is controlled by the balance between the electrostatic force and the elastic restoring force in membrane legs which is described by the second-order linear differential equation as (Ochoa, 2007):

$$\frac{d^2 y}{dx^2} = \frac{M(x)}{EI}$$

(1)

Where E is Young's modulus (Pa), and I is the moment of inertia (m^4). EI product is known as the flexural rigidity of the beam. $M(x)$ represents the bending moment (N.m^2).

Here L, t and w are flexure's length, thickness and width, respectively. For small deflections, the electrostatic force obeys the Hook's law (F=k$_{total}$ d), so for 4 flexures:

$$k_{total} = \frac{4Etw^3}{L^3}$$

(2)

If the reverse bias voltage is applied across the over-lapping electrode areas, the electrostatic force is given by:

$$F = \frac{A\varepsilon_0 V^2}{2g^2}$$

(3)

Where A is the overlapping electrode area, ε_0 is the permittivity of free space. V is the voltage across the electrodes, $g = (h- d)$, h is the initial air-gap thickness, and d is the deflection of the upper electrode toward the lower electrode. Solving above equations for V gives:

$$V = (h-d)\sqrt{\frac{L^3 h}{2Etw^3\varepsilon_0 A}}$$

(4)

Due to the elastic movement, there is no hysterisis in the wavelength-tuning curve. So, the membrane returns to its original position, where, the voltage is removed. Although increased the air gap leads to a wider tuning range, a larger air gap means a longer cavity length which results in a narrower FSR, and therefore a shorter tuning range. Thus to achieve the maximum tuning range these two parameters must be optimized. The calculated displacement versus voltage for a 150 × 150 µm piston micro mirror with four 100 µm flexures, and a 1.4 µm starting air gap is 470 nm for a 19 volt membrane bias. In this simulation the flexure material is 1.5 µm thick gold (Au) with a Young's modulus of E = 79 GPa. A key feature of this calculation is the expected "snap-down" of top n-DBR which is about 1/3 of the starting air gap.

Coupled thermal equation and rate equation

Dynamic of the VCSELs is primarily governed by the following space and time dependent rate equations coupled with thermal equation as follows (Jungo and Erni, 2003; Samal, 2004):

$$\frac{dN(r,t)}{dt} = \frac{\eta_i J(r,t)}{eV} - \frac{N(r,t)}{\tau_n} + CN^3(r,t) + D_n \nabla^2 N(r,t)$$
$$- v_g \sum_m G_m(r,t)S_m(t)$$

(5)

$$\frac{dS_m(t)}{dt} = \Gamma_m \beta_m \frac{\iint N(r,t)dr}{\pi R^2 \tau_n} + \frac{\Gamma_m v_g S_m(t)}{\pi R^2} \iint G_m(r,t)dr - \frac{S_m(t)}{\tau_s}$$

(6)

$$C_{th}\frac{dT}{dt} = (P_{el} - P_{opt}) - \frac{T - T'_0}{R_{th}}$$

(7)

where N is the carrier density, J is injection current, η_i is the injection efficiency, τ_n is the carrier recombination lifetime, V is the active-region volume, C is Auger recombination coefficient, D_n is electron diffusion coefficient, v_g is group velocity, S_m is mth mode the photon density, τ_s is the photon lifetime and β_m is the spontaneous emission coupling coefficient in mth transverse mode, Γ_m is mth mode of optical confinement, R is radius of cavity, P_{opt} and P_{el} are the injected electrical power and generated optical power R_{th} is the thermal resistance of each regions, and C_{th} is thermal capacitance, respectively. G_m is the mth modal gain, which can be approximated as a linear function of the carrier density.

$$G_m(r,t) = g_0 \frac{N(r,t) - N_{tr}}{1 + \varepsilon S_m(t)}|\Psi_m(r,t)|^2$$

(8)

Here g_0 is differential gain, ε is gain compression coefficient and

Table 2. The parameters used in the simulation.

Symbol	Value	Units
η_i	1	
D_n	12	cm.^{2}s^{-1}
β	1.5×10^{-4}	
C	7×10^{-31}	cm^6.s^{-1}
τ_s	2.2	ps
τ_n	2.5	ns
N_{tr}	2×10^{18}	cm^{-3}
g_0	1500	cm^{-1}
ε	1.4×10^{-16}	cm^3
N_{th}	5×10^{18}	cm^{-3}
R	12	μm
R_{ox1}	10	μm
R_{ox2}	20	μm
R_{th}	3000	K/W
R_{top}	0.997	
R_{bottom}	0.9985	
P_0	-0.9×10^7	
P_1	0.6×10^{-12}	cm^3
P_2	0.25×10^{-14}	cm^3 k^{-1}
P_3	-0.25×10^{26}	cm^{-3}
J_{l0}	7×10^{-4}	kA/cm^2
T_{ref}	139	$^\circ$k

$|\Psi_m(r,t)|$ is the normalized field amplitude of the mth mode. The modification of Equation (5) begins with separation of the time and space variables. This is done for photon density by describing as azimuthal and radial components of the optical field so-called c and s modes (Jungo and Erni, 2003; Samal, 2004):

$$S_m(\rho,\varphi,t) = S_m^c(t)|\Psi_m(\rho)|^2 \cos^2(l\varphi) + S_m^s(t)|\Psi_m(\rho)|^2 \sin^2(l\varphi) \tag{9}$$

where the index m indicates mode's order and the azimuthal order l is a function of m. The carrier density profile can be expanded in a time dependent orthogonal series:

$$N(\rho,\varphi,t) = \sum_i N_i(\varphi,t) J_0(\frac{\gamma_i\rho}{R}) = \sum_i J_0(\frac{\gamma_i\rho}{R}) \sum_q \left[N_{iq}^c(t)\cos(q\varphi) + N_{iq}^s(t)\sin(q\varphi) \right] \tag{10}$$

Hence, the carrier profile is modelled by describing azimuthal and radial components of the carrier so-called c and s, the carrier profile radial components of the carrier so-called c and s, the carrier profile is expanded in an orthogonal series with time dependent expansion coefficients. Where γ_i is the ith root of the first-order Bessel function of the first kind. The functions is a family of functions with vanishing slope at $\rho=0$ and $\rho=R$ that R is effective cavity radius. The number of required Bessel terms is a function of the effective cavity radius, number of modes, oxide apertures radius, ambipolar diffusion coefficient, and current profile. In this case, 10 terms Bessel series expansion has been taken.

Current confinement and spreading in the cavity is controlled by the size and position of the oxide apertures. Current injection profile engineering in device provides single mode operation. In the

structure, smaller aperture is positioned far away from the active region in the p-mirror (R_{ox1}) and a larger aperture is positioned near to the active region in the bottom n-mirror (R_{ox2}). So, the current profile J (ρ, φ, t) can be written as:

$$J(\rho,\varphi,t) = J(t)\rho_c(\rho) \tag{11}$$

where the normalized function $\rho_c(\rho)$ describes the current injection profile, including spreading effects. The electrical model for current injection can be described as:

$$\rho_c(\rho) = 1, \quad \rho \le R_{ox1}$$

$$\rho_c(\rho) = \exp\left[-\frac{(\rho - R_{ox1})^2}{2\gamma^2} \right]$$

$$\gamma = r_s + R_{ox2} \tag{12}$$

Where r_s is the current spreading coefficient.

Small active volume of VCSELs, poor heat dissipation and large resistance introduced by DBRs can exhibit strong thermally dependent behavior. Therefore, for accurate modeling of the device, thermal effects have to be taken into consideration (Jungo and Erni, 2003; Samal, 2004). In our simulations two dominant thermal effects, namely, gain detuning and thermionic emission are considered. Before evaluating these effects, the cavity temperature must be computed. In Equation (8) the dependence of gain on temperature is taken by assuming a parabolic gain spectrum.

$$g_0(\lambda,T) = g_0(1 - \frac{T-T_0}{T_{ref}})\left[1 - 2(\frac{\lambda(T) - \lambda_P(T)}{\Delta\Lambda})^2 \right] \tag{13}$$

where T is the cavity temperature, T_0 is the room temperature, T_{ref} is a fitting parameter and $\Delta\Lambda$ is the characteristic width of the parabolic gain approximation. Current leakage due to thermionic emission is also considered as (Jungo and Erni, 2003; Samal, 2004):

$$J_l(N,T) = J_{l0}\exp(-\frac{p_0}{T} + \frac{P_1N}{T} + P_2N + \frac{P_3}{T(N+N_{th})}) \tag{14}$$

J_{l0} is reference leakage current density and P_0, P_1, P_2 and P_3 are leakage parameters. The numerical values of parameters are summarized in Table 2.

RESULTS AND DISCUSSION

To enhance the performance of MEMS-VCSEL, we propose two new structures with n-aperture fixed at the first mirror-pair of the n-DBR whose size is twice the diameter of the p-aperture. In the first structure (new structure-1) the p-aperture is placed in the first mirror-pair of the p-DBR, in the second structure (new structure-2) the p-aperture is placed in the third mirror-pair of the p-DBR as compared with conventional VCSEL (with a 5 μm single oxide aperture in bottom n-DBR). Spatial-current distribution versus radius distance of conventional VCSEL and the proposed structures are shown in Figure 2.

The relative size and location of the dual asymmetric

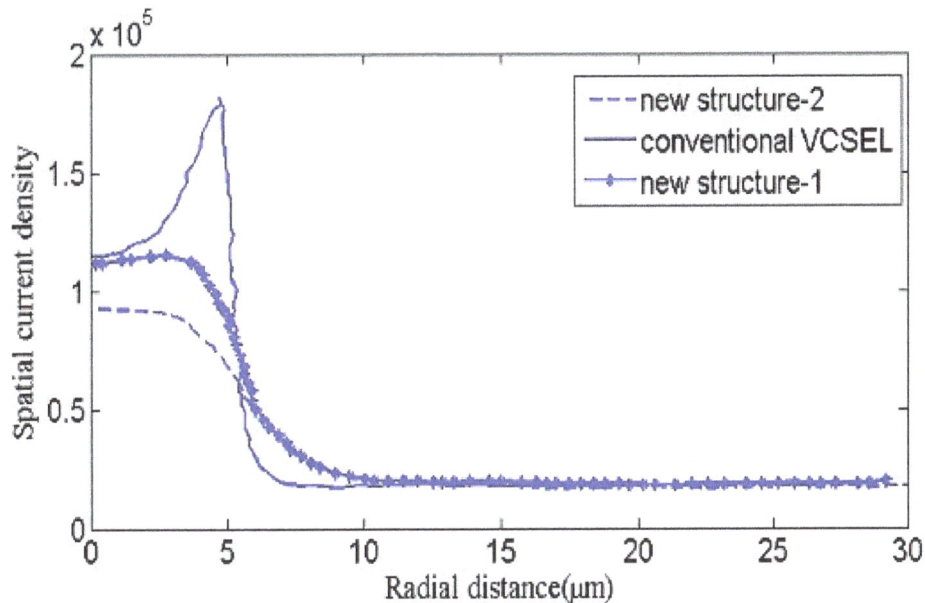

Figure 2. Spatial-current distribution vs. radial distance in conventional (solid), New structure-1(star dash) and new structure-2 (dash).

oxide apertures and doping profiles of the DBR mirrors, affect the shape of spatial injection current profile in active region. The spatial current density distribution shows a ring shape injection profile for a conventional VCSEL with a maximum at the periphery and a minimum at the center of the device. The new structure-1 and new structure-2 show an improvement in the profile over the conventional structure when the p-aperture is moved farther away from the active region. The new designs show bell shape injection profiles, which make them suitable for single mode operation. A bell shape current injection profile is favourable for a LP01 mode of a VCSEL laser. However, the new structure-2 shows a better performance for single mode operation and we use this structure in our simulations.

The steady-state output power is shown in Figure 3. Because of higher leakage current and gain detuning in conventional VCSEL, the thermal rollover appears above 10 mA compared with new structure which occurs at 15 mA. It can be seen that new structure has lower temperature rise versus conventional structure. This MEMS-VCSEL can be less temperature sensitive than conventional VCSEL. In addition, for a constant injection current the temperature difference goes to zero far from the VCSEL center. This lower temperature sensitivity enables us to stabilize the emission wavelength of a tunable VCSEL within one nanometer over a broad range of operation temperature.

3D distribution of carrier density and photon intensity for conventional and the new structures are shown in Figure 4a and b. The conventional structure exhibits several higher order modes, but the single mode

operation of the new structure is evident in Figure 4b. Figures show the dominant modes for conventional and new structure designs. The conventional design shows LP41 is the most dominant mode and the fundamental LP01 mode more than 40 dB lower than the dominant mode. New structure design shows LP01 as the dominant mode with the next higher mode more than 40 dB lower in this case.

The large signal responses of devices are compared in Figure 5. Calculated results show multi transverse modes operation for conventional VCSEL and high power output with dominant LP01 mode in the new structure. Tuning range of 40 nm for a 19 volt membrane bias is shown in Figure 6. The continuous repeatable and hysteresis-free tuning range can be observed clearly, which makes it suitable for DWDM systems.

Conclusions

We have designed a novel MEMS-VCSEL with asymmetric double oxide aperture and highly strained GaInNAsSb GaNAs quantum wells for DWDM optical communication systems. We have demonstrated a comprehensive model including electrostatic membrane equation coupled with thermal, spatial and temporal rate equations considering SRH, Auger and carrier diffusion effects. The coupled equations are solved numerically using FDM to find self-consistent solutions for tunable, single mode and high power of operation. The electrically-pumped MEMS-VCSEL can be directly modulated with wide tuning range. We have illustrated

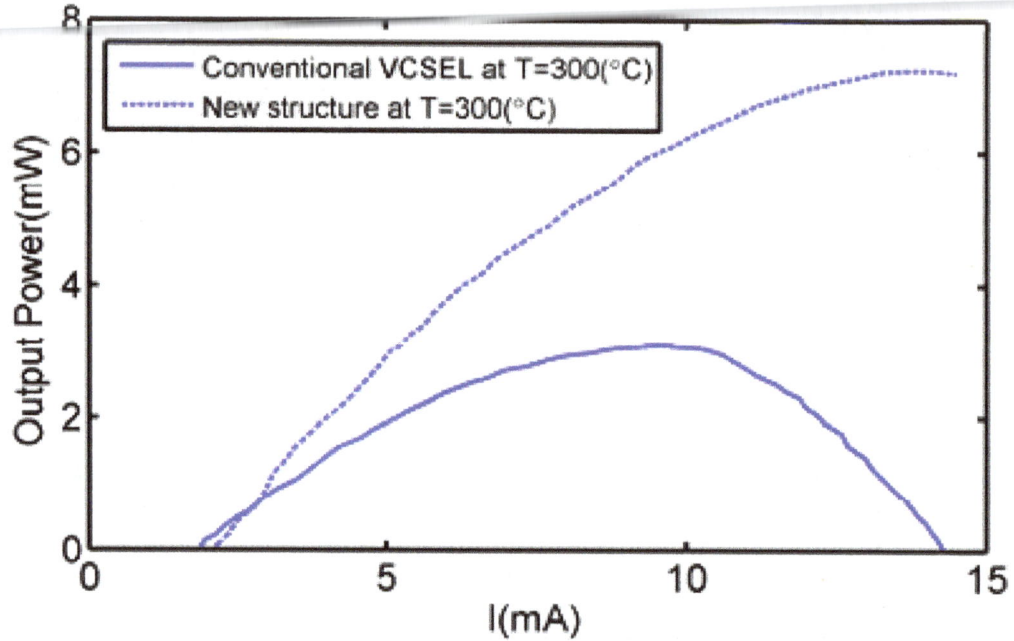

Figure 3. Thermal rollover in the new structure and conventional design at room temperature.

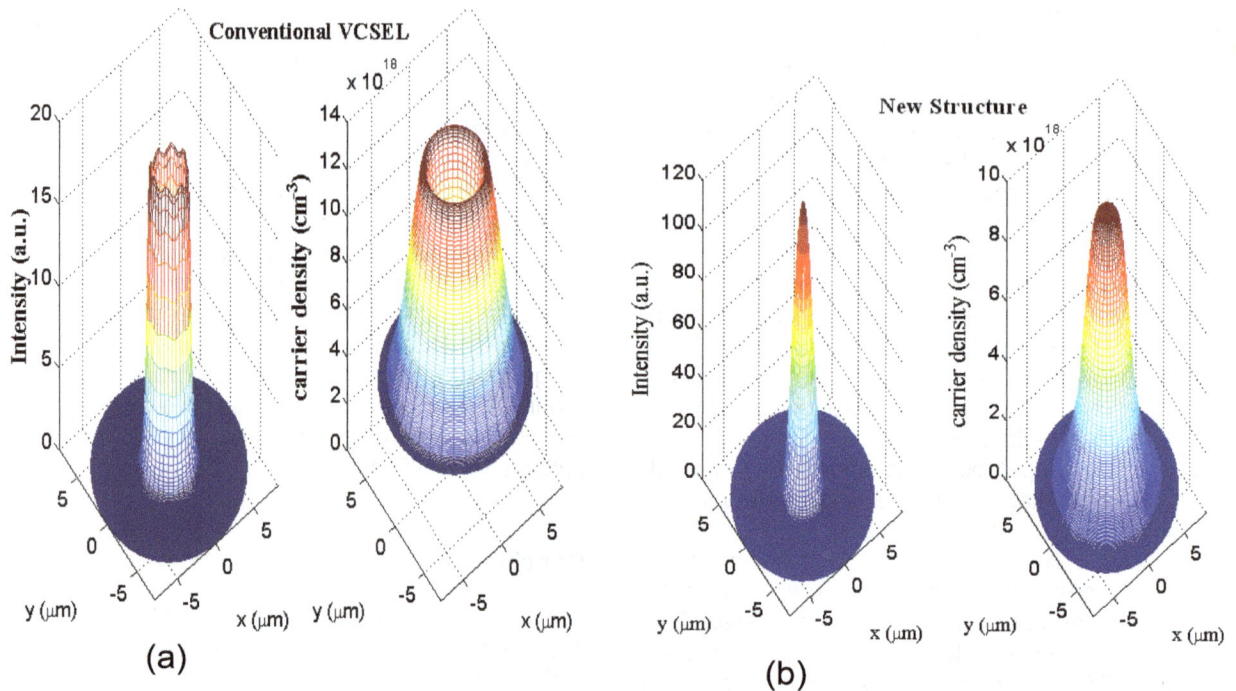

Figure 4. Carrier density and photon intensity of (a) conventional design and (b) new structure.

the large and small signal response of new structure with 40 nm tuning range. Single mode operation is achieved by asymmetric double oxide apertures. The thermal effects as gain detuning and carrier leakage are investigated. The new structure design, demonstrated a room temperature CW single-mode output power of more than 7 mW with a side mode suppression ratio greater than 25 dB for an active device size more than 12 µm in

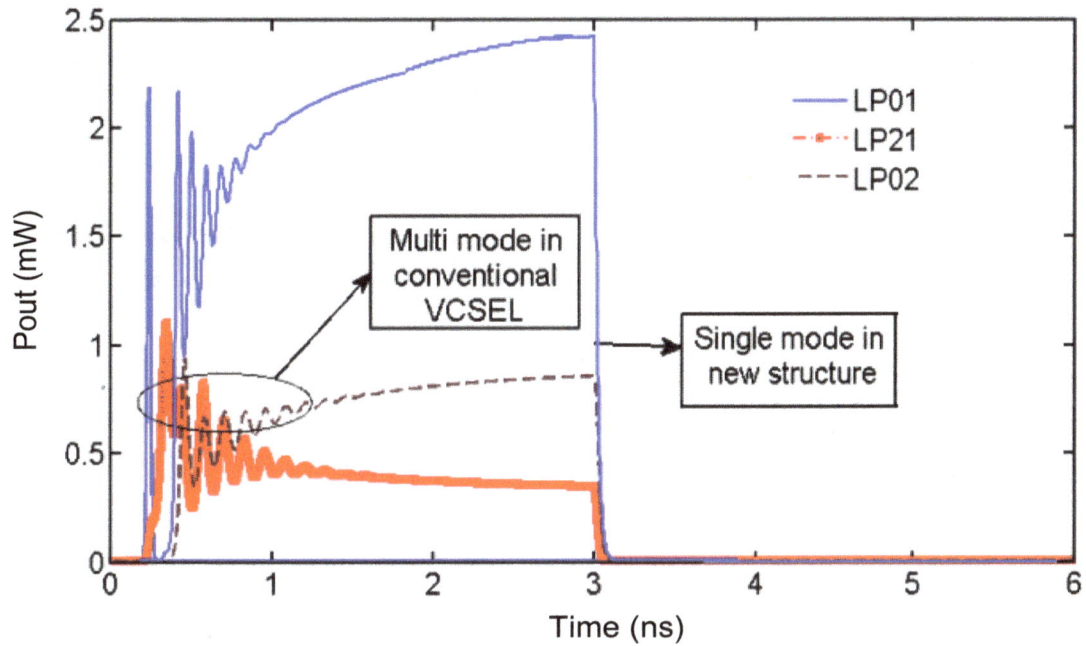

Figure 5. Small signal response in conventional (solid) and the new structure (dash).

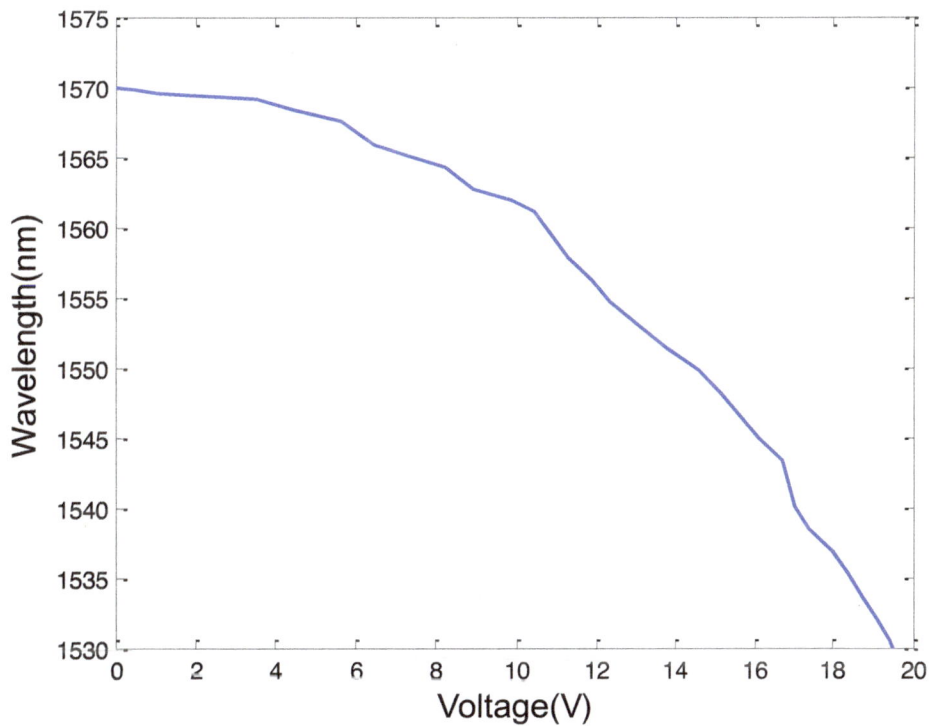

Figure 6. Tuning range in the new structure.

radius. It has been shown that thermal rollover occurs at higher injection current as compared to with conventional VCSEL.

REFERENCES

Chang-Hasnain CJ (2000). Tunable VCSEL. IEEE J. Selected Topic Quantum Elect. 6:978-987.

Guan B, Guo X, Deng J, Shen G (2009). Investigation on Tunable Wavelength and Modal Characteristics of MEMS Tunable Vertical Cavity Surface Emitting Lasers, Asia Communications and Photonics Conference and Exhibition (ACP), China.

Gutowski K, Sarzala RP (2008). Computer simulation of tuned and detuned GaInNAsSb QW VCSELs, Mate. Sci. Poland. pp. 26-45.

Jungo M, Erni D (2003). VISTAS: A Comprehensive System Oriented Spatio temporal VCSEL Model. IEEE J. Selected Topics Quantum Elect. 9(3):939-948.

Kogel B, Debernardi P, Westberg P, Haglund A, Gustavsson J, Bengtsson J, Haglund E, Larsson A (2011). Single mode tunable VCSELs with integrated MEMS technology, Proceedings of the European Conference on Laser and Electro-Optics (CLEO/Europe), Paper CB8.5-WED. Germany.

Martin WA (2001). Micro machined Tunable VCSELs for Wavelength-Division Multiplexing Systems, Ph.D. Thesis, Stanford University.

Nakwaski W, Sarzala R (1998). Transverse-Modes Control of Vertical-Cavity Surface-Emitting Lasers. Opt. Commun. 148:63-69.

Ochoa EM (2007). Hybrid Micro Electro Mechanical Tunable Filter Ph.D. Thesis, Air Force Institute of Technology University.

Samal N (2004). Ph.D. Thesis. High-power Single-mode Vertical Cavity Surface Emitting Lasers, Arizona State University.

Samal N, Johnson SR, Ding D (2005). High-power Single-mode Vertical-Cavity Surface-Emitting Lasers. J. Appl. Phys. Lett. 87(22):161108-161124.

Tayebati P, Wang P, Azimi M, Maflah L, Vakhshoori D (1999). Half-Symmetric Cavity Micro electro mechanically Tunable Vertical Cavity Surface Emitting Lasers with Single Spatial Mode Operating Near 950 nm. IEEE Photon. Technol. Lett. 75(7):897-898.

Zhou D, Mawst LJ (2002). High-power single-mode anti resonant reflecting optical waveguide-type vertical-cavity surface-emitting lasers. IEEE J. Quantum. 38:1599-1606.

10

Multiband fractal based reconfigurable antenna with introduction of RF MEMS switches for next generation devices

Paras Chawla and Rajesh Khanna

Department of Electronics and Communication Engineering, Thapar University, Patiala, Punjab, India.

Next generation network's main requirements are compact size and also support to heterogeneous network. Fractal antennas have useful applications in cellular telephone and microwave communications. Video conferencing, streaming video are main applications that are included in next generation networks and requirements for these applications are high data rates require to have high bandwidth. But as size of antenna reduces, bandwidth support also reduces. So it is required to have small size with high bandwidth. Fractal antennas are proposed in this paper for different applications as it is compact and also has good bandwidth. In this paper, we propose a reconfigurable fractal shape antenna for *Ku* and *Ka* band. Its performance analysis shows these antennas are multiband antennas support *Ku* and *Ka* bands and also support mobile terminal applications which are sight to be used in next generation networks. Different number of iteration is performed for its support to different bands. Re-configurability in antenna is another important concept and in this work; the reconfigurability is achieved by using radio frequency microelectromechanical system (RF MEMS) switches.

Key words: Meander, RF MEMS switch, reconfigurable antenna, fractal antenna.

INTRODUCTION

Rapid developments in the wireless communication industry continue to drive the requirements for small, compatible, and affordable reconfigurable antennas and antenna arrays. Reconfigurable multiband antennas are attractive for many military and commercial applications where it is desirable to have a single antenna that can be dynamically reconfigured to transmit and/or receive on multiple frequency bands (Brown, 1998; Weedon et al., 1999; Kiriazi et al., 2003). Such antennas find applications in space-based radar, unmanned aerial vehicles, satellites, electronic intelligence aircraft and many other communications and sensing applications. The technology of design and fabrication of MEMS for RF circuits had a major positive impact on reconfigurable antennas (Cetiner et al., 2003; Chiao et al., 1999). Next

generation network aims at high data rates requires wide bandwidth. Antennas performance in a communication system requires being highly efficient. But today's miniaturization technologies need antennas size to be less as well. Both requirements at a time first for small size and wider bandwidth can be fulfilled by using antenna fractal shape (Song et al., 2004; Wu et al., 2007; Werner and Ganguly, 2003). Re-configurability antenna can be further used to improve overall system performance.

Pattern-reconfigurable antennas can either increase capacity or extend radio coverage by increasing the carrier-to-interference ratio. In this paper, we analyze antennas which explore their potentials for different emerging frequency bands.

Puente et al. (1996; 1998) introduces fractal multiband antenna based on the Sierpinski gasket. Reconfigurability to these fractal antennas will enhance the performance and this reconfigurability is achieved by using RF MEMS. Anagnostou et al. (2006) defines an RF MEMS based self similar shaped reconfigurable antenna which has potential of multiband antenna with small size and reconfigurability. In this paper, proposed antenna with RF MEMS support shows an improvement in insertion loss and isolation loss offered by switches. In this paper, we propose a switch which uses meander shape beam rather than cantilever beam for improvement in scattering parameters. It has wide potential with multiband support for different applications like Ku and Ka band which is to be sight for different satellite communication (Pacheco et al., 2001). It is also supposed to support mobile terminal applications.

ANTENNA AND RF MEMS SWITCH DESIGN

Reconfigurablity in antennas offers us tunability of frequency, so that we can use single antenna for different range of frequencies. Electric switching proved useful in assuring re-configurability for frequency tuning in antenna. MEMS switches are used in this paper to achieve frequency tuning of antenna. As size and bandwidth requirement is achieved by fractal shape, we propose here fractal shaped reconfigurable antenna which is reconfigured by RF MEMS switch and it supports bands including Ku, Ka band and also mobile terminal applications.

RF MEMS switch

MEMS are the integration of mechanical elements, sensors, actuators, and electronics on a common silicon substrate through micro-fabrication technology. The micro-fabrication process normally involves a lithography-based micromachining, fabricated on batch basis, which offers great advantages of low cost when manufacturing in large volume (Rebeiz, 2003). MEMS are also built on high-resistivity gallium arsenide (GaAs) wafers, and quartz substrates using semiconductor micro fabrication technology. From a mechanical point of view, MEMS switches can be a thin metal cantilever, air bridge, or diaphragm; from RF circuit configuration point of view, it can be series connected or parallel connected with an RF transmission line. The contact condition can be capacitive (metal–insulator–metal) or resistive (metal-to-metal), polar ceramics such as (Ba,Sr)TiO3 - BST and designed to open the line or shunt it to ground upon actuation of the MEMS switch. Each type of switch has certain advantages in performance or manufacturability. Tunable ferroelectrics have great potential applications in tunable microwave devices. Main mechanical operations of RF MEMS switches depends mainly on spring constant of

material used, that is, k. We always require to have less k, that is, less stiff material because the deflection of beam depends on spring constant k and we need more deflection with given force for a given RF MEMS switch. In this paper, we have used Serpentine flexure type meander shape to lower the k value.

Calculation for spring constant for meander shaped beam (Rebeiz, 2003), is given as:

Serpentine flexure

$$k \approx \frac{48GJ}{l_a^2 \left(\frac{GJ}{EI_x} l_a + l_b \right) n^3} \quad n >> \frac{3l_b}{\frac{GJ}{EI_x} l_a + l_b} \quad \text{for} \qquad (1)$$

where n is the number of meanders in the serpentine flexure, $G = E / 2(1+v)$ is the torsion modulus, $I_x = wt^3 / 12$ is the moment of inertia, E = elasticity, v = poison ratio and the torsion constant is given by

$$J = \frac{1}{3} t^3 w \left(1 - \frac{192}{\pi^5} \frac{t}{w} \sum_{i=1,iodd}^{\infty} \frac{1}{i^5} \tanh \left(\frac{i\pi w}{2t} \right) \right) \qquad (2)$$

For the case where $l_a >> l_b$, the spring constant of the serpentine flexure becomes

$$k \approx 4Ew\left(t / (nl_a)^3 \right) \qquad (3)$$

DESIGN OF A COPLANAR WAVEGUIDE

Coplanar waveguide (CPW) is a one-sided three-conductor transmission line. Coplanar waveguide have two grounds in the same plane of center conductor, reducing the coupling effects and allows for easy inclusion of series and shunt elements. Since microwave integrated circuits are basically coplanar in structure, coplanar waveguide lines are used widely as circuit elements and as interconnecting lines. At millimeter-wave frequencies, coplanar waveguide offers the potential of lower conductor and radiation losses as compared to microstrip lines (Payne and Weedon, 2000). Coplanar waveguide also allows for varying the dimensions of the transmission line without changing the characteristic impedance.

An approximate formula, for the characteristic impedance of the coplanar waveguide, assuming t is small, 0<k<1, and h>>w, is

$$Z_o = \frac{30\pi^2}{\sqrt{(\varepsilon_r + 1)/2}} [\ln(2 \frac{1+\sqrt{k}}{1-\sqrt{k}})]^{-1} \text{ohms} \qquad (4)$$

where $k = \frac{w}{w+2s}$ and w = center strip width; s = slot width; ε_{re} = relative dielectric constant of the dielectric substrate.

Figure 1. Serpentine flexure beam based RF MEMS switch.

An empirical equation for effective relative dielectric constant ε_{re} is given as

$$\varepsilon_{re} = \frac{\varepsilon_r + 1}{2}\left[\tanh\left(1.785\log\left(\frac{h}{w} + 1.75\right) + \frac{kw}{h}\left(0.04 - 0.7k + (1 - 0.1\varepsilon_r)\frac{(0.25 + k)}{100}\right)\right)\right] \quad (5)$$

As shown in Figure 1, the silicon substrate of $\varepsilon_r = 11.8$ was chosen, and CPW center conductor width w = 2 μm and ground spacing s = 0.8 μm, nearly thickness is 0.5 mm and nearly infinite compared to width. The outer CPW ground wave conductor width is 14 μm. The effective dielectric constant is 4.4. Microwave parameters that should be optimized for any RF switch are the insertion loss, isolation, and switching frequency and return loss (Reid, 1999; Smith, 1999). The insertion loss is due to mismatch the characteristic impedance of the line and switch. The contact resistance and beam metallization loss will also contribute to the insertion loss. Switch influence on these microwave parameters will be discussed subsequently.

ACTIVATION MECHANISM FOR ELECTROSTATIC ACTUATION

When the voltage is applied between a fixed-fixed or cantilever beam and the pull down electrode, an electrostatic force is induced on the beam, the electrostatic force applied to the beam is found by considering the power delivered to a time-dependent capacitance and is given by,

$$F_e = \frac{1}{2}V^2\frac{dC(g)}{dg} = -\frac{1}{2}\frac{\varepsilon_0 WwV^2}{g^2} \quad (6)$$

where V is the voltage applied between the beam and electrode. W is the electrode area.

The electrostatic force is approximated as being distributed evenly across the beam section above the electrode. Equating the applied electrostatic force with the mechanical restoring force due to the stiffness of the beam (F= kx),

$$\frac{1}{2}\frac{\varepsilon_0 WwV^2}{g^2} = k(g - g_0) \quad (7)$$

where g_0 is the zero-bias bridge height. At ($2/3g_0$), the increase in the electrostatic force is greater than the increase in the restoring force, resulting in the beam position becoming unstable and collapse of the bam to the down-state position. The pull-down (also called pull-in) voltage is found to be

$$V_p(V) = V\left(\frac{2g_0}{3}\right) = \sqrt{\frac{8k}{27\varepsilon_0 W.w}go^3} = \sqrt{\frac{8k}{27\varepsilon_0 A}go^3} \quad (8)$$

As shown in Equation (8), the pull down voltage depends on the spring constant of beam structure, and, beam gap g_0 and electrode area A. To reduce the actuation voltage, the key is beam structure of low spring constant k. The pull-in voltage was investigated in terms of beam structure (different k), beam thickness, gap and beam materials.

Table 1. Parameters for capacitive RF MEMS shunt switch.

Parameter	Value	Parameter	Value
Length (µm)	160	Actuation area (µm^2)	20 * 80
Width (µm)	80	Actuation voltage (V)	21.5 V (20-30 V)
Height (µm)	2	C_d (pF)	50
Membrane type	Aluminium (film)	Poision ratio	3.0e-001
Thickness (µm)	0.5	Young's modulus	7.7e+004
Residual stress (MPa)	0	Isolation (dB) > 20 GHz	20 dB
Spring constant (Nm)	10-50	Isolation (dB) 20-55 GHz	35 dB
Holes (µm)	No	Loss (dB) (0.2-55 GHz)	0.2-0.4
Sacrificial layer	BPSG (2 µm)	C_d (pF)	50
Dielectric (Å)	SiN (0.2 µm)		

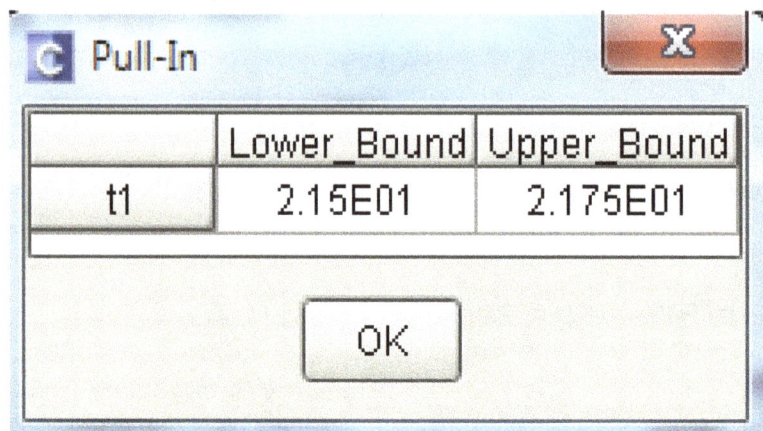

Figure 2. Pull-in voltage for capacitive MEMS switch.

FRACTAL ANTENNA

From the antenna side, microstrip antennas are well known for their features such as low profile, light weight, low cost, conformability to planar and non-planar surfaces, rigid, and easy installation. They are most commonly incorporated into mobile communications devices because of low cost and versatile designs. An emphasis has been given in microstrip antenna structures and reconfigurable aperture (recap), in order to achieve multiple octave tunability. Reconfigurable multiband slot antennas are receiving a lot of attention lately due to the emergence of RF-MEMS switches. In recent years neural networks and genetic algorithms are being used extensively for new reconfigurable multiband antenna designs.

The feature of self-similarity of a fractal antenna can also provide a basis for the design of multiple-frequency antennas. These antennas have the advantage that they radiate similar patterns in a variety of frequency bands. The major predecessor is the widely studied Sierpinski gasket (Puente et al., 1996, 1998). Benoit Mandelbrot,

the pioneer of classifying this geometry, first coined the term "fractal" in 1975 from the Latin word fractal, which means "broken or irregular fragments". The various fractals shape that posse self-similarity have been applied to multi-band or miniaturized antenna design. Many fractal geometries such as Sierpinski gasket, Sierpinski carpet, Koch island, Hilbert curve and Minkwoski etc has been used in fractal antennas.

RESULTS AND DISCUSSION

RF MEMS design and analysis

Table 1 shows the various parameters calculated and measured for meander based RF MEMS switch. Since coventorware software could synthesize the multiply factors, such as electrostatic-forces, pull-down voltages, Young's modulus, and other vector values could are obtained, and the result is intuitionistic. So, we also used coventore software to know the relationship shape, material and actuated voltage of switch. Figure 2 shows

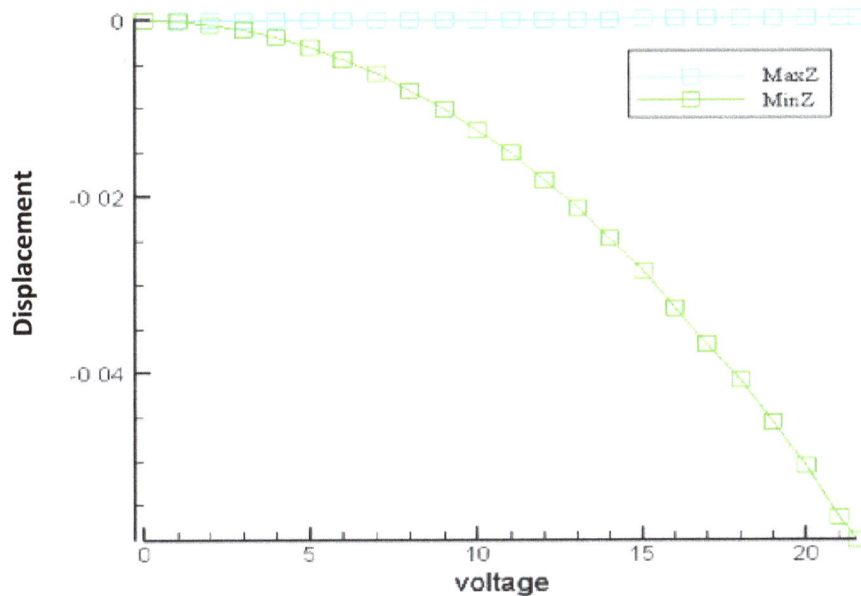

Figure 3. Displacement versus voltage graph for capacitive MEMS switch.

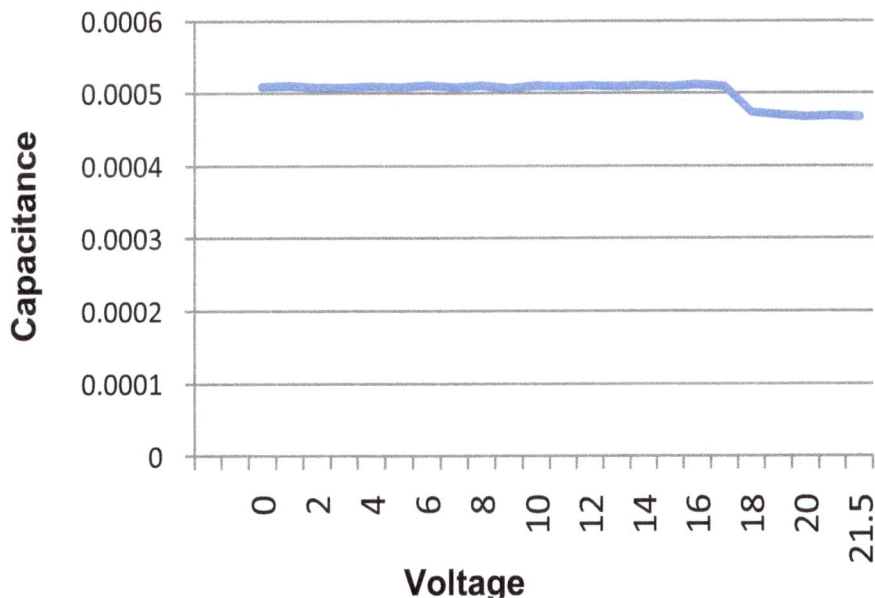

Figure 4. Voltage versus capacitance graph for meander based RF MEMS switch.

the pull-in voltage for RF MEMS switch which ranges from 21.5 to 21.7 V. Variation of displacement, that is, from maximum to minimum z-direction with respect to voltage is shown next in Figure 3, when the switch is electrostatically actuated. The next graph (in Figure 4) shows change in capacitance value with different values of voltages. In off–state, its value is 509.09 µF and at pull-in voltage, that is, in on-state comes out to be 467.9 µF.

FRACTAL ANTENNA DESIGN AND ANALYSIS

We have design Sierpinski gasket shaped patch on FR4 substrate having dielectric constant is equal to 4.4 and thickness is 1.6 mm. The size of substrate has dimensions of 26 × 30 mm2. Figure 5 shows the geometry of designed planar antenna in finite element based electromagnetic solver HFSS software. Figure 6 shows the result of return loss, which show designed

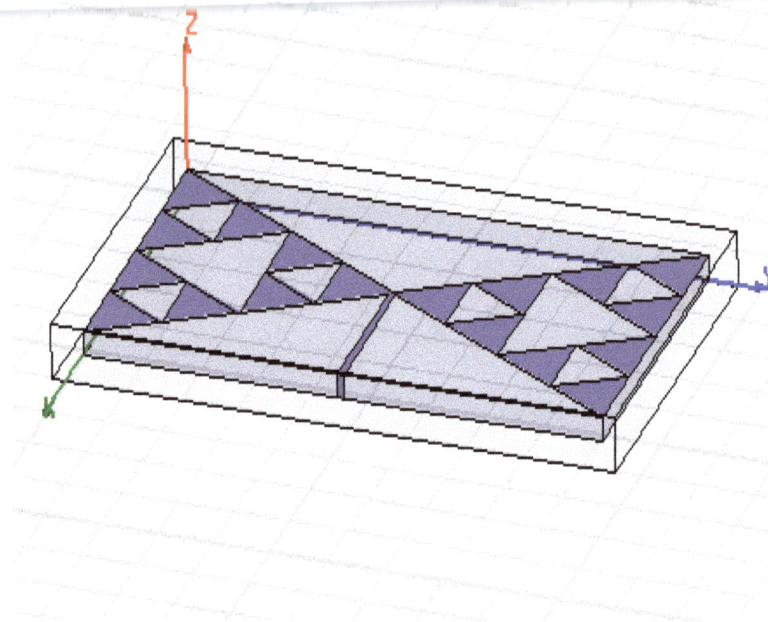

Figure 5. Fractal Sierpinski gasket shape for antenna.

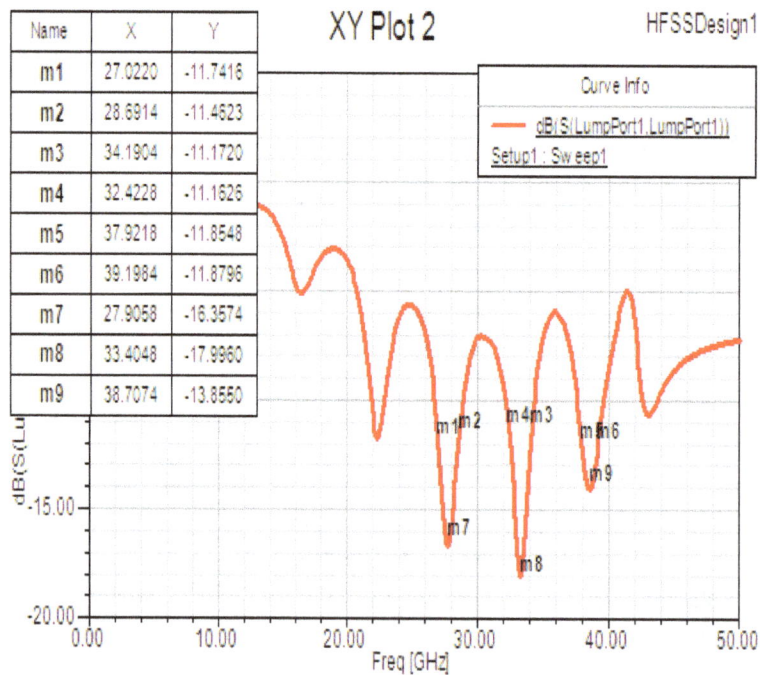

Figure 6. Return loss (different mark points to measure bandwidth and resonant frequencies).

antenna are resonating on three bands. Resonant frequencies are marked as 7, 8, 9 m, that is, 27.9056, 33.4048 and 38.7074 GHz, respectively.

The aforementioned multi-band antenna having bandwidth (in GHz) are given as:

m2 – m1 = 28.6914-27.0220 = 1.6694
m3 – m4 = 34.1904-32.3228 = 1.8676
m6 – m5 = 39.1984-37.9218 = 1.2766

Further, as expected from geometry (Figure 7) show that

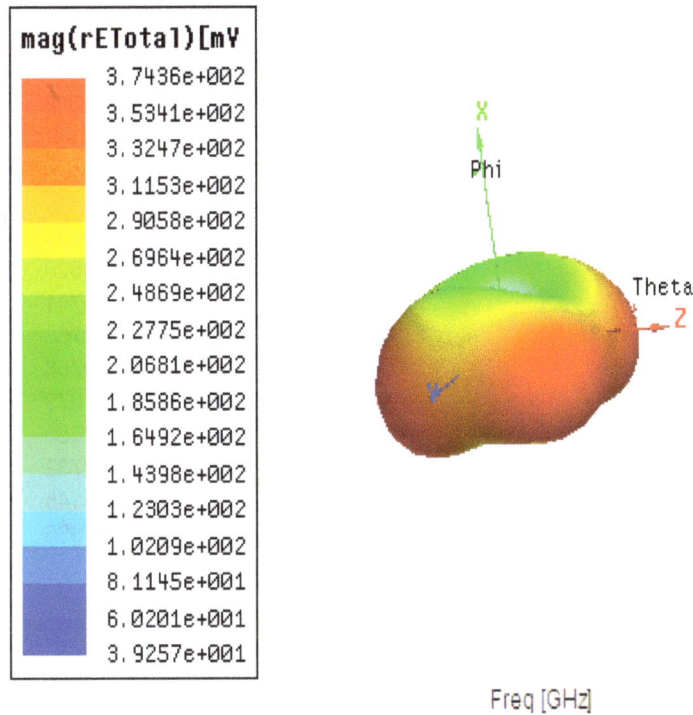

Figure 7. Radiation pattern and polar plot for proposed antenna.

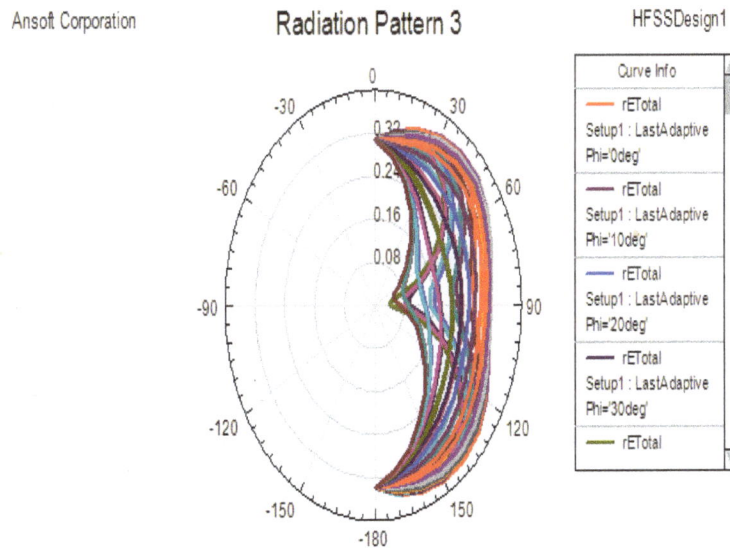

Figure 8. Simulated total normalized electric field intensities of the designed antenna.

antenna radiates equally in all directions and defining the properties of omnidirectional radiation pattern. There is not much difference in radiation patterns of all cases namely, with/without switches of designed antennas. Simulated result of Figure 8 show the total normalized electric field intensities of the designed antenna for all possible values of phi and theta. A Table 2 result signifies that at resonant frequencies, the voltage standing wave ratio (VSWR) is less than 2.

Figure 9 to 14 show the geometries and results of return loss for 3rd to 1st iteration. Figure 15 shows the fabricated structure of all antennas for proof of concept.

Table 2. VSWR at different selected frequencies shows value less than 2.

S/N	Frequency in (GHz)	VSWR (less than 2)
1	27.218437	1.544088
2	27.316633	1.480224
3	27.414830	1.427252
4	27.513026	1.386671
5	27.611222	1.359630
6	27.709419	1.346528
7	27.807615	1.346741
8	27.905812	1.358768
9	28.004008	1.380758
10	28.102204	1.411079
11	28.200401	1.448609
12	28.298597	1.492732
13	28.396794	1.543145
14	28.494990	1.599635

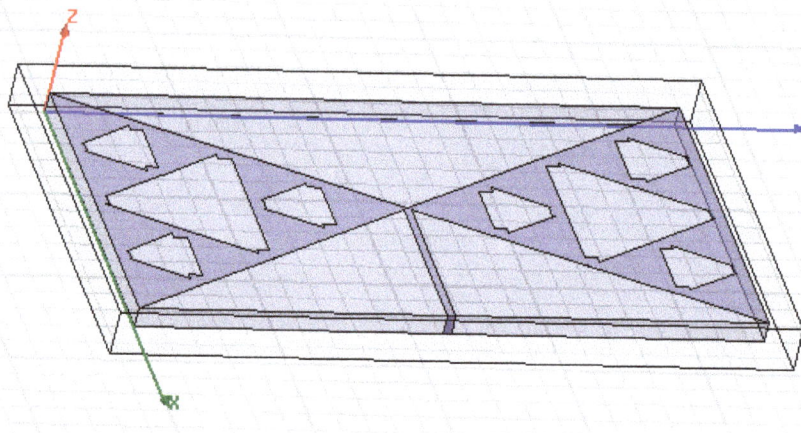

Figure 9. 3rd iteration Sierpenski fractal antenna.

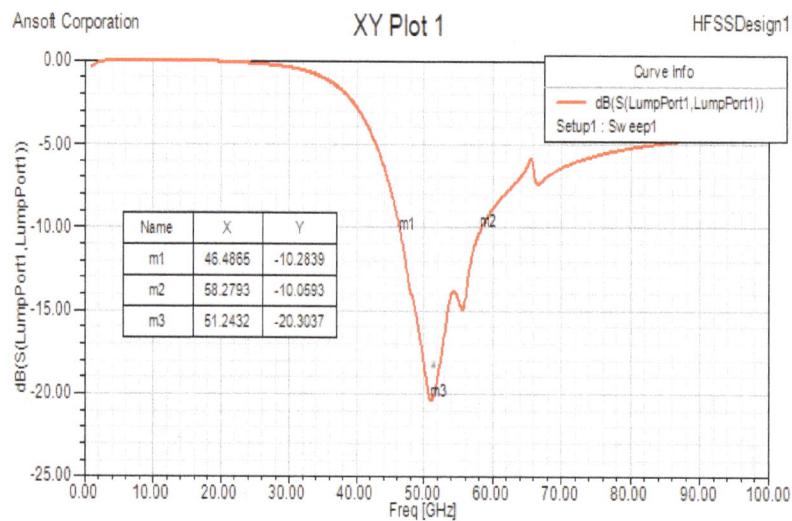

Figure 10. Different mark points to measure bandwidth and resonant frequencies.

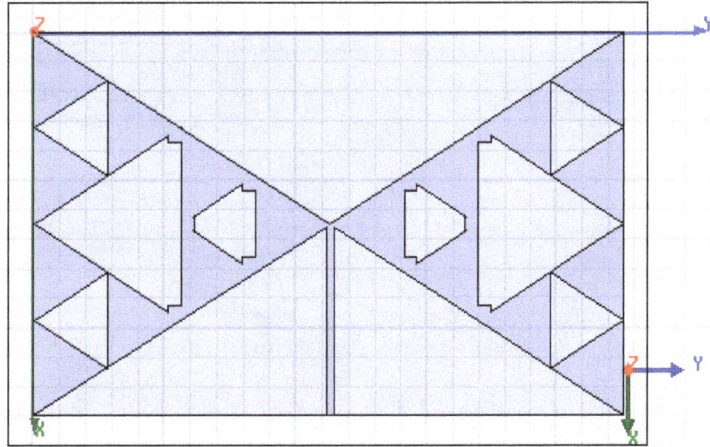

Figure 11. 2nd iteration Sierpenski fractal antenna.

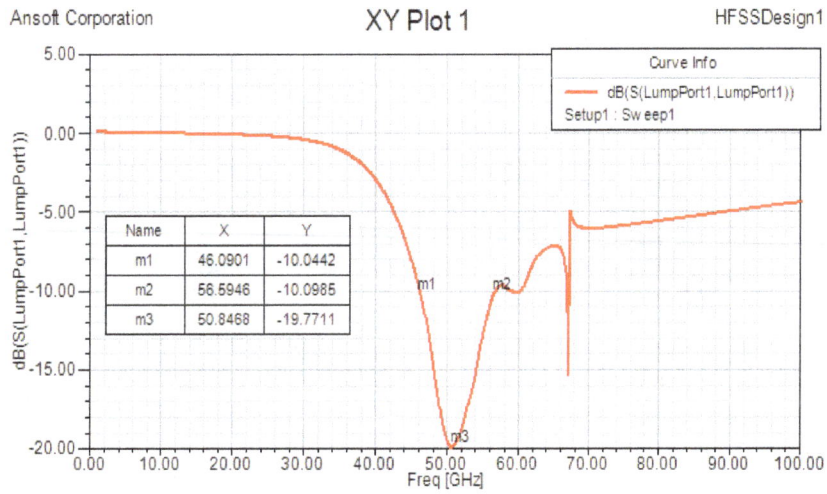

Figure 12. Different mark points to measure bandwidth and resonant frequencies.

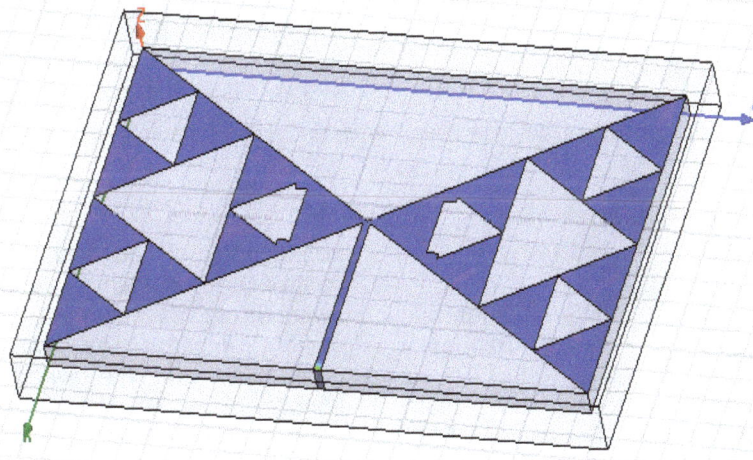

Figure 13. 1st iteration Sierpenski fractal antenna.

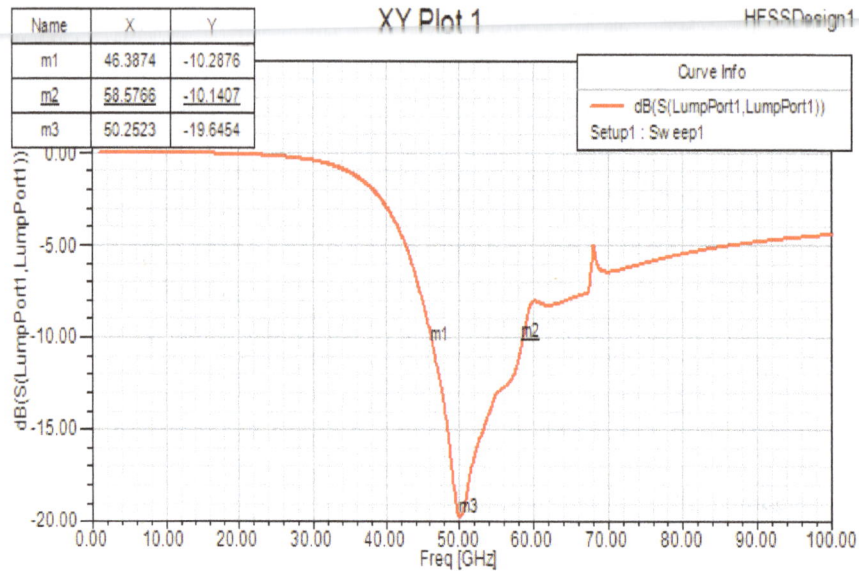

Figure 14. Different mark points to measure bandwidth and resonant frequencies.

Figure 15. Fabricated 1st to 3rd iteration Sierpenski Fractal Antenna.

Conclusion

In this paper, meander based RF MEMS switches has been designed and simulated for a multiple-frequency fractal antenna. Further, the frequency tunability is achieved by the reconfiguring the antenna electronically or changing the order of iteration. By using MEMS switches, the losses are kept to a minimum. This is very important to obtain high reconfigurability and because the connection/ disconnection of a switch may lead to mismatched systems. Also, the integration of the switches to the antenna imposes restrictions on the system design. The antenna structure should be compatible with the switch structure and made with specific materials. The dimensions of the switches often cannot be altered and thus put a low-bound in the size of the antenna system and a high-bound in the power the system can handle. This technology can be applied to many other devices, including tunable filters, other antenna geometries, signal splitters or military applications.

Further in this paper, fractal shaped microstrip antenna is proposed with its return loss, radiation pattern and voltage standing wave ratio (VSWR) as shown in the result. As can be analyze from the result, it shows good characteristics to work for many wireless communication applications. It is offering return loss to be less than -10 dB for many frequencies so supporting multiband applications. For proof of concept open and short path are provided instead of switches. Radiation pattern being omnidirectional and VSWR is also less than 2. As it is microstrip antenna, its size is small. So requirements for multiband support and small size are met with the proposed antenna.

REFERENCES

Anagnostou DE, Zheng G, Chryssomallis MT, Lyke JS, Ponchak GE, Papapolymerou J, Christodoulou CG (2006). Design, fabrication, and measurements of an RF-MEMS-based self-similar Reconfigurable Antenna. IEEE Trans. Antennas Propag. 54(2):422-432.

Brown ER (1998). RF-MEMS switches for reconfigurable integrated circuits. IEEE Trans. Microw. Theory Tech. 46(11):1868-1880.

Cetiner BA, Qian JY, Chang HP, Bachman M (2003). Monolithic integration of RF MEMS switches with a diversity Antenna on PCB substrate. IEEE Trans. Microw. Theory Tech. 51:332-335.

Chiao JC, Fu Y, Chio IM, Lisio MD, Lin LY (1999). MEMS reconfigurable Vee antenna.Proceeding IEEE MTT-S Int. Microw. Symp. Dig. 4:1515-1518.

Kiriazi J, Ghali H, Ragaie H, Haddara H (2003). Reconfigurable dual band dipole Antenna on Silicon using series MEMS switches. Proc. Antennas Propag. Soc. Int. Symp. 1:403-406.

Pacheco SP, Peroulis D, Katehi LPB (2001). MEMS single-pole double-throw (SPDT) X and K-band switching circuits. Microwave Theory Tech -S Int. Microw. Symp. 1:165-168.

Payne WJ, Weedon WH (2000). Stripline feed networks for reconfigurable patch antennas. Proceedings for the Antenna Applications Symposium. Allerton Park Monticello, Illinois.

Puente C, Romeu J, Pous R, Garcia X, Benltez F (1996). Fractal multiband antenna based on the Sierpinski gasket. Electr. Lett. 32(1):1-2.

Puente C, Romeu J, Pous R, Cardama (1998). On the Behavior of the Multiband Sierpinski Fractal Antenna. IEEE Trans. Antennas propagation.

Rebeiz GM (2003). RF MEMS Theory, Design, and Technology. New Jersey: John Wiley & Sons Publication.

Reid JR (1999). An Overview of micro-electro-mechanical systems (MEMS).Tutorial Session on MEMS for antenna applications antenna applications symposium. Allerton Park Monticello, Illinois.

Smith JK (1999). MEMS and advanced radar. Tutorial session on MEMS for antenna applications.antenna applications symposium.Allerton Park Monticello, Illinois.

Song CTP, Hall PS, Ghafouri-Shiraz H (2004). Shorted Fractal Sierpinski Monopole Antenna. IEEE Trans. Antennas Propag. 52 (10).

Wu W, Wang BZ, Yang XS, Zhang Y (2007). A Pattern- Reconfigurable Planar Fractal Antenna and its Characteristic-Mode Analysis. IEEE Antennas Propag. Mag. 49(3).

Werner DH, Ganguly S (2003). An Overview of Fractal Antenna Engineering Research. IEEE Antennas Prop. Mag. 45(1):38-57.

Weedon WH, Payne WJ, Rebeiz GM, Herd JS, Champion M (1999). MEMS-Switched reconfigurable multi-band antenna: Design and modeling. Proceedings for the 1999 Antenna Applications Symposium. Allerton Park Monticello, Illinois.

11

A Mode-I crack problem for two-dimensional problem of a fiber-reinforced thermoelastic with normal mode analysis

Kh. Lotfy[1] and Wafaa Hassan[2,3]

[1]Department of Mathematics, Faculty of Science, Zagazig University, Zagazig P. O. Box 44519 Egypt.
[2]Department of Mathematics and Physics, Faculty of Engineering, Port Said Branch of Suez Canal University, Port Said, Egypt.
[3]Department of Mathematics, Faculty of Science and Arts, Al-mithnab, Qassim University, P.O. Box 931, Buridah 51931, Al-mithnab, Kingdom of Saudi Arabia.

The aim of the present work is to investigate the influence of the Mode-I crack on the plane waves in a linearly fiber-reinforced. A general model of the equations of coupled theory (CD) and Lord-Shulman (L-S) theory with one relaxation time are applied to study the influence of reinforcement on the total deformation for an infinite space weakened by a finite linear opening Mode-I crack is solving. The material is homogeneous isotropic elastic half space. The crack is subjected to prescribed temperature and stress distribution. The normal mode analysis is used to obtain the exact expressions for the displacement components, force stresses and temperature. The variations of the considered variables with the horizontal distance are illustrated graphically. Comparisons are made with the results in two theories. A comparison also is made between the two theories for different depths.

Key words: Fiber-reinforced, Lord-Shulman theory, Mode-I crack, normal mode analysis, thermoelasticity.

INTRODUCTION

Fiber-reinforced composites are widely used in engineering structures, due to their superiority over the structural materials in applications requiring high strength and stiffness in lightweight components. A continuum model is used to explain the mechanical properties of such materials. A reinforced concrete member should be designed for all conditions of stresses that may occur and in accordance with principles of mechanics. The characteristic property of a reinforced concrete member is that its components, namely concrete and steel, act together as a single unit as long as they remain in the elastic condition that is, the two components are bounded together so that there can be no relative displacement between them. In the case of an elastic solid reinforced by a series of parallel fibers, it is usual to assume transverse isotropy. In the linear case, the associated constitutive relations, relating infinitesimal stress and strain components have five material constants. In the last three decades, the analysis of stress and deformation of fibre-reinforced composite materials has been an important research area of solid mechanics. A reinforced concrete member shall be designed for all conditions of stresses that may occur and accordance with principle of mechanics. Fiber-reinforced composites are used in a variety of structures due to their low weight and high strength. The characteristic property of a reinforced composite is that its components act together as single anisotropic units as long as they remain in the elastic condition. The waves propagation in a reinforced media plays a very interesting role in civil engineering and geophysics. The studies of propagation, reflection and transmission of waves are of great interest to seismologists. Such studies help them to obtain knowledge about the rock structures as well as their elastic properties and at the same time information regarding minerals and fluids present inside the earth. The idea of introducing a continuous self reinforcement at every point of an elastic solid was given by Belfield et al. (1983). The model was later applied to the rotation of a tube by Verma (1986), who has also discussed the magneto elastic shear waves in self-reinforced bodies. Singh (2002) showed that, for wave propagation in fibre-reinforced anisotropic media, this decoupling cannot be achieved by the introduction of the displacement potential. Sengupta and Nath (2001) discussed the problem of the surface waves in fibre-reinforced anisotropic elastic media. Hashin and Rosen (1964) gave the elastic moduli for fibre-reinforced materials. The problem of reflection of plane waves at the free surface of a fibre-reinforced elastic half-space was discussed by Singh and Singh (2004). Chattopadhyay and Choudhury (1990) have discussed the problem of propagation, reflection and transmission of magneto elastic shear waves in a self–reinforced medium. The reflection and transmission of plane SH wave through a self-reinforced elastic layer sandwiched between two homogeneous visco-elastic solid half-spaces has been studied by Chaudhary et al. (2004). Chattopadhyay and Choudhury (1995) studied the propagation of magneto-elastic shear waves in an infinite self-reinforced plate. Pradhan et al. (2003) studied the dispersion of Loves waves in a self-reinforced layer over an elastic non-homogenous half space. The propagation of plane waves in fibre-reinforced media is discussed by Chattopadhyay et al. (2002). The theory of couple thermo-elasticity was extended by Lord and Şhulman (1967) and Green and Lindsay (1972) by including the thermal relaxation time in constitutive relations.

Othman and Song (2007) showed the effect of initial stress, thermoelastic parameter and thermal boundary condition upon the reflection amplitude ratios. The problem of magneto-elastic transverse surfaces waves in self-reinforced elastic solid was studied by Verma et al. (1988). The problem of wave propagation in thermally conducting linear fibre-reinforced composite materials was discussed by Singh (2006).

Othman and Lotfy (2009) studied two-dimensional problem of generalized magneto-thermoelasticity under the effect of temperature dependent properties. Othman et al. (2009) studied transient disturbance in a half-space under generalized magneto-thermoelasticity with moving internal heat source. Othman and Lotfy (2010) studied the plane waves in generalized thermo -microstretch elastic half-space by using a general model of the equations of generalized thermo-microstretch for a homogeneous isotropic elastic half space. Othman and Lotfy (2009) studied the generalized thermo-microstretch elastic medium with temperature dependent properties for different theories. Othman and Lotfy (2010) studied the effect of magnetic field and inclined load in micropolar thermoelastic medium possessing cubic symmetry under three theories. The normal mode analysis was used to obtain the exact expression for the temperature distribution, thermal stresses, and the displacement components.

The investigation of interaction between a magnetic field, stress, and strain in a thermoelastic solid is very important due to its many applications in diverse field, such as geophysics (for understanding the effect of the Earth's magnetic field on seismic waves), damping of acoustic waves in a magnetic field, designing machine elements like heat exchangers, boiler tubes where the temperature induced elastic deformation occurs, biomedical engineering (problems involving thermal stress), emissions of the electromagnetic radiations from nuclear devices, development of a highly sensitive super conducting magnetometer, electrical power engineering plasma physics etc. (Lotfy, 2012a, b). The problem has been solved numerically using the normal mode analysis and many works in generalized magneto-thermo elasticity with effect of rotation and other fields can be found in (Othman and Saied, 2012a, b; 2013a, b; Othman and Atwa, 2012) for a half space fiber-reinforced. Numerical results for the conductive temperature, thermodynamic temperature, displacement components and the stresses are represented graphically and the results are analyzed.

In the recent years, considerable efforts have been devoted the study of failure and cracks in solids. This is due to the application of the latter generally in industry and particularly in the fabrication of electronic components. Most of the studies of dynamical crack problem are done using the equations of coupled or even uncoupled theories of thermoelasticity (Dhaliwal, 1980; Hasanyan et al., 2005; Ueda, 2003; Elfalaky and Abdel-Halim, 2006). This is suitable for most situations where long time effects are sought. However, when short time are important, as in many practical situations, the full system of generalized thermoelastic equations must be used (Lord and Şhulman, 1967).

In the present work we shall formulate the fiber-reinforced two-dimensional problem of thermally for an infinite space weakened by a finite linear opening Mode-I crack is solving for the considered variables. The normal mode method is used to obtain the exact expressions for the considered variables. The distributions of the considered variables are represented graphically. A comparison is carried out between the temperature, stresses and displacements as calculated from the generalized thermoelasticity L-S and coupled theories for half space fiber-reinforced in two problems. A comparison also is made between the two theories for different depths.

FORMULATION OF THE PROBLEM AND BASIC EQUATIONS

We shall consider the problem of a homogeneous, isotropic and linearly fiber-reinforced thermoelastic half-space ($x \geq 0$). The constitutive equations for a fibre-reinforced linearly elastic anisotropic medium with respect to the reinforcement direction 'a' are (Belfield et al., 1983):

$$\sigma_{ij} = \lambda e_{kk}\delta_{ij} + 2\mu_T e_{ij} + \alpha(a_k a_m e_{km}\delta_{ij} + a_i a_j e_{kk}) + 2(\mu_L - \mu_T)(a_i a_k e_{kj} + a_j a_k e_{ki}) + \beta a_k a_m e_{km} a_i a_j - \gamma(T-T_0)\delta_{ij}.$$

(1)

Where σ_{ij} are components of stress; e_{ij} are the components of strain; λ, μ_T are elastic constants; $\alpha, \beta, (\mu_L - \mu_T)$ are reinforcement parameters, $\gamma = (3\lambda + 2\mu)\alpha_t$, α_t thermal expansion coefficient and $\underline{a} \equiv (a_1, a_2, a_3)$, $a_1^2 + a_2^2 + a_3^2 = 1$. We choose the fiber-direction as $\underline{a} \equiv (1,0,0)$. The strains can be expressed in terms of the displacement u_i as:

$$e_{ij} = \frac{1}{2}(u_{i,j} + u_{j,i}).$$

(2)

For plane strain deformation in the xy-plane [displacement $\underline{u} = (u,v,0)$] $\frac{\partial}{\partial z} = 0$, $w = 0$. Equation (1) then yields

$$\sigma_{xx} = A_{11}u_{,x} + A_{12}v_{,y} - \gamma(T-T_0),$$

(3)

$$\sigma_{yy} = A_{22}v_{,y} + A_{12}u_{,x} - \gamma(T-T_0),$$

(4)

$$\sigma_{zz} = A_{12}u_{,x} + \lambda v_{,y} - \gamma(T-T_0),$$

(5)

$$\sigma_{xy} = \mu_L(u_{,y} + v_{,x}), \qquad \sigma_{zx} = \sigma_{zy} = 0.$$

(6)

Where

$$A_{11} = \lambda + 2(\alpha + \mu_T) + 4(\mu_L - \mu_T) + \beta, \qquad A_{12} = \alpha + \lambda,$$
$$A_{22} = \lambda + 2\mu_T.$$

(7)

The equation of motion in a rotating frame of reference in the context of Lord-Shulman's(LS) theory is

$$\rho \ddot{u}_i = \sigma_{ij,j}, \ (i,j = 1,2,3).$$

(8)

The heat conduction equation

$$kT_{,ii} = \left(\frac{\partial}{\partial t} + \tau_0 \frac{\partial^2}{\partial t^2}\right)(\rho C_E T + \gamma T_0 u_{i,i}).$$

(9)

Where ρ is the density, k is the thermal conductivity, C_E is specific heat at constant strain and T is temperature above reference temperature T_0.

Using the summation convection. From Equations (3) to (6), we note that the third equation of motion in Equation (8) identically satisfied and first two equations become:

$$\rho\left(\frac{\partial^2 u}{\partial t^2}\right) = A_{11}\frac{\partial^2 u}{\partial x^2} + B_2\frac{\partial^2 v}{\partial x \partial y} + B_1\frac{\partial^2 u}{\partial y^2} - \gamma\frac{\partial T}{\partial x},$$

(10)

$$\rho\left(\frac{\partial^2 v}{\partial t^2}\right) = A_{22}\frac{\partial^2 v}{\partial y^2} + B_2\frac{\partial^2 u}{\partial x \partial y} + B_1\frac{\partial^2 v}{\partial x^2} - \gamma\frac{\partial T}{\partial y}.$$

(11)

Where $B_1 = \mu_L$, $\qquad\qquad B_2 = \alpha + \lambda + \mu_L$.

For convenience, the following non-dimensional variables are used:

$$x' = c_1\eta x, \quad y' = c_1\eta y, \quad u' = c_1\eta u, \quad v' = c_1\eta v, \quad t' = c_1^2\eta t, \quad \tau_0' = c_1^2\eta\tau_0,$$

$$\Omega' = \frac{\Omega}{c_1^2\eta}, \quad \theta = \gamma\frac{(T-T_0)}{\lambda + 2\mu_T}, \quad \sigma_{ij}' = \frac{\sigma_{ij}}{\mu_T}. \quad i,j = 1,2.$$

(12)

$$\eta = \frac{\rho C_E}{k}, \qquad\qquad c_1^2 = \frac{\lambda + 2\mu_T}{\rho}.$$

Where

In terms of non-dimensional quantities defined in Equation (12), the above governing equations reduce to (dropping the dashed for convenience)

$$\frac{\partial^2 u}{\partial t^2} = h_{11}\frac{\partial^2 u}{\partial x^2} + h_2\frac{\partial^2 v}{\partial x \partial y} + h_1\frac{\partial^2 u}{\partial y^2} - \frac{\partial\theta}{\partial x},$$

(13)

$$\frac{\partial^2 v}{\partial t^2} = h_{22}\frac{\partial^2 v}{\partial y^2} + h_2\frac{\partial^2 u}{\partial x \partial y} + h_1\frac{\partial^2 v}{\partial x^2} - \frac{\partial\theta}{\partial y},$$

(14)

$$\frac{\partial^2\theta}{\partial x^2} + \frac{\partial^2\theta}{\partial y^2} = \left(\frac{\partial}{\partial t} + \tau_0\frac{\partial^2}{\partial t^2}\right)\left(\theta + \varepsilon\left(\frac{\partial u}{\partial x} + \frac{\partial v}{\partial y}\right)\right).$$

(15)

Where

$$(h_{11}, h_{22}, h_1, h_2) = \frac{(A_{11}, A_{22}, B_1, B_2)}{(\lambda + 2\mu_T)}, \qquad \varepsilon = \frac{\gamma^2 T_0}{\rho C_E(\lambda + 2\mu_T)}.$$

$$\mu_T\sigma_{xx} = A_{11}u_{,x} + A_{12}v_{,y} - (\lambda + 2\mu_T)\theta,$$

(16)

$$\mu_T\sigma_{yy} = A_{22}v_{,y} + A_{12}u_{,x} - (\lambda + 2\mu_T)\theta,$$

(17)

$$\mu_T\sigma_{zz} = A_{12}u_{,x} + \lambda v_{,y} - (\lambda + 2\mu_T)\theta,$$

(18)

$$\mu_T\sigma_{xy} = \mu_L(u_{,y} + v_{,x}), \qquad \sigma_{zx} = \sigma_{zy} = 0.$$

(19)

NORMAL MODE ANALYSIS

The normal mode analysis gives exact solutions without any assumed restrictions on temperature, displacement and stress distributions. It is applied to a wide range of problems in different branches. It can be applied to boundary-layer problems, which are described by the linearized Navier-Stokes equations in electro hydrodynamics. The normal mode analysis is, in fact, to look for the solution in Fourier transformed domain. Assuming that all the field quantities are sufficiently smooth on the real line such that normal mode analysis of these functions exists. The solution of the considered physical variable can be decomposed in terms of normal modes as the following form:

$$[u,v,\theta,\sigma_{ij}](x,y,t) = [u^*(x),v^*(x),\theta^*(x),\sigma_{ij}^*(x)]\exp(\omega t + iay). \quad (20)$$

Where ω is the (complex) time constant. $i=\sqrt{-1}$, a is the wave number in the $y-$ direction and $u^*(x),v^*(x),\theta^*(x)$ and $\sigma_{ij}^*(x)$ are the amplitude of the field quantities. Using Equation (20), then Equations (13) to (19) take the form

$$[h_{11}D^2 - A_1]u^* + [iah_2D]v^* = D\theta^*, \quad (21)$$

$$[h_1D^2 - A_2]v^* + [iah_2D]u^* = ia\theta^*, \quad (22)$$

$$A_4Du^* + iaA_4]v^* = [D^2 - A_3]\theta^*, \quad (23)$$

$$\mu_T\sigma_{xx}^* = A_{11}Du^* + iaA_{12}v^* - (\lambda + 2\mu_T)\theta^*, \quad (24)$$

$$\mu_T\sigma_{yy}^* = A_{12}Du^* + iaA_{22}v^* + -(\lambda + 2\mu_T)\theta^*, \quad (25)$$

$$\mu_T\sigma_{zz}^* = A_{12}Du^* + ia\lambda v^* - (\lambda + 2\mu_T)\theta^*, \quad (26)$$

$$\mu_T\sigma_{xy}^* = \mu_L(iau^* + Dv^*), \qquad \sigma_{zx}^* = \sigma_{zy}^* = 0. \quad (27)$$

Where

$$A_1 = \omega^2 + h_1a^2, \qquad\qquad A_2 = \omega^2 + h_{22}a^2,$$

$$A_3 = a^2 + \omega + \omega^2\tau_0, \qquad A_4 = \omega\varepsilon + \varepsilon\omega^2\tau_0, \qquad D = \frac{d}{dx}$$

Eliminating $\theta^*(x)$ and $v^*(x)$ between Equations (21) to (23), we obtain the ordinary differential equation satisfied by $u^*(x)$.

$$[D^6 - AD^4 + BD^2 - C]u^*(x) = 0. \quad (28)$$

where

$$A = \frac{1}{h_1h_{11}}\{h_{11}A_2 + h_1A_1 + h_1A_4 + h_1h_{11}A_3 - h_2^2a^2\}, \quad (29)$$

$$B = \frac{1}{h_1h_{11}}\{A_1A_2 + h_{11}A_2A_3 + h_1A_1A_3 + h_{11}a^2A_4 + A_2A_4 - h_2^2a^2A_3 - 2h_2a^2A_4\}, \quad (30)$$

$$C = \frac{1}{h_1h_{11}}\{A_1A_2A_3 + A_1A_4a^2\}. \quad (31)$$

In a similar manner, we get

$$[D^6 - AD^4 + BD^2 - C]\{v^*(x),\theta^*(x)\} = 0. \quad (32)$$

The above equation can be factorized as

$$(D^2 - k_1^2)(D^2 - k_2^2)(D^2 - k_3^2)u^*(x) = 0, \quad (33)$$

where, $k_n^2 (n=1,2,3)$ are the roots of the following characteristic equation

$$k^6 - Ak^4 + Bk^2 - C = 0.. \quad (34)$$

The solution of Equation (33) which is bounded as $x \to \infty$, is given by

$$u^*(x) = \sum_{n=1}^{3} M_n(a,\omega)\exp(-k_nx) \quad (35)$$

Similarly

$$v^*(x) = \sum_{n=1}^{3} M_n'(a,\omega)\exp(-k_nx) \quad (36)$$

$$\theta^*(x) = \sum_{n=1}^{3} M_n''(a,\omega)\exp(-k_nx) \quad (37)$$

Where M_n, M_n' and M_n'' are some parameters depending on a and ω.

Substituting from Equations (35) to (37) into Equations (21) to (23), we have

$$M_n'(a,\omega) = H_{1n}M_n(a,\omega), \qquad n=1,2,3. \quad (38)$$

$$M_n''(a,\omega) = H_{2n}M_n(a,\omega), \qquad n=1,2,3. \quad (39)$$

Where

$$H_{1n} = \frac{-iah_{11}k_n^2 + iaA_1 + iah_2k_n^2}{h_2k_na^2 + h_1k_n^3 - A_2k_n}, \qquad n=1,2,3. \quad (40)$$

$$H_{2n} = \frac{-A_4k_n + iaA_4H_{1n}}{k_n^2 - A_3}, \qquad n=1,2,3. \quad (41)$$

Thus, we have

$$v^*(x) = \sum_{n=1}^{3} H_{1n} M_n(a,\omega)\exp(-k_n x)$$

(42)

$$\theta^*(x) = \sum_{n=1}^{3} H_{2n} M_n(a,\omega)\exp(-k_n x)$$

(43)

Substitution of Equations (35), (42) and (43) into Equations (24) to (27), we get

$$\mu_T \sigma_{xx}^* = \sum_{n=1}^{3} H_{3n} M_n(a,\omega)\exp(-k_n x)$$

(44)

$$\mu_T \sigma_{yy}^* = \sum_{n=1}^{3} H_{4n} M_n(a,\omega)\exp(-k_n x)$$

(45)

$$\mu_T \sigma_{zz}^* = \sum_{n=1}^{3} H_{5n} M_n(a,\omega)\exp(-k_n x)$$

(46)

$$\mu_T \sigma_{xy}^* = \sum_{n=1}^{3} H_{6n} M_n(a,\omega)\exp(-k_n x)$$

(47)

where

$$H_{3n} = -A_1 k_n + iaA_{12}H_{1n} - (\lambda + 2\mu_T)H_{2n} \qquad n=1,2,3.$$

(48)

$$H_{4n} = -A_{12} k_n + iaA_{22}H_{1n} - (\lambda + 2\mu_T)H_{2n} \qquad n=1,2,3.$$

(49)

$$H_{5n} = -A_{12} k_n + ia\lambda H_{1n} - (\lambda + 2\mu_T)H_{2n} \qquad n=1,2,3.$$

(50)

$$H_{6n} = \mu_L(ia - k_n H_{1n}) \qquad n=1,2,3.$$

(51)

The normal mode analysis is, in fact, to look for the solution in Fourier transformed domain. Assuming that all the field quantities are sufficiently smooth on the real line such that normal mode analysis of these functions exists.

APPLICATION

Problem 1 (A Mode-I crack)

The plane boundary subjects to an instantaneous normal point force and the boundary surface is isothermal, the boundary conditions (Figure 1) at the vertical plan $y=0$ and in the beginning of the crack at $x=0$ are

$$\sigma_{yy}(x,y,t) = -p(y,t), \quad |x| < a$$

(52)

$$\theta(x,y,t) = f(y,t) \quad |x| < a \text{ and } \frac{\partial\theta}{\partial y} = 0 \quad |x| > a$$

(53)

$$\sigma_{xy}(x,y,t) = 0 \quad -\infty < x < \infty$$

(54)

Substituting the expressions of the variables considered into the above boundary conditions, we get

$$\sum_{n=1}^{3} H_{3n} M_n(a,\omega) = -p^*(y,t)$$

(55)

$$\sum_{n=1}^{3} H_{2n} M_n(a,\omega) = f^*(y,t)$$

(56)

$$\sum_{n=1}^{3} H_{6n} M_n(a,\omega) = 0$$

(57)

Invoking the boundary conditions (55) to (57) at the surface $x=0$ of the plate, we obtain a system of three equations. After applying the inverse of matrix method, we have the values of the three constants $M_j, j=1,2,3$. Hence, we obtain the expressions of displacements, temperature distribution and another physical quantities of the plate the muscles.

$$\begin{pmatrix} M_1 \\ M_2 \\ M_3 \end{pmatrix} = \begin{pmatrix} H_{41} & H_{42} & H_{43} \\ H_{21} & H_{22} & H_{23} \\ H_{61} & H_{62} & H_{63} \end{pmatrix}^{-1} \begin{pmatrix} -p^* \\ f^* \\ 0 \end{pmatrix}$$

(58)

Problem 2

A time-dependent heat punches across the surface of semi-infinite thermo-elastic half space. In the physical problem, we should suppress the positive exponential that are unbounded at infinity. The constants M_1, M_2 and M_3 have to be chosen such that the boundary conditions on the surface x = 0 take the form

(1) Thermal boundary condition that the surface of the half-space subjected to a

$$\theta(0,y,t) = 0 \quad \text{on} \quad x=0,$$

(59)

(2) Thermal boundary condition that the surface of the half-space subjected to a

$$\sigma_{xx}(0,y,t) = -p_1(y,t) \quad \text{on} \quad x=0,$$

(60)

$$\sigma_{xy}(0,y,t) = 0 \quad \text{on} \quad x=0.$$

(61)

where p is given function of y and t. Invoking the boundary conditions (59 to 61) at the surface $x=0$ of the plate, we obtain a system of three equations. After applying the inverse of matrix method, we have the values of the three constants $M_j, j=1,2,3$.

$$\begin{pmatrix} M_1 \\ M_2 \\ M_3 \end{pmatrix} = \begin{pmatrix} H_{31} & H_{32} & H_{33} \\ H_{21} & H_{22} & H_{23} \\ H_{61} & H_{62} & H_{63} \end{pmatrix}^{-1} \begin{pmatrix} 0 \\ -p_1 \\ 0 \end{pmatrix}$$

(62)

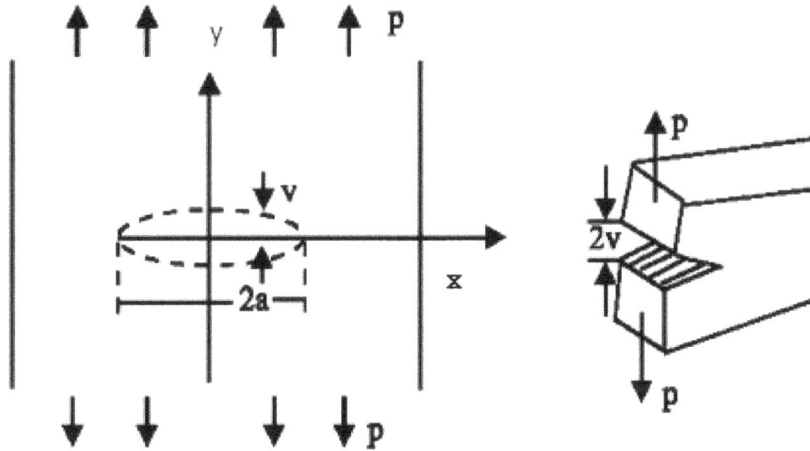

Figure 1. Displacement of an external Mode-I crack.

From this matrix we obtain the values of the three constants M_1, M_2 and M_3. Hence, we obtain the expressions of displacements, force stress, coupled stress and temperature distribution for generalized thermoelastic medium.

NUMERICAL RESULTS

In order to illustrate the theoretical results obtained in preceding section, to compare these in the context of various theories of thermoelasticity and reinforcement on wave propagation. We now present some numerical results for the physical constants. We now present some numerical results for the physical constants (Singh and Singh, 2004).

$\lambda = 7.59 \times 10^9 \, N/m^2$, $\mu_T = 1.89 \times 10^9 \, N/m^2$, $\mu_L = 2.45 \times 10^9 \, N/m^2$,

$\alpha = -1.28 \times 10^9 \, N/m^2$, $\beta = 0.32 \times 10^9 \, N/m^2$, $\rho = 7800 \, kg/m^2$,

$\alpha_t = 1.78 \times 10^{-5} \, N/m^2$, $k = 386$, $C_E = 383.1$, $\tau_0 = 0.02$,

$a = 1$

$T_0 = 293k$ $f^* = 1$, $p^* = 2$, $\omega = \omega_0 + i\xi$ $\omega_0 = 2$,

$\xi = 1$

$C_E = 383.1 \quad J/(kgk)$, $\mu = 3.86 \times 10^{10} \quad kg/ms^2$

The computations were carried out for a value of time $t = 0.1$. The numerical technique, outlined above, was used for the distribution of the real part of the thermal temperature θ, the displacement u and v, the stresses $\sigma_{xx}, \sigma_{yy}, \sigma_{zz}$ and σ_{xy} distribution for the problem. The field quantities, temperature, displacement components u, v and stress components $\sigma_{xx}, \sigma_{yy}, \sigma_{zz}$

and σ_{xy} depend not only on space x and time t but also on the thermal relaxation time τ_0. Here all the variables are taken in non dimensional forms.

The results are shown in Figures 1 to 14. The graph shows the two curves predicted by different theories of thermoelasticity. In these figures, the solid lines represent the solution in the coupled theory; the dashed lines represent the solution in the generalized Lord and Shulman theory. We notice that the results for the temperature, the displacement and stresses distribution when the relaxation time is including in the heat equation are distinctly different from those when the relaxation time is not mentioned in heat equation, because the thermal waves in the Fourier's theory of heat equation travel with an infinite speed of propagation as opposed to finite speed in the non-Fourier case. This demonstrates clearly the difference between the coupled and the theory of thermoelasticity (LS).

Problem 1

For the value of y, namely y = -1, were substituted in performing the computation. It should be noted (Figure 2) that in this problem I, the crack's size, x is taken to be the length in this problem so that $0 \le x \le 1.2$, $y = 0$ represents the plane of the crack that is symmetric with respect to the y-plane. It is clear from the graph that θ has a maximum value at the beginning of the crack ($x = 0$), it begins to fall just near the crack edge ($x \approx 1.2$), where it experiences sharp decreases (with a maximum negative gradient at the crack's end). The value of temperature quantity converges to zero with increasing the distance x.

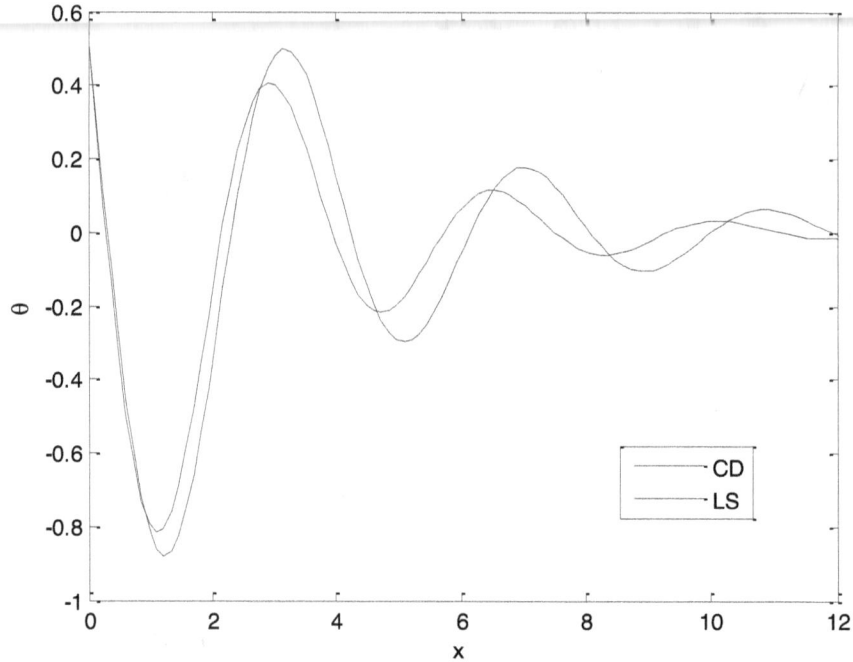

Figure 2. The temperature distribution for problem 1.

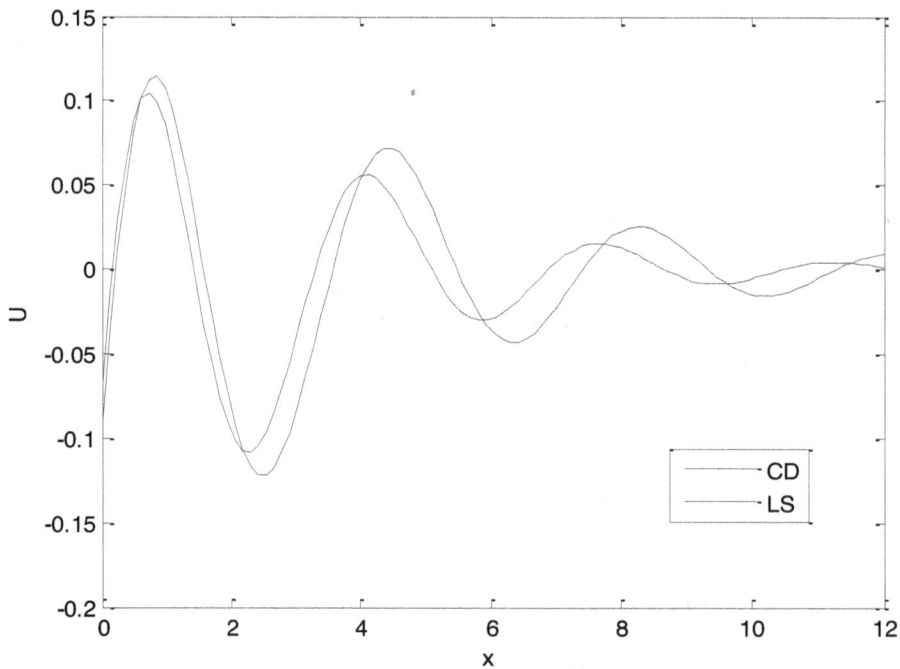

Figure 3. Horizontal displacement distribution u for problem 1.

Figure 3 the horizontal displacement, u begins with increases near the crack edge ($x \approx 1.2$), then smooth decreases again to reach its minimum magnitude just at the crack end. Beyond it u falls again to try to retain zero at infinity. Figure 4, the vertical displacement V, we see that the displacement component V always starts from the negative value and terminates at the zero value. Also, at the crack end to reach minimum value, beyond

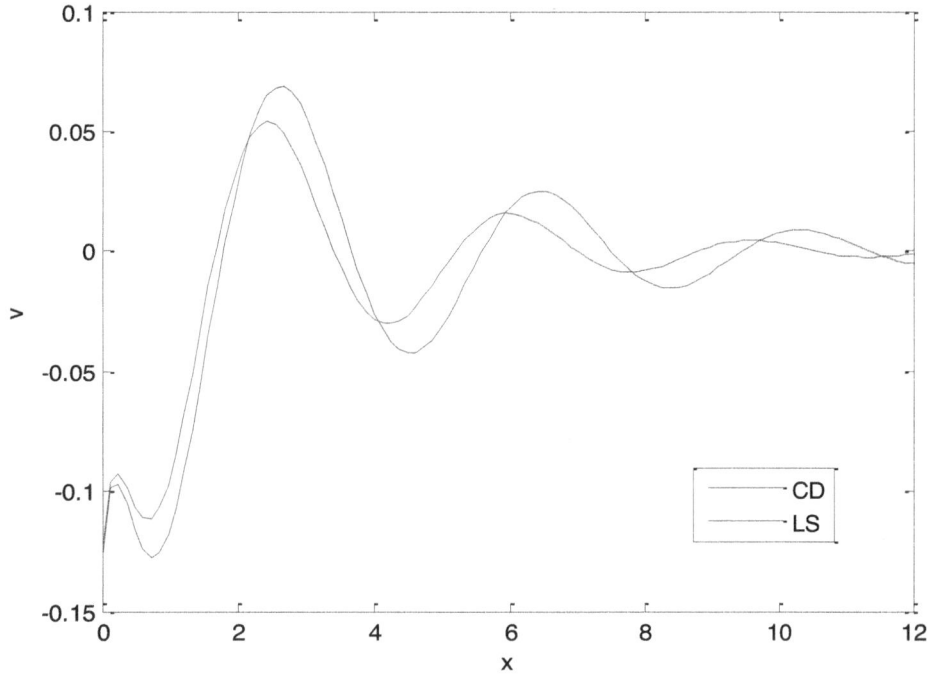

Figure 4. Vertical displacement distribution v for problem 1.

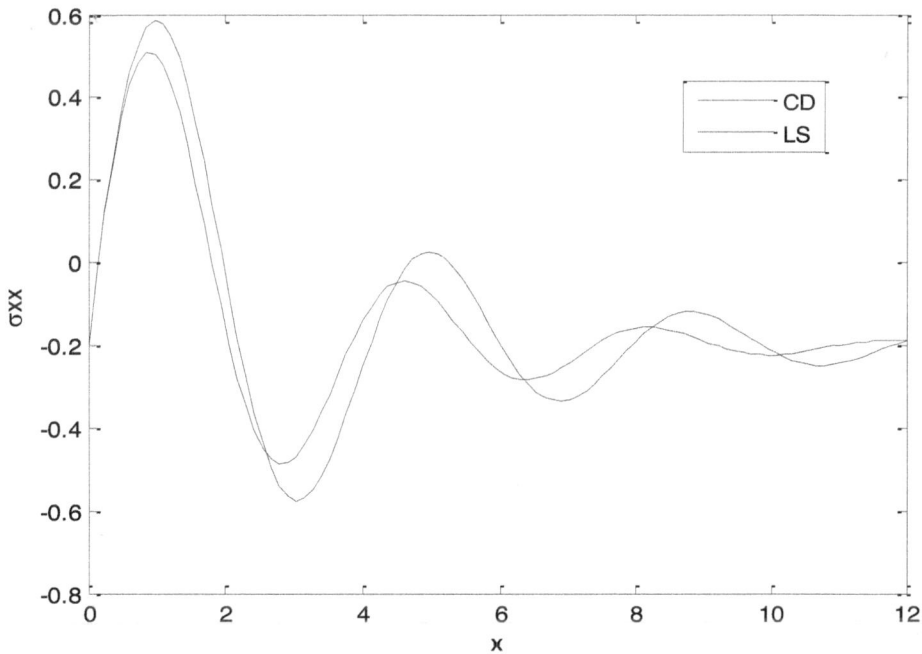

Figure 5. The distribution of stress component σ_{xx} for problem 1.

reaching zero at the three double of the crack size (state of particles equilibrium). The displacements u and v show different behaviours, because of the elasticity of the solid tends to resist vertical displacements in the problem under investigation. Both of the components show different behaviours, the former tends to increase to maximum just before the end of the crack. Then it falls to a minimum with a highly negative gradient. Afterwards it

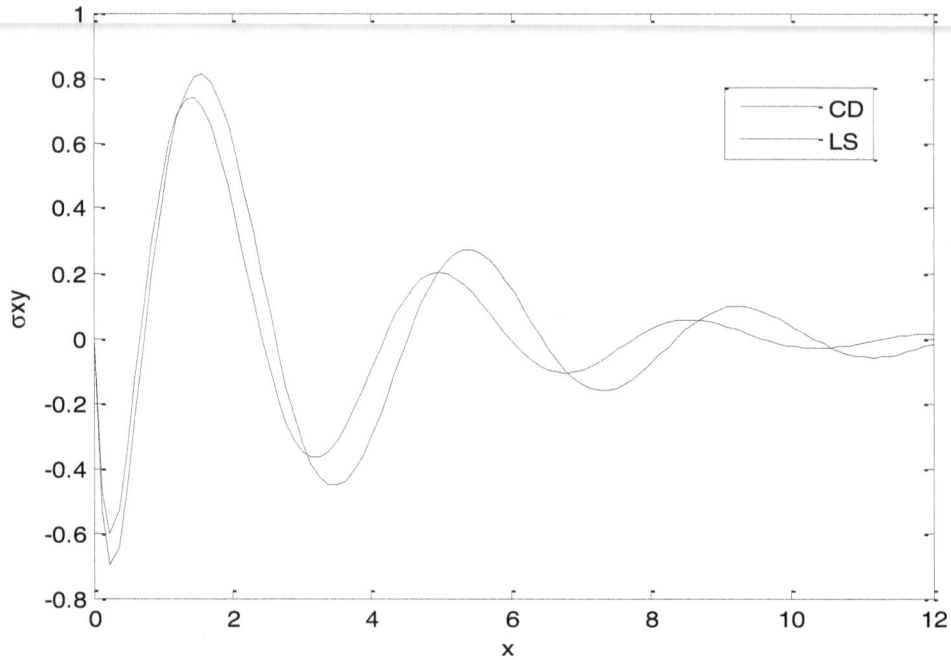

Figure 6. The distribution of stress component σ_{xy} for problem 1.

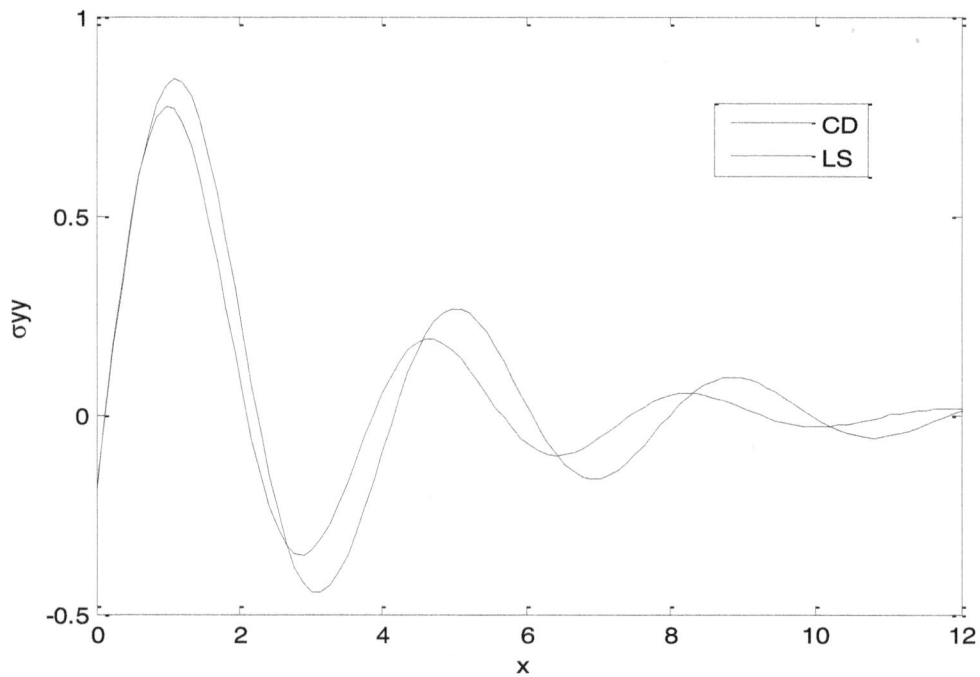

Figure 7. The distribution of stress component σ_{yy} for problem 1.

rises again to a maximum beyond about the crack end.

The stress component, σ_{xx} reach coincidence with negative value (Figure 5) and satisfy the boundary condition at $x = 0$, reach the maximum value near the end of crack ($x \approx 1.2$) and converges to zero with increasing the distance x, also Figure 7 (satisfy the

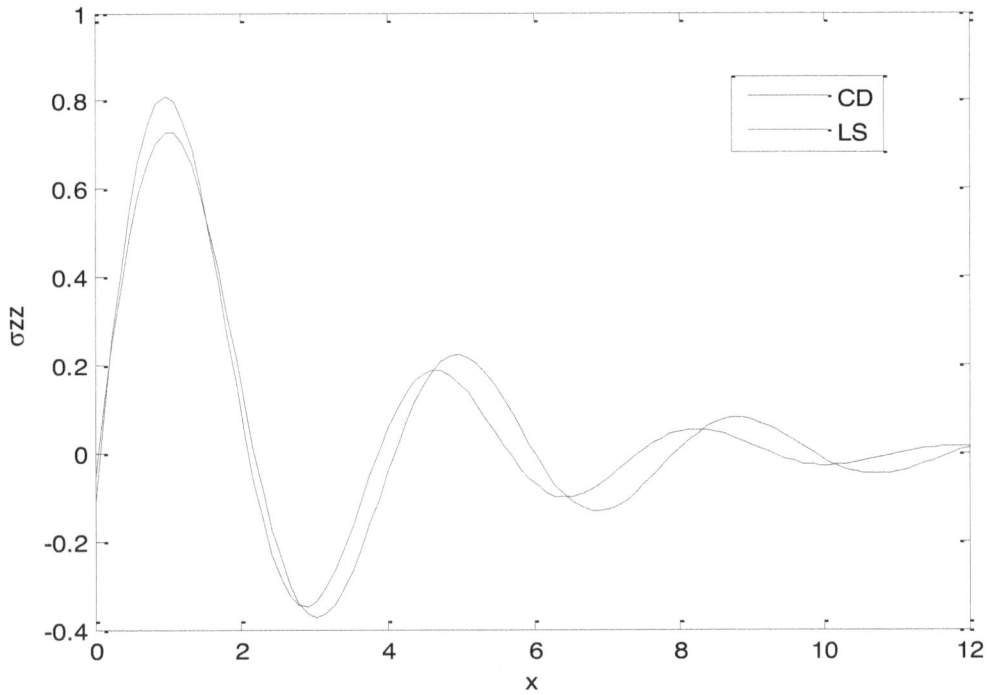

Figure 8. The distribution of stress component σ_{zz} for problem 1.

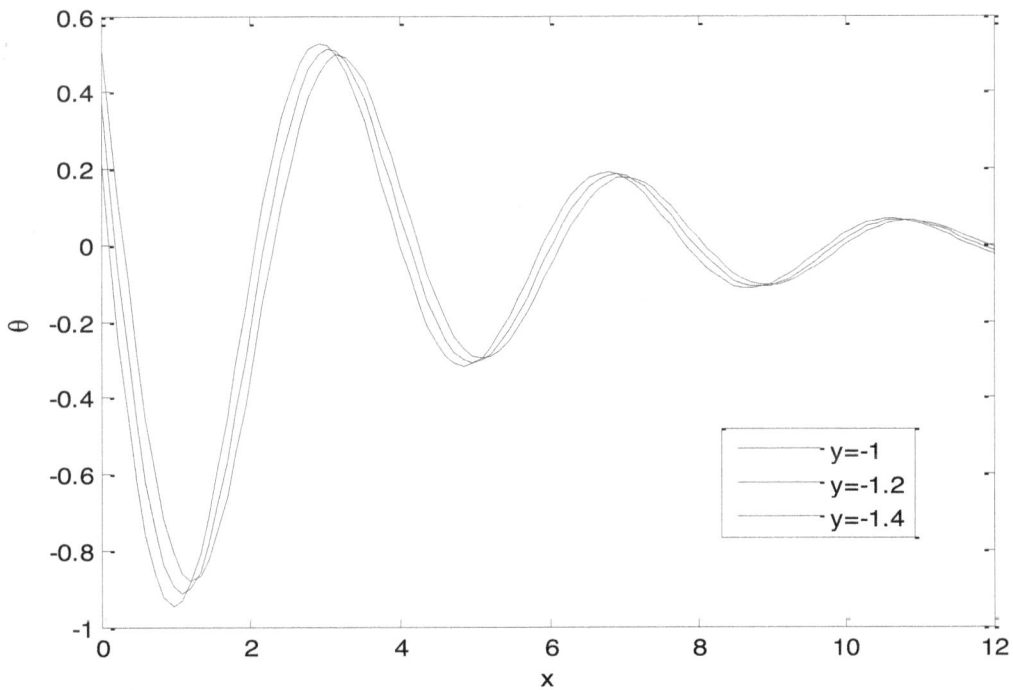

Figure 9. Temperature distribution with Variation of distances under LS theory (problem 1).

boundary condition at $x=0$) and 7, take the same behavior of Figure 5. Figure 6, shows that the stress component σ_{xy} satisfies the boundary condition at $x=0$ and had a different behaviour. It decreases in the start and start increases (maximum) in the context of the two theories until reaching the crack end. These trends

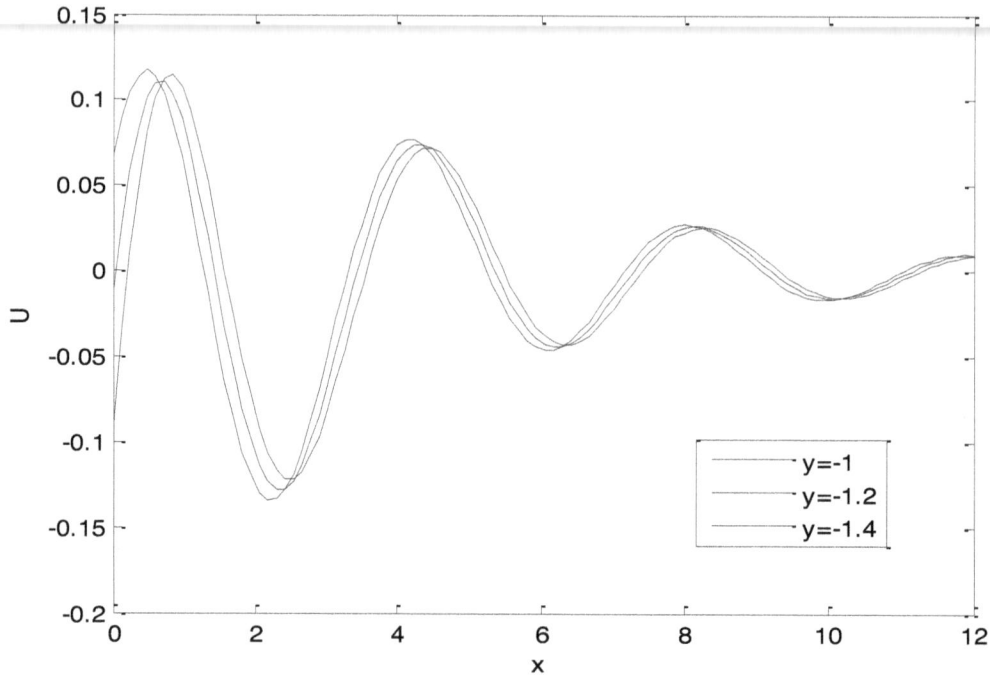

Figure 10. Displacement distribution u with variation of distance under L-S theory(problem 1).

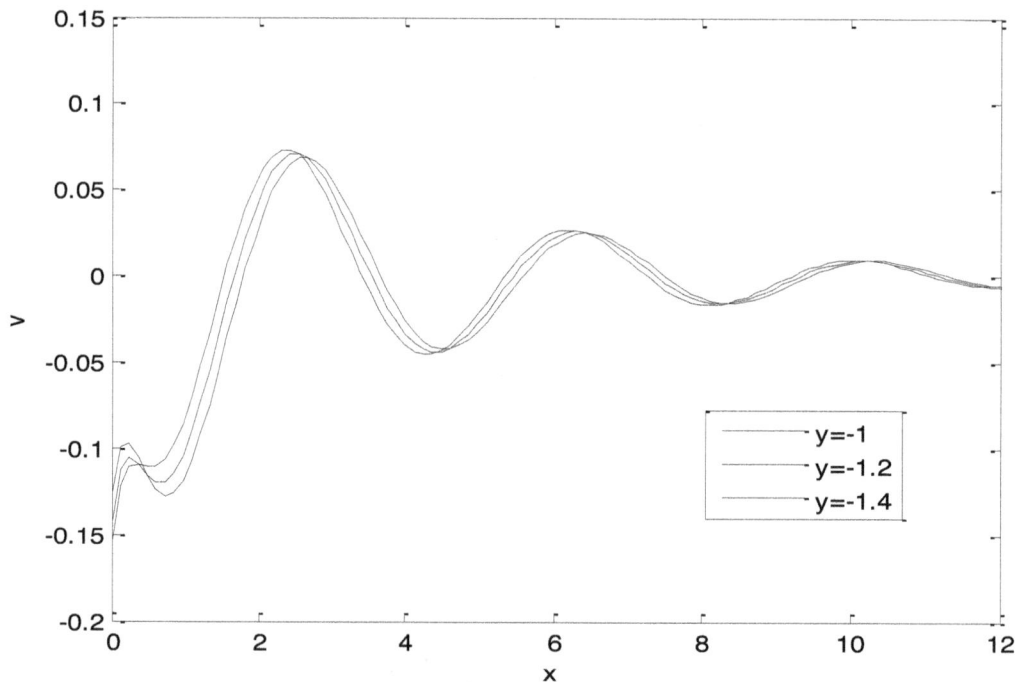

Figure 11. Displacement distribution v with variation of distance under LS theory(problem 1).

obey elastic and thermoelastic properties of the solid under investigation.

Figure 9-14 show the comparison between the temperature θ, displacement components u, v, the force stresses components σ_{xx}, σ_{yy} σ_{zz} and σ_{xz}, the case of different three values of y, (namely y= -1, y=-1.2 and y=-1.4) under (LS) theory. It should be noted (Figure 9) that in this problem. It is clear from the graph that θ

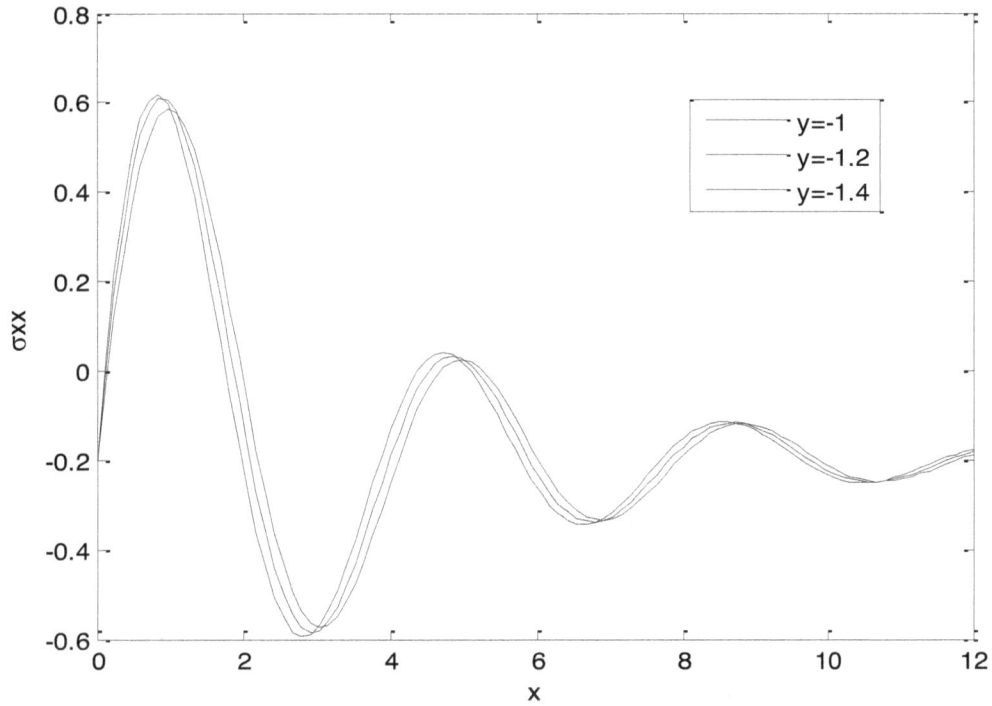

Figure 12. Stress distribution σ_{xx} with variation of distance under LS theory(problem 1).

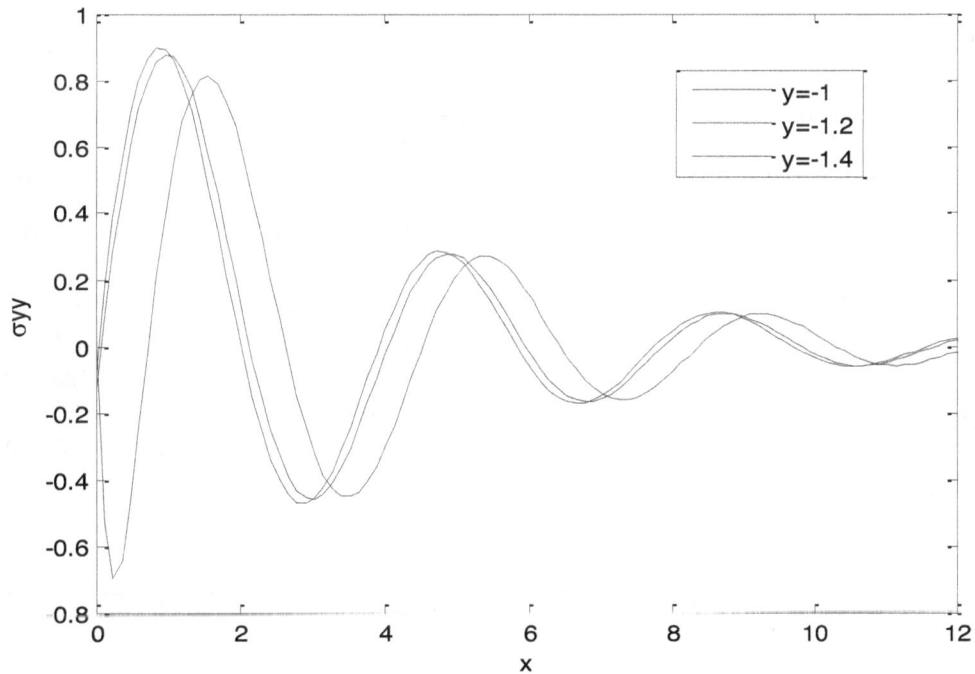

Figure 13. Stress distribution σ_{yy} with Variation of distance under LS theory (problem 1).

has minimum value at the beginning of the crack ($x=0$), it begins to fall just near the crack edge ($x \approx 1.2$), where it experiences sharp increases (with maximum positive gradient at the crack's end). Graph

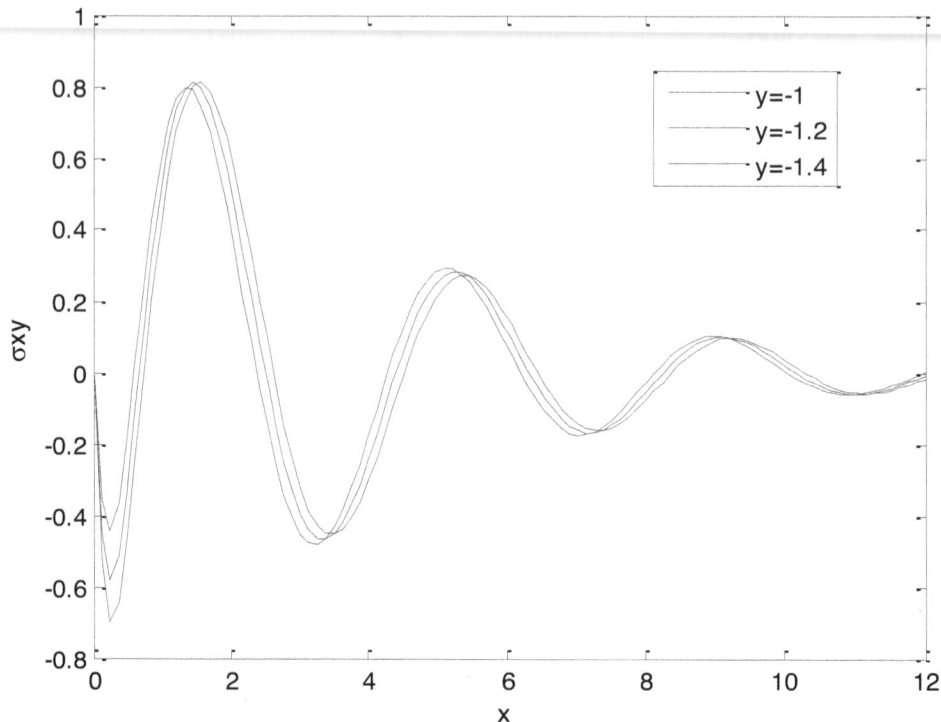

Figure 14. Stress distribution σ_{xy} with variation of distance under LS theory(problem 1).

lines for both values of y show different slopes at crack ends according to y-values. In other words, the temperature line for y =- 1 has the highest gradient when compared with that of y = -1.2 and y= -1.4 at the first of the range. In addition, all lines begin to coincide when the horizontal distance x is beyond the three double of the crack size to reach the reference temperature of the solid. These results obey physical reality for the behaviour of fiber as a polycrystalline solid.

Figure 10 the horizontal displacement u, despite the peaks (for different vertical distances y= -1, y=-1.2 and y=-1.4) occur at equal value of x, the magnitude of the maximum displacement peak strongly depends on the vertical distance y. it is also clear that the rate of change of u decreases with increasing y as we go farther apart from the crack. On the other hand, Figure 11 shows atonable increase of the vertical displacement V, near the crack end to reach minimum value beyond $x \approx 1.2$ reaching zero at the three double of the crack size (state of particles equilibrium). Figure 12, the horizontal stresses σ_{xx} Graph lines for both values of y show different slopes at crack ends according to y-values. In other words, the σ_{xx} component line for y = -1.4 has the highest gradient when compared with that of y = -1.2 and y= -1.4 at the edge of the crack. In addition, all lines begin to coincide when the horizontal distance x is

beyond the three double of the crack size to reach zero after their relaxations at infinity. Variation of y has a serious effect on both magnitudes of mechanical stresses. These trends obey elastic and thermoelastic properties of the solid under investigation.

Figure 13, shows that the stress component σ_{yy} ,satisfy the boundary condition at $x = 0$, the line for y = -1.4 has the highest gradient when compared with that of y= -1 and y= -1.2 in the range $0 \leq x \leq 1.6$ (near the crack edge), the line for y = -1 has the highest gradient when compared with that of y=-1.2 and y=-1.4 in the range (near the crack end) $1.6 \leq x \leq 3$ and converge to zero when $x > 12$. These trends obey elastic and thermoelastic properties of the solid. Figure 14, shows that the stress component σ_{xy} satisfy the boundary condition, it decreases in the start and start increases (maximum) in the context of the three values of y until reaching the crack end, the line for y=-1.4 has the highest gradient when compared with that of y=-1.2 and y=-1 in the range $0 \leq x \leq 1.7$, the line for y=-1 has the highest gradient when compared with that of y=-1.2 and y=-1.4 in the range $1.7 \leq x \leq 3.5$ and converge to zero when $x > 12$. These trends obey elastic and thermoelastic properties of the solid.

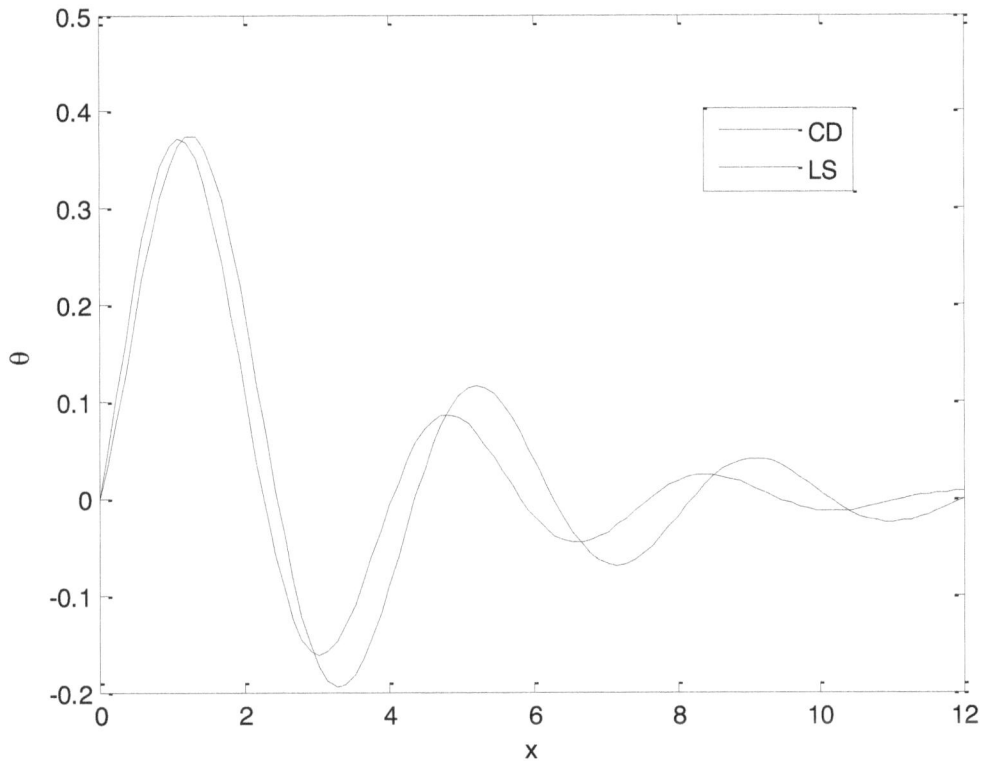

Figure 15. The temperature distribution for problem 2.

Problem 2

Figure 15 described the values of temperature θ under two theories. It indicts that the values of temperature θ increasing in the ranges $0 \leq x \leq 1.7$ and $3.2 \leq x \leq 5.2$ with fibre-reinforced and then decreases in the ranges $1.7 \leq x \leq 3.2$ and $5.2 \leq x \leq 6.5$. The values of temperature θ converge to zero with increasing the distance x.

In Figure 16, the horizontal displacement, u, begins with sharp decreases near the ($x \approx 1.4$), then smooth increases again to reach it's a maximum value just at near $x \approx 2.8$. Beyond it u falls again to try to retain zero at infinity. In Figure 17, the vertical displacement v, we see that the displacement component v always starts from the positive value and terminates at the zero value at the infinity (state of particles equilibrium). The displacements u and v show different behaviours, because of the elasticity of the solid tends to resist vertical displacements in the problem under investigation. Both of the components show different behaviours.

In Figure 18 the stress component, σ_{xx} reach coincidence with negative value and satisfy the boundary

condition at $x = 0$, reach the maximum value near ($x \approx 3.3$) and converges to zero with increasing the distance x, also Figures 19-21 take the same behavior.

Figure 19, shows that the stress component σ_{xy} satisfy the boundary condition at $x = 0$ and had a different behaviour. It sharp increases in the start and start decreases (minimum) in the context of the two theories until reaching the crack end. These trends obey elastic and thermoelastic properties of the solid under investigation.

Conclusions

In the present study, the normal mode analysis is used to study the effect of the cracks of the problem under consideration at the free surface of a fiber-reinforced thermoelastic half-space based on the CD and L-S theory of thermoelasticity. According to the analysis above, we can conclude the following points:

(1) The curves in the context of the (CD) and (L-S) theories decrease exponentially with increasing x, this indicate that the thermoelastic waves are unattenuated and non dispersive, where purely thermoelastic waves undergo both attenuation and dispersion.

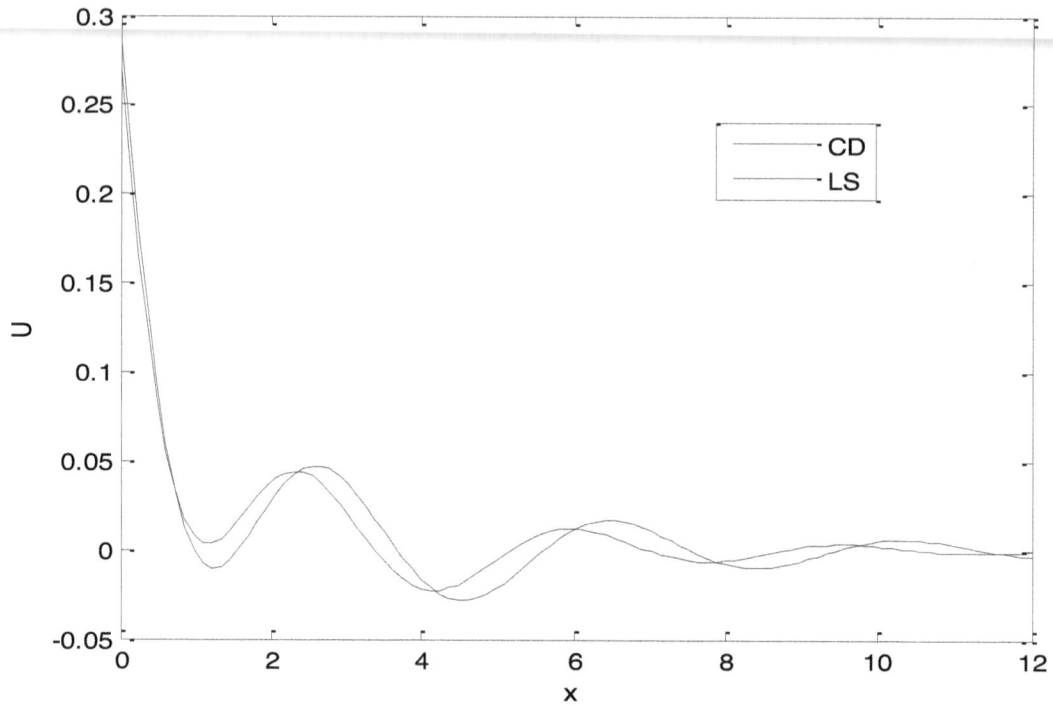

Figure 16. Horizontal displacement distribution u for problem 2.

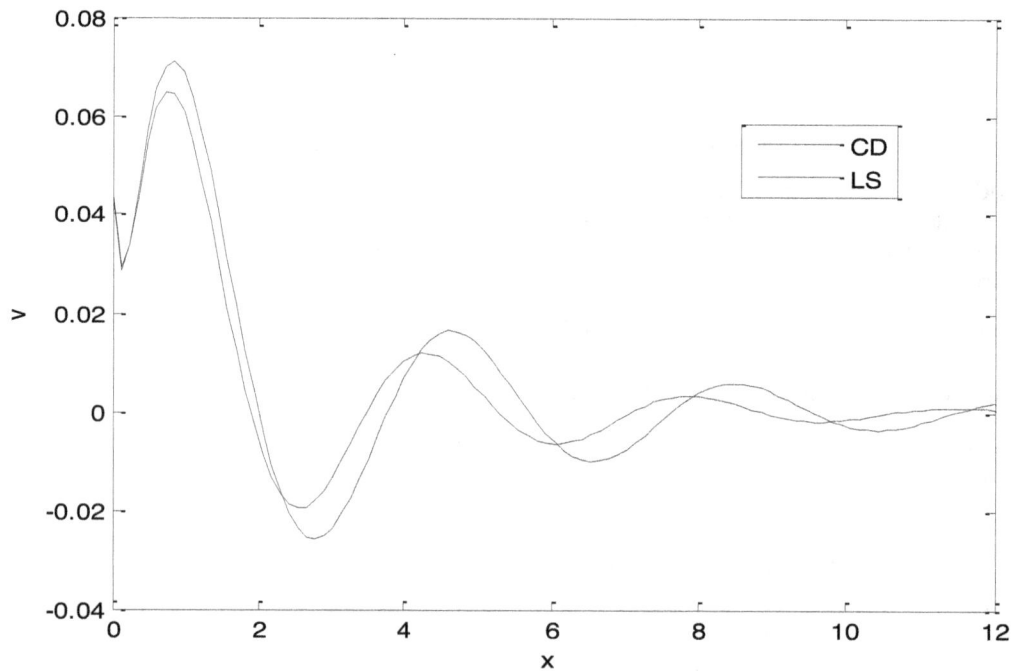

Figure 17. Vertical displacement distribution v for problem 2.

(2) The curves of the physical quantities with (CD) theory in most of figures are lower in comparison with those under (L-S) theory, due to the relaxation times.

(3) Analytical solutions based upon normal mode analysis for themoelastic problem in solids have been developed and utilized.

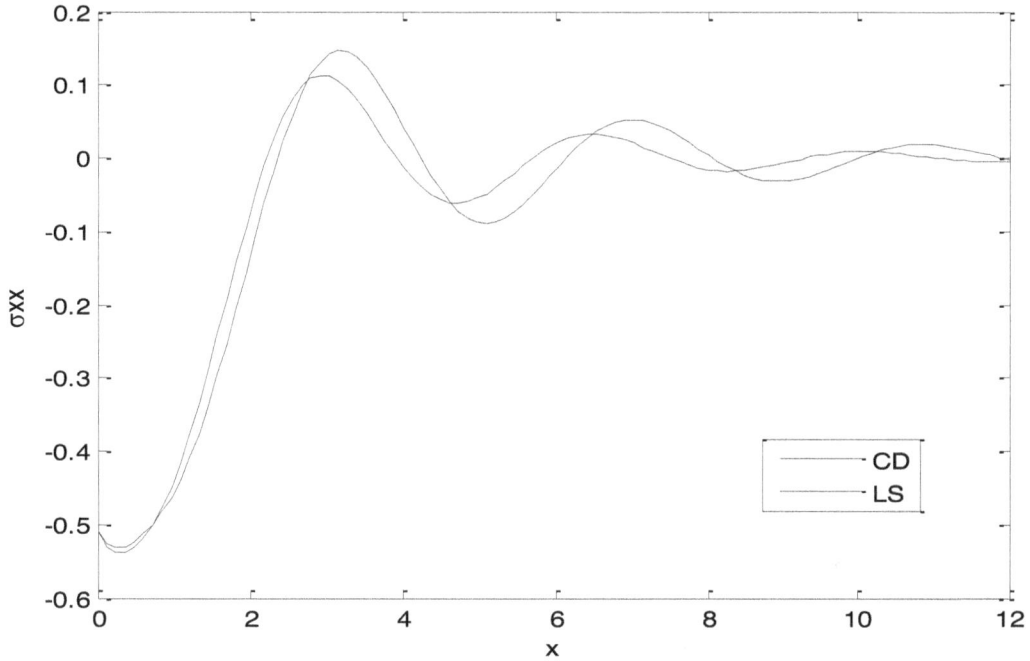

Figure 18. The distribution of stress component σ_{xx} for problem 2.

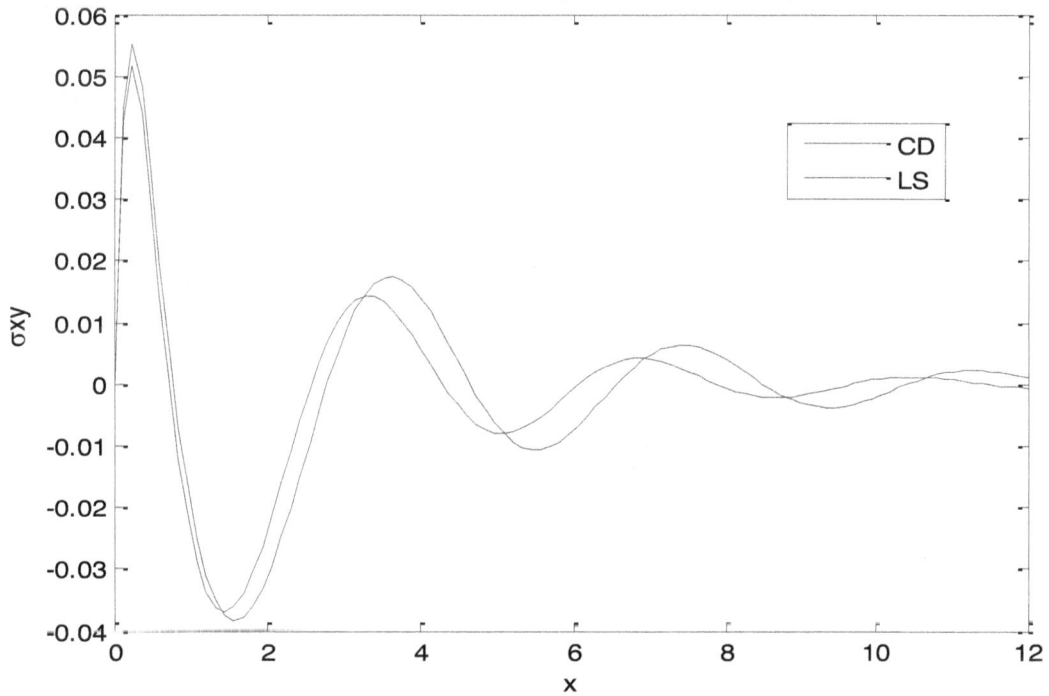

Figure 19. The distribution of stress component σ_{xy} for different problem 2.

(4) A linear opening mode-I crack has been investigated and studied for copper solid.

(5) Temperature, radial and axial distributions were estimated at different distances from the crack edge.

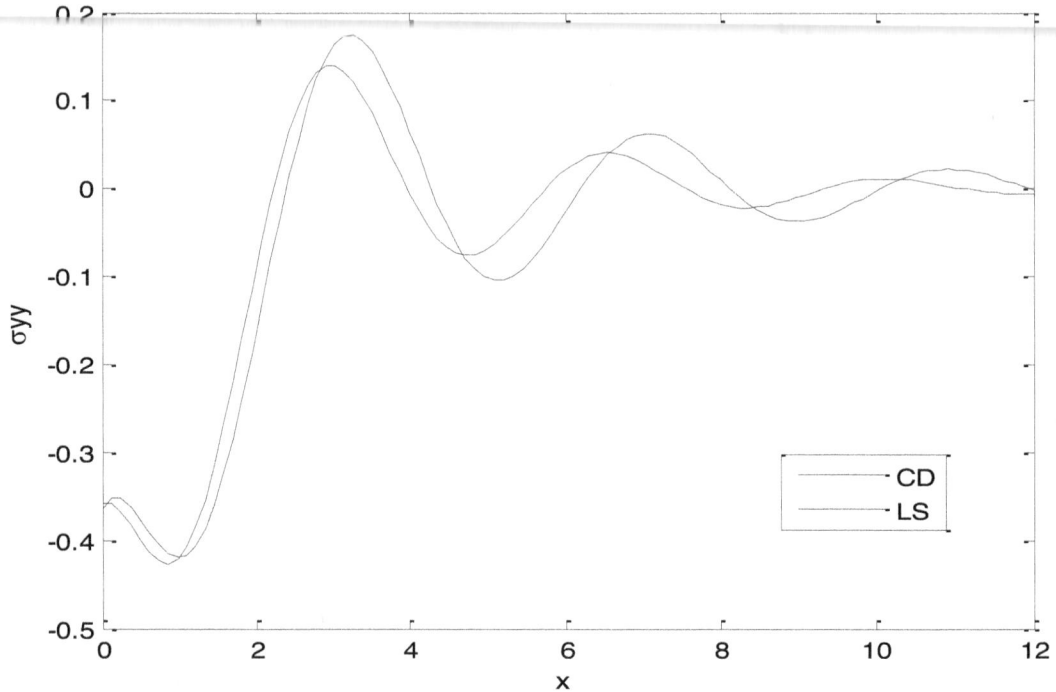

Figure 20. The distribution of stress component σ_{yy} for different problem 2.

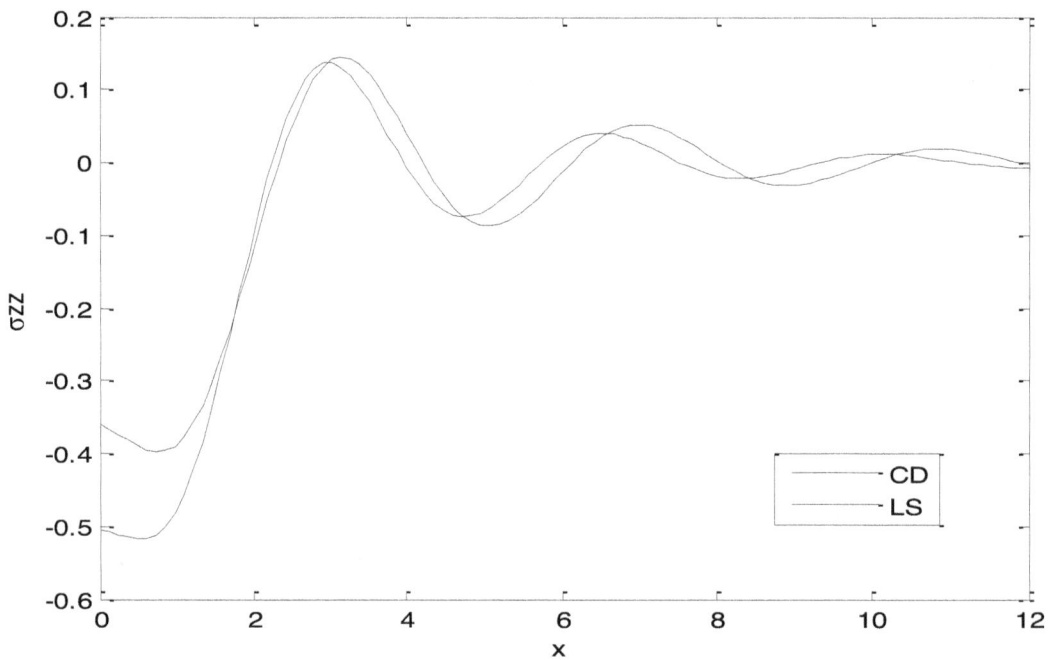

Figure 21. The distribution of stress component σ_{zz} for problem 2.

(6) The stresses distributions, and temperature were evaluated as functions of the distance from the crack edge.

(7) Crack dimensions are significant to elucidate the mechanical structure of the solid.

(8) Cracks are stationary and external stress is demanded

to propagate such cracks.

(9) It can be concluded that a change of volume is attended by a change of the temperature while the effect of the deformation upon the temperature distribution is the subject of the theory of thermoelasticity.

(10) The value of all the physical quantities converges to zero with an increase in distance y and All functions are continuous.

(11) The fibre-reinforced has an important role on the distribution of the field quantities.

(12) The method which used in the present article is applicable to a wide range of problems in thermodynamics and thermoelasticity.

(13) Deformation of a body depends on the nature of the applied force as well as the type of boundary conditions

(14) It is clear from all the figures that all the distributions considered have a non-zero value only in a bounded region of the half-space. Out side of this region, the values vanish identically and this means that the region has not felt thermal disturbance yet.

(15) The results presented in this paper should prove useful for researchers in material science, designers of new materials, low temperature physicists, as well as for those working on the development of a theory of hyperbolic thermoelasticity. The introduction of the crack to the generalized fiber-reinforced thermoelastic medium provides a more realistic model for these studies.

REFERENCES

Belfield AJ, Rogers TG, Spencer AJM (1983). Stress in elastic plates reinforced by fiber lying in concentric circles, J. Mech. Phys. Solids, 31:25-54.

Chattopadhyay A, Choudhury S (1990). Propagation, reflection and transmission of magneto-elastic shear waves in a self-reinforced medium, Int. J. Eng. Sci. 28:485-495.

Chattopadhyay A, Choudhury S (1995). Magnetoelastic shear waves in an infinite self-rein-forced plate, Int. J. Num. Anal. Methods Geomech. 19:289-304.

Chattopadhyay A, Venkateswarlu RLK, Saha S (2002). Reflection of quasi-P and quasi-SV waves at the free and rigid boundaries of a fibre-reinforced medium, Sādhanā, 27:613-630.

Chaudhary S, Kaushik VP, Tomar SK (2004). Reflection/transmission of plane wave through a self-reinforced elastic layer between two half-spaces, Acta Geophysica Polonica. 52:219-235.

Dhaliwal R (1980). External Crack due to Thermal Effects in an Infinite Elastic Solid with a Cylindrical Inclusion. Thermal Stresses in Server Environments Plenum Press, New York and London, pp. 665-692.

Elfalaky A, Abdel-Halim AA (2006). A Mode-I Crack Problem for an Infinite Space in Thermoelasticity, J. Appl. Sci. 6:598-606.

Green AE, Lindsay KA (1972). Thermoelasticity. J. Elasticity 2:1-7.

Hasanyan D, Librescu L, Qin Z, Young R (2005). Thermoelastic Cracked Plates Carrying nonstationary Electrical Current. J. Thermal Stresses, 28:729-745.

Hashin Z, Rosen WB (1964). The elastic moduli of fibre-reinforced materials, J. Appl. Mech. 31:223-232.

Lord HW, Şhulman YA (1967). Generalized dynamical theory of thennoelasticity. J. Mech. Phys. Solid, 15:299-306.

Lotfy KH (2012a). Mode-I crack in a two-dimensional fibre-reinforced generalized thermoelastic problem, Chin. Phys. B, 21:1-014209.

Lotfy KH (2012b). The effect of a magnetic field on a 2D problem of fibre-reinforced thermoelasticity rotation under three theories, Chin. Phys. B; 21:6- 064214

Othman M, Atwa S (2012). Generalized Magneto-thermoelasticity in a Fibre- reinforced Anisotropic Half-space with Energy Dissipation, Int. J. Thermophys. 33(6):1126-1142.

Othman M, Lotfy KH (2009a). Effect of Magnetic Field and Inclined Load in Micropolar Thermoelastic Medium Possessing Cubic Symmetry, Int. J. Ind. Math. 1(2):87-104.

Othman M, Lotfy KH (2009b). Two-dimensional Problem of Generalized Magneto-thermoelasticity under the Effect of Temperature Dependent Properties for Different Theories, MMMS, 5:235-242.

Othman M, Lotfy KH (2010a). Generalized Thermo-microstretch Elastic Medium with Temperature Dependent Properties for Different Theories, Engineering Analysis with Boundary Elements, 34:229-237.

Othman M, Lotfy KH (2010b). On the Plane Waves in Generalized Thermo-microstretch Elastic Half-space, International Communication in Heat and Mass Transfer, 37:192-200.

Othman M, Lotfy KH (2013). The effect of magnetic field and rotation of the 2-D problem of a fiber-reinforced thermoelastic under three theories with influence of gravity, Mech. Mater. 60:120-143.

Othman M, Lotfy KH, Farouk RM (2009). Transient Disturbance in a Half-space under Generalized Magneto-thermoelasticity due to Moving Internal Heat Source, Acta Physica Polonica A, 116:186-192.

Othman M, Saied S (2012a). The Effect of Mechanical Force on Generalized Thermoelasticity in a Fiber-reinforced under Three Theories, Int. J. Thermophys. 33(6):1082-1099.

Othman M, Saied S (2012b). The Effect of Rotation on Two-dimensional Problem of a Fibre-reinforced Thermoelastic with One Relaxation Time, Int. J. Thermophys. 33(2):160-171.

Othman M, Saied S (2013). Two-Dimensional Problem of Thermally Conducting Fiber-reinforced under Green-Naghdi Theory, J. Thermoelasticity. 1 (1):13-20

Othman MIA, Song YQ (2007). Reflection of plane waves from an elastic solid half-space under hydrostatic initial stress without energy dissipation, Int. J. Sol. Struct. 44:5651-5664.

Pradhan A, Samal SK, Mahanti NC (2003). "Influence of anisotropy on the love waves in a self- reinforced medium", Tamkang J. Sci. Eng. 6(3):173-178.

Sengupta PR, Nath S (2001). Surface waves in fibre-reinforced anisotropic elastic media, Sādhanā, 26:363-370.

Singh B (2006). Wave propagation in thermally conducting linear fibre-reinforced composite materials, Arch. Appl. Mech.75:513-520.

Singh B, Singh SJ (2004). "Reflection of planes waves at the free surface of a fibre- reinforced elastic half-space", Sādhanā, 29:249-257.

Singh SJ (2002). Comments on "Surface waves in fibre-reinforced anisotropic elastic media" by Sengupta and Nath [Sādhanā, 2001 26, 363-370]. Sādhanā. 27:1-3.

Ueda S (2003). Thermally induced fracture of a piezoelectric Laminate with a crack Normal to Interfaces, J. Thermal Stresses, 26:311-323.

Verma PDS (1986). Magnetoelastic shear waves in self-reinforced bodies, Int. J. Eng. Sci. 24:1067-1073.

Verma PDS, Rana OH, Verma M (1988). Magnetoelastic transverse surfaces waves in self-reinforced elastic bodies, Ind. J. Pure Appl. Math. 19:713-716.

12

Discrete singular convolution for Lennard-Jones potential using Shannon kernel

Mrittunjoy Guha Majumdar

St. Stephen's College, Delhi, India.

In this paper, the idea of discrete singular convolution (DSC) as a viable computation method for analyzing physical systems has been underlined. Discrete singular convolution has been used for solving the Schrödinger equation for water molecules, subject to the Lennard Jones potential, and the DSC differentiator has been used for obtaining the energy eigen-states of water, for a given grid size for discretization, along with the Shannon kernel for approximation of the singular delta-type kernel in the problem.

Key words: Discrete singular convolution, computational methodology, Schrödinger equation, Lennard Jones potential.

INTRODUCTION

In the world of computational methodology, one has global and local methods for analysis. Global methods are more accurate and localized than local methods. However, local methods are more useful for handling particular kinds of problems, especially those involving certain boundary conditions.

Discrete singular convolution (DSC) is a potential numerical approach for solving computational problems. Wei worked on the application of this method for problems such as the use of discrete singular convolution for solving the Fokker-Planck equation (Wei, 1999) analyzing the nonlinear dynamic response of laminated plates (Civalek, 2013) and free vibration analysis of multiple-stepped beams (Duan and Xinwei, 2013). The purpose of this paper is to use the DSC algorithm for the numerical solution of the Schrodinger equation for solving for the energy eigen values of the Schrodinger equation for the Lennard Jones potential in water using the Shannon kernel. This is along the lines drawn by *Wei* in his seminal paper on the topic (Wei, 2000).

The Lennard Jones potential has been a subject of interest among physicists with the associated computational methods used to analyze it. Huacuja et al. (2007) have analyzed the molecular configurations which minimize the Lennard Jones potential. Barr et al. (1995) and McGeoch (1996) have established the plinth of computational analysis in terms of statistical and experimental aspects of algorithms. As per one of the main criteria for computational methods, one needs to optimize the resources available in one's disposal for the evaluation or analysis of a problem. For studying problems involving either sample size approaching to infinity or a discontinuity in any variable, one needs to devise a way to circumvent this problem using the tools available. Barr et al. (1995), McGeoch (1996), Pattengale et al. (2009) and Terán-Villanueva et al. (2013) have proposed a set of principles that allow us to extract scientific knowledge from these methods. This includes relevant computational experiments, as per the literature related to the given subject, which must be reproducible and should relate with other studies carried out in the relevant field.

Solving the Schrödinger equation is a non-deterministic polynomial (NP) type of problem. This means that if the problem's solution can be guessed in some polynomial time, but no rule is followed on how to make the guess for the solution Arkady (2013). None of the instances of the equation's guessed solutions have yielded an algorithm for exactly solving this equation in a polynomial or reasonable number of steps for any quantum system. It is in this context that the idea of discretizing the Schrödinger equation for solutions becomes worth noting. For this one could use the idea of convolution, as in the case of discrete singular convolution (DSC), explained in the following section. More importantly, going by McGeoch's formulation, one needs to test the relevance of Wei's DSC algorithm for solving for the Schrodinger equation for more complex systems than the Morse potential for iodine.

DISCRETE SINGULAR CONVOLUTION

Discrete singular convolution is a general approach for numerically solving problems involving singular convolution such as in Hilbert transforms and Abel Transforms. Using the appropriate kernel, one can use the formalism to solve a number of physical problems. The discrete singular convolution approach is based on the theory of distributions. Let W be a distribution and $f(x)$ be an element of the space of test functions for the kernel. If W is a singular kernel, we can define what is known as the singular convolution:

$$F(x) = W * f(x) = \int_{-\infty}^{\infty} W(x-y)f(y)dy$$

One such kernel is the singular kernel of the delta type (δ). One must remember that the Dirac Delta function is a generalized function itself and not strictly a function, as per the conditions satisfied by functions. Since these are singular kernels, one cannot use them in computation methods. To overcome this problem, one defines a sequence of functions that are approximations of the distribution:

$$\lim_{\alpha \to \alpha_0} W_\alpha(x) \to W(x)$$

Here α_0 is the generalized limit. For a singular kernel of the delta type, one has a delta sequence.

Delta sequence

Any sequence of functions $f_n(x)$ which converges to the Delta function for $n \to \infty$. One can define the approximation to the convolution as:

$$F_\alpha(x) = \sum_j W_\alpha(x-x_j) \times f(x_j)$$

Here, $\{x_j\}$ represents the set of points for which the algorithm is defined. One must note that this algorithm is valid only for smooth approximations of the Kernel. One kernel of the Delta type is the Shannon kernel, which is defined as:

$$\frac{\sin(\alpha(x-x_0))}{(x-x_0)}$$

One often defines an algorithm sampling element or the *Nyquist* frequency for discrete singular convolution. It can be defined as:

$$\alpha = \frac{\pi}{\Delta}$$

One can then define the Shannon kernel as:

$$\frac{\sin(\frac{\pi}{\Delta}(x-x_0))}{\frac{\pi}{\Delta}(x-x_0)}$$

DSC approach for Schrodinger equation

We have a grid representation for the coordinate so that the potential part of the Hamiltonian is made diagonal. One can represent the Hamiltonian in the matrix form:

$$H(x_i, x_j) = -\frac{\hbar^2}{2m}\delta_\alpha''(x_i - x_j) + V(x_i)\delta_{ij}$$

where

$$\delta_\alpha''(x_i - x_j) = \frac{d^2}{dx^2}[\delta_\alpha(x-x_j)]|_{x=x_i}$$

and

$$\delta_\alpha(x-x_j) = \frac{\sin(\frac{\pi}{\Delta}(x-x_j))}{\frac{\pi}{\Delta}(x-x_j)}$$

in this case.

Lennard Jones potential

Lennard-Jones potential describes the interaction between uncharged molecules. It is mildly attractive as two molecules approach each another from a distance,

Table 1. Energy Eigenvalues *(in Hartree)*

1.005E+03	8.074E-06	-2.730E-05	2.890E-06
2.294E-01	2.477E-06	-2.730E-05	1.282E-05
3.142E-04	-2.913E-06	-2.661E-05	2.156E-05
-3.876E-04	-7.982E-06	-2.523E-05	2.890E-05
-2.592E-04	-1.259E-05	-2.340E-05	3.509E-05
-2.122E-04	-1.668E-05	-2.106E-05	3.991E-05
-1.906E-04	-2.018E-05	-1.821E-05	4.335E-05
-1.755E-04	-2.294E-05	-1.505E-05	4.542E-05
-1.619E-04	-2.523E-05	-1.161E-05	4.633E-05
-1.484E-04	-2.661E-05	-7.982E-06	4.587E-05
-1.344E-04	2.913E-06	-4.312E-06	4.450E-05
-1.202E-04	6.285E-06	-6.400E-07	4.220E-05
-1.055E-04	9.381E-06	-6.124E-05	3.876E-05
-9.060E-05	1.216E-05	-4.679E-05	3.486E-05
-7.569E-05	-3.326E-05	-2.025E-05	3.005E-05
1.379E-05	1.945E-05	-8.166E-06	2.500E-05

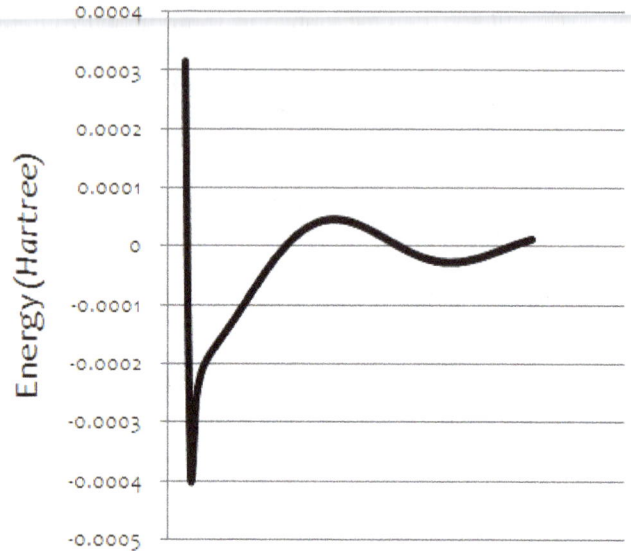

Graph 1. Plot of eigenvalues obtained.

but is strongly repulsive when they are close to each other. The 12:6 Lennard Jones potential form is given as:

$$V = 4\varepsilon\left(\left(\frac{\sigma}{r}\right)^{12} - \left(\frac{\sigma}{r}\right)^{6}\right)$$

METHODOLOGY

In this work, discrete singular convolution methodology was used to analyze the system of water molecules, subject to the conditions:

$$\sigma = 0.3166 \times 10^{-9} m$$
$$\varepsilon = 1.08 \times 10^{-21} J/mol$$

To begin with, the Hamiltonian matrix was obtained for the system, using the DSC-Hamiltonian matrix formulation (A), using the Shannon kernel, for the parameters,

$$\Delta = 1 \times 10^{-12} m$$
$$r = 0.942 \times 10^{-10} \text{ m.}$$

the mean bond length of water.

Thereafter, the diagonalization of the matrix was carried out to obtain the eigen vectors of the given system, subject to the Lennard Jones potential.

RESULTS

For 64 grid-points for the DSC evaluation, the following eigen values were found in Table 1. Magnifying Graph 1, one finds these interesting characteristics. The representation for the entire range of eigenvalues gives us a graph similar to that for the Wigner Distribution (Graph 2). Wigner distribution is the probability

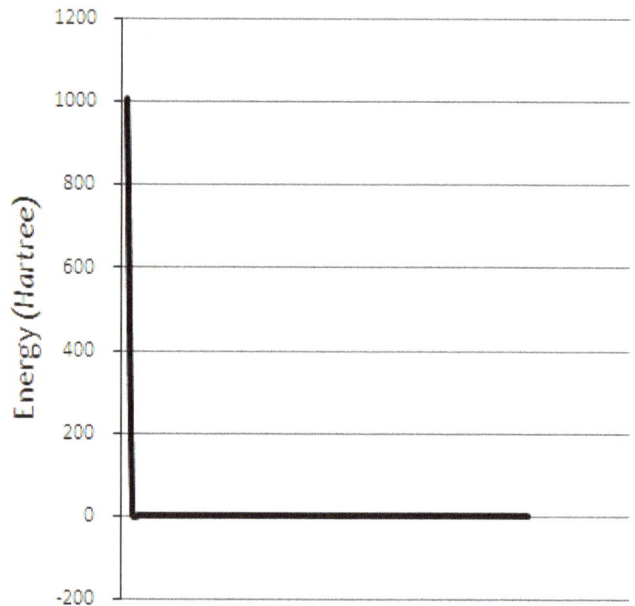

Graph 2. Wigner Distribution.

distribution on the interval $[-R, R]$ whose probability density function is a semicircle of radius R centered at $(0, 0)$, and given by:

$$f(x) = \begin{cases} \frac{2}{\pi R^2}\sqrt{R^2 - x^2}, & -R \leq x \leq R \\ 0, & R < |x| \end{cases}$$

This distribution arises as the limiting distribution of eigen values of symmetric matrices whose size approaches infinity. The given data complies with this distribution, and increasingly so as the number of grid points are increased from 16 to 64.

Conclusion

The discrete singular convolution method was studied and used for finding the eigen values of water molecules under the influence of the Lennard Jones potential.

REFERENCES

Arkady B (2013). Computational Complexity and the Interpretation of a Quantum State Vector, arXiv:1304.0508v1 [quant-ph] 2 Apr 2013.

Barr RS, Golden BL, Kelly JP, Resende MGC, Stewart WR (1995). "Designing and reporting on computational experiments with heuristic methods." J. Heuristics. 1(1):9-32.

Civalek Ö (2013). "Nonlinear dynamic response of laminated plates resting on nonlinear elastic foundations by the discrete singular convolution-differential quadrature coupled approaches." *Composites Part B: Eng.* 50:171-179.

Duan G, Xinwei W (2013). "Free vibration analysis of multiple-stepped beams by the discrete singular convolution." Appl. Mathe. Computation. 219(24):11096-11109.

Huacuja HJF, Vargas DR, Valdez GC, Andrade CAC, Valdez GC, Flores JAM (2007). "Experimental Analysis for the Lennard-Jones Problem Solution."Innovations in Hybrid Intelligent Systems. Springer Berlin Heidelberg, 44:239-246.

McGeoch CC (1996). "Toward an experimental method for algorithm simulation."INFORMS J. Computing. 8(1):1-15.

Pattengale ND, Alipour M, Bininda-Edmonds ORP, Moret BME, Stamatakis A (2009). "How many bootstrap replicates are necessary?." Res. Computational Molecular Biol. Springer Berlin Heidelberg, 2009. pp. 184-200.

Terán-Villanueva JD, Huacuja HJF, Juan MCV, Rodolfo APR, Héctor JPS, José AMF (2013). "Cellular Processing Algorithms." Soft Computing Applications in Optimization, Control, and Recognition. Springer Berlin Heidelberg. 294:53-74.

Wei GW (1999). "Discrete singular convolution for the solution of the Fokker–Planck equation." The J. Chem. Phys.110: 8930.

Wei GW (2000). "Solving quantum eigenvalue problems by discrete singular convolution." J. Phys. B: Atomic, Mol. Optical Phys. 33(3):343.

13

Fuzzy cell mapping on dynamical systems

Y. Song[1], D. Edwards[2] and V. S. Manoranjan[3]

[1]Savannah College of Art and Design, P. O. Box 3146, Savannah, GA 31402-3146, USA.
[2]Department of Mechanical Engineering, University of Idaho, Moscow, ID 83844-0902, USA.
[3]Department of Mathematics, Washington State University, Pullman, WA 99164-3113, USA.

Fuzzy cell mapping is a novel computational technique that combines fuzzy logic and a simple cell mapping method. In a simple cell mapping method, the information about mapping locations of image cells is never incorporated into the method. This limits the usage of a simple cell mapping method. In our fuzzy cell mapping method, we account for the mapping locations of image cells and incorporate the information by employing triangular membership functions. This paper demonstrates the use of fuzzy cell mapping on nonlinear dynamical systems. Our results indicate that fuzzy cell mapping can give good estimates on the global behavior of autonomous dynamical systems that posses limit cycles and strange attractors.

Key words: Fuzzy cell mapping, nonlinear dynamical systems, limit cycle, strange attractor.

INTRODUCTION

Nonlinear systems appear in many scientific disciplines such as engineering, physics, chemistry, biology, economics, and demography. Therefore, methods of analysis of nonlinear systems, which can provide a good understanding of their behavior, have wide applications. Although there are several analytical methods (Hsu, 1987), determining the global behavior of strongly nonlinear systems is still a substantially difficult task. The direct approach of numerical integration is a viable method. However, such an approach is sometimes prohibitively time consuming even with the powerful present-day computers.

In an attempt to find more efficient and practical ways of determining the global behavior of strongly nonlinear systems, methods such as cell-to-cell mapping were proposed and developed in the 1970's and 1980's (Hsu, 1987; Lind and Marcus, 1995). The basic idea behind the cell-to-cell mapping method is to consider the state space not as a continuum but rather as a collection of a large number of state cells, with each cell taken as a state entity. In the past, only two types of cell mappings, that is, simple cell mapping (SCM) and generalized cell mapping (GCM), have been investigated. A few years ago, Edwards and Choi (1997) proposed a new type of cell mapping method, namely fuzzy cell mapping (FCM). This method combines fuzzy logic and cell mapping techniques in a way to maintain the simplicity of SCM and achieve the results of GCM. Edwards and Choi (1997) were able to demonstrate the usefulness of their method by applying it to the Duffing equation, a nonlinear dynamical system where the time variable appears explicitly (non-autonomous system). In this paper, we extend the work of Edwards and Choi (1997) and show FCM's versatility by applying the FCM to various autonomous systems (nonlinear dynamical systems where the time variable *does not* appear explicitly) with simple steady states, limit cycles and strange attractors. Frequently, many real world mechanical, biological and

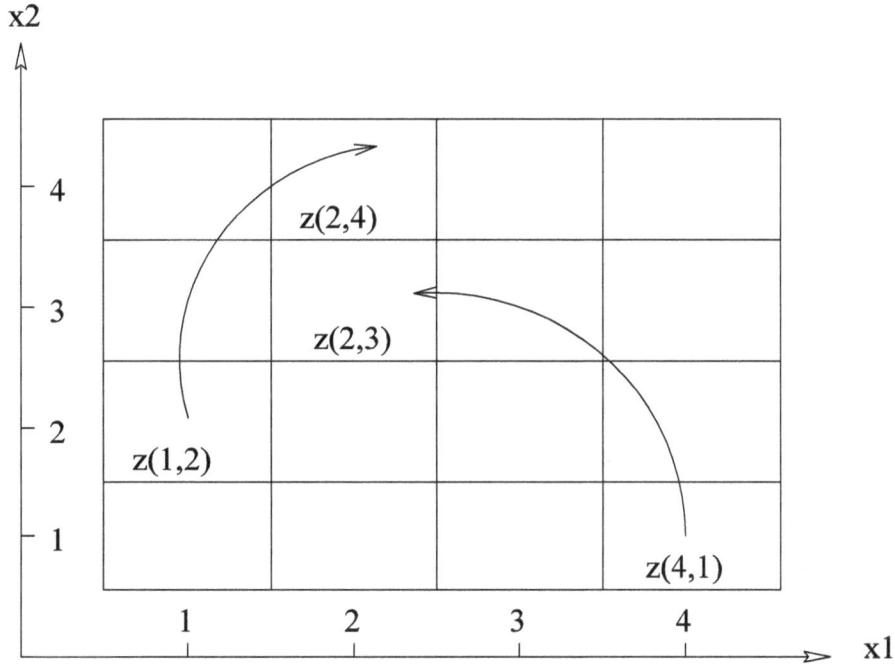

Figure 1. Cell structure and cell mapping illustration in two dimensions.

geometrical problems are formulated in terms of nonlinear autonomous dynamical systems (Jordan and Smith, 1989; Manoranjan et al., 2006; Manoranjan et al., 2008). So, there is an increasing need to consider and study methods that can provide a better understanding of such nonlinear dynamical systems.

Next are brief reviews on the different cell mapping methods; thereafter results of our study with FCM are presented.

SIMPLE CELL MAPPING AND GENERALIZED CELL MAPPING

In a cell-to-cell mapping, as mentioned before, the state space is considered as a collection of state cells, with each cell taken as a state entity. There are many ways to construct a cell structure over a given Euclidean state space. The simplest way is to construct a cell structure consisting of rectangular parallelepipeds of uniform size. Let x_i, $i = 1, 2, ..., N$ be the state variables of the state space. Let the coordinate axis of a state variable x_i be divided into a large number, say m, of intervals of uniform interval size h_i. The interval z_i along the x_i-axis is defined to be one which contains all x_i (Figure 1) satisfying:

$$(z_i - \frac{1}{2})h_i \leq x_i < (z_i + \frac{1}{2})h_i,$$

(1)

Here, by definition z_i is an integer, such that $z_i = 1, 2, ...,$

m. An N-tuple $(z_1, z_2, ..., z_N)$ is then called a cell vector and is denoted by z. A point x with components x_i belongs to a cell z with components z_i if and only if x_i and z_i satisfy Equation (1) for all i.

Practically, for all physical problems, only a finite region of a cell state space is of interest to us. Thus, all the cells outside of this finite region can be lumped together into a single cell, a *sink cell*, which will be assumed to map into itself in the mapping scheme. With this definition of the sink cell, one can assume m is finite, as is the total number of cells, m^N.

Let n represent the state of the system at $t = nT$, where T is the sampling time interval, which can be of any duration, but is usually associated with the period of the forcing function or the natural period of the system. The transition for a cell mapping method from the state at $\xi(n)$ to the state at $\xi(n+1)$ can be represented in a matrix notation as:

$$\xi(n+1) = C\xi(n)$$

(2)

where, C is the transition matrix of dimensions $m^N \times m^N$ and ξ is a m^N-tuple obtained by rearranging the indices of the cell vector z into one dimension such that ξ_k, the kth component of ξ, represents the kth cell for $k = 1, 2, ..., m^N$.

In SCM, the system of equations is integrated for the

mapping from cell j to cell i

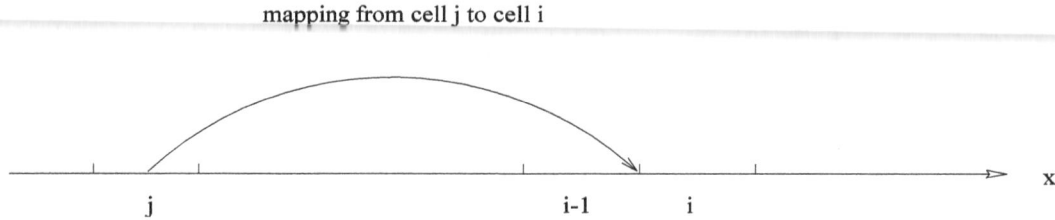

Figure 2. SCM in one dimension.

fixed time period T using the center point of each cell as the initial condition. After the integration, each cell (represented by the center point of the cell), called the original cell or the domain cell, is mapped into another cell, the image cell (regardless of the actual image location in the cell). Mapping (2) becomes $\xi(n+1) = C_S \xi(n)$, where, the elements $c_{ij}(n)$ in the transition matrix C_S are defined as the mapping from cell j at $t = nT$ to cell i at $t = (n + 1)T$. The subscript on C_S denotes SCM.

The matrix defines a Markov chain since only one element of each column is nonzero and which is equal to one. Though such a mapping, which results in a type of Poincare map, usually provides a good approximation to the global behavior of most systems, one cannot expect the method to disclose any structural details of the system behavior at a scale which is comparable to the cell size, as for a chaotic system, since only one point within the cell is used. One way to improve the power of the cell mapping method is to incorporate more system dynamics into the mapping. This leads to the generalized cell mapping method.

In GCM, the original cell is divided into subcells and each subcell is mapped, in the same fashion as in SCM, into an image cell. The probability that the original cell will map into an image cell is then based on where each subcell maps. Here, $\xi(n)$, instead of representing the cell vector as in SCM, represents the cell probability vector. The transition matrix becomes the transition probability matrix C_G. The elements, $c_{ij}(n)$, in C_G map the probability that the state is in cell j at $t = nT$ to the probability that the state will be in cell i at $t = (n+1)T$.

Since each subcell represents a fixed, equal percentage of the original cell, the elements of a column in C_G will sum to one with the number of nonzero elements being equal to the number of subcells. Once the probability matrix is determined, the iteration can be carried out by simple matrix-vector multiplications:

$$\xi(n + k) = C_G^{k-1} \xi(n) \tag{3}$$

This mapping process is also a Markov chain just like in SCM. The method is well suited, for instance, to find

strange attractors of chaotic systems and the statistical properties associated with these systems. The principal drawback with this method is that it requires much more computations than the SCM method.

FUZZY CELL MAPPING METHOD

The GCM is needed because the SCM method does not provide an accurate estimate of the global properties for certain systems. This limitation exists because of the crisp way in which the mapping is allocated to the image cell. For example, in the one dimensional case, even when a mapping lands right in the middle of cell i - 1 and cell i, where it is at equal distance from the center point of the two cells, by definition (1), cell i is still the only image cell, though the mapping could equally belong to cell i - 1 and cell i, (Figure 2). So, the SCM ignores an important piece of information, the actual mapping location.

In order to make use of all the information from SCM, FCM associates each cell with a membership function such that each cell, ξ_k, is described as a fuzzy set and the center point of the cell, $x_{\xi_k}, (k = 1, 2, \dots m^N)$, as a fuzzy number. There are various ways to choose the membership functions. A simple, natural choice is:

$$F_{\xi_k}(x) = \begin{cases} \prod_{i=1}^{i=N}\left(1 - \dfrac{|x_i - (x_{\xi_k})_i|}{h_i}\right) & if \ |x_i - (x_{\xi_k})_i| \le h_i, i = 1, 2, \dots, N. \\ 0 & elsewhere \end{cases} \tag{4}$$

Obviously, $F_{\xi_k}(x_{\xi_k}) = 1$, and $F_{\xi_k}(x_{\xi_j}) = 0$, for $k \ne j$, and $F_{\xi_k}(x)$ is symmetric about x_{ξ_k} for all k. Intuitively, these are a set of triangular shaped functions in one-dimension (Figure 3), and pyramid-like functions in two-dimensions.

Let x be the actual location of a mapping from an original cell. $\sum_{k=1}^{k=m^N} F_{\xi_k}(x) = 1$, and the number of ξ_k such that $F_{\xi_k}(x) \ne 0$ is at most 2^N.

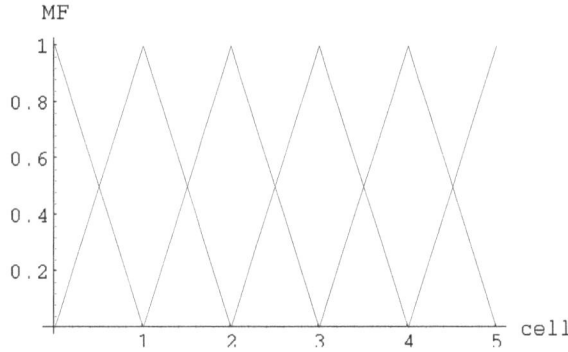

Figure 3. Triangular membership functions (MF) in one dimension.

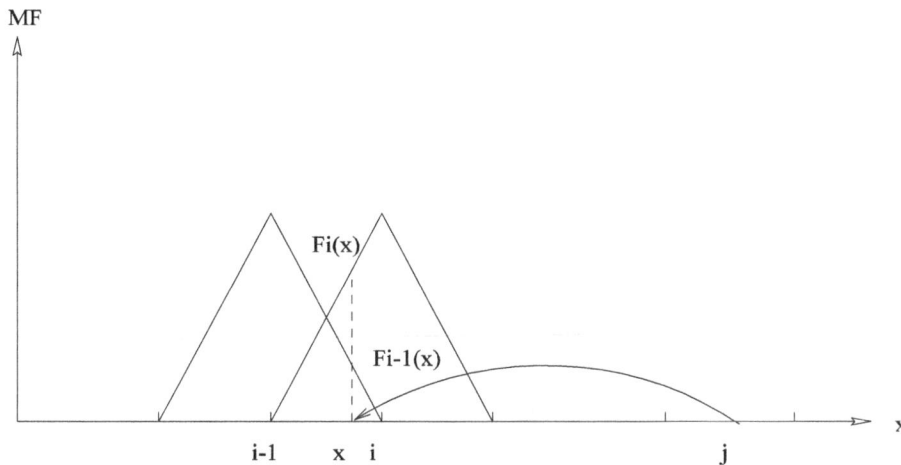

Figure 4. Mapping location and membership functions (MF)

$$\sum\nolimits_{k=1}^{m} F_k(x) = F_{i-1}(x) + F_i(x) = 1.$$

With such a choice of the membership functions, we define the transition matrix in (2) as C_F such that the elements of C_F, $c_{ij} = F_{\xi_i}(x)$, where x is the mapping location of the *j-th* cell (Figure 4).

So, FCM can be described by the following formula:

$$\xi(n+k) = C_F^{k-1}\xi(n). \tag{5}$$

Again, the mapping process is a Markov chain.

The real advantage of FCM is that it uses only the information from SCM and no further integration is required.

NUMERICAL EXPERIMENTS USING FCM

We apply the FCM method on a variety of nonlinear dynamical systems. In the case of dynamical systems that possess either a limit cycle or a strange attractor, we solve the system using the FCM and Euler's methods. The results are compared pictorially.

Use of FCM to determine stable equilibrium points

Consider a two-species Lotka-Volterra competition model (Hofbauer and Sigmund, 1998),

$$\dot{N}_1 = N_1(1 - N_1 - 2N_2)$$
$$\dot{N}_2 = 2N_2(1 - N_2 - 1.5N_1) \tag{6}$$

Here, N_1 and N_2 are the respective population densities of the competing species. This system has four equilibria, (0, 0), (0.5, 0.25), (1, 0), and (0, 1). The equilibria (0, 0)

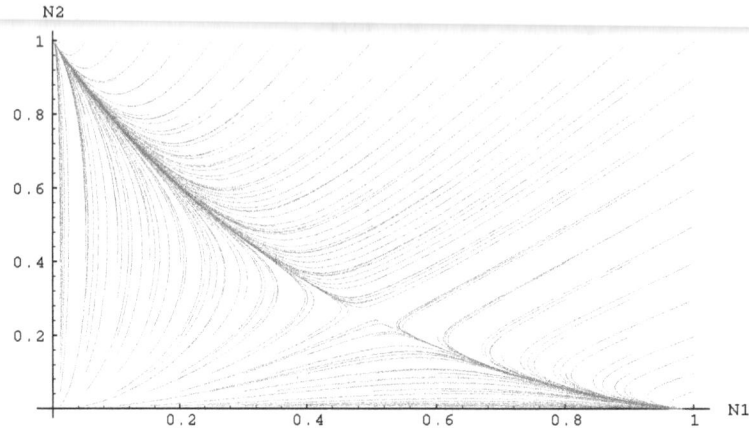

Figure 5. Phase trajectories for the competition model (6).

Table 1. Use of FCM on the competition model (6).

FCM converged to	Percentage of points that converged
(0, 0)	0.01
(0, 1)	69.85
(0.5, 0.25)	0.15
(1, 0)	29.99

and (0.5, 0.25) are *unstable*; whereas, (1, 0) and (0, 1) are *stable* states with each having a domain of attraction separated by a separatrix that passes through (0.5, 0.25), the *saddle point* (Figure 5).

Table 1 presents the results obtained in solving the problem using the FCM method. Here, we uniformly divide the unit square in the phase plane into 101 × 101 = 10201 cells, and the center points were taken as the initial conditions for the integration.

The system was integrated for a time period of $T = 6$. The process stopped when $\epsilon = \ \| \xi(n+1) - \xi(n) \|_2 \le 10^{-3}$ and it took 7 iterations. The one point that converged to the unstable equilibrium (0, 0) is (0, 0) itself and the points that converged to (0.5, 0.25); the saddle point, started on the separatrix.

Use of FCM to determine limit cycles

Consider the van der Pol equation (Seydel, 1988):

$$\ddot{u} - \lambda(1 - u^2)\dot{u} + u = 0, \tag{7}$$

with $\lambda = 0.5$. This equation originally arose as an idealization of a spontaneously oscillating, or self-excited, valve circuit. This has a nice property where regions of

positive and negative damping are separated by a closed loop (or path) known as a *limit cycle*.

Figure 6 shows the limit cycle solution for the van der Pol equation (7) obtained numerically by the Euler's method, which integrated the system for a time period of $T = 10^3$, and by the FCM. For the FCM, the phase plane is divided into 303 × 303 = 91809 uniform cells over the region

$$-3 \le u_1 \le 3,$$
$$-3 \le u_2 \le 3, \tag{8}$$

where, $u_1 = u$, and $u_2 = \dot{u}$. Here we choose $T = 2\pi$, the period of the limit cycle. The figure presents the limit cycle determined by the FCM after 3 iterations. The FCM solution and the Euler solution practically overlap on each other.

FCM can be used to determine strange attractors for chaotic systems

Edwards and Choi (1997) gave such an example which was the Duffing equation, a non-autonomous system. Here, we apply the FCM method to two different

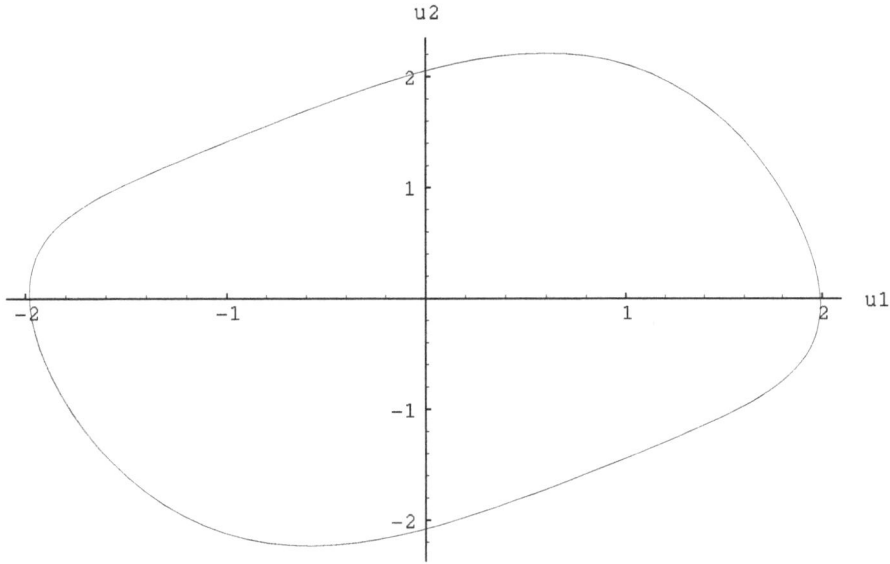

Figure 6. Limit cycle solution for the van der Pol Equation (7) (Solid line - Euler's method, Points - FCM).

nonlinear autonomous systems.

First we consider the Lorenz equations (Lorenz, 1963):

$$\dot{x}_1 = 10(-x_1 + x_2)$$
$$\dot{x}_2 = 18x_1 - x_2 - x_1 x_3$$
$$\dot{x}_3 = -\frac{8}{3}x_3 + x_1 x_2 \tag{9}$$

These equations arose in a model for convective motion in the atmosphere. The system (9) is known to possess a *chaotic* (or *strange*) attractor which results in a bistable pole-reversing behavior that is aperiodic, even without any *random* input.

The strange attractor determined by the Euler's method, which integrated the system for a time period of $T = 10^4$, is shown in Figure 7. Figure 8 presents the same attractor determined by the FCM. The region,

$$-20 \leq x_1 \leq 20,$$
$$-20 \leq x_2 \leq 20,$$
$$0 \leq x_3 \leq 40, \tag{10}$$

in the phase space is divided into $61 \times 61 \times 61 = 226981$ cells. For 3 × 3 systems like this one, choosing the smallest cell size was limited by our computing resources. Here $T = 10, \epsilon = 10^{-5}$, and it took 85 iterations for the process to come to a stop.

We also applied the FCM to the chaotic system given in Beltrami (1987):

$$\dot{x}_1 = x_1 x_2 - \mu x_1$$
$$\dot{x}_2 = (x_3 - \gamma)x_1 - \mu x_2$$
$$\dot{x}_3 = 1 - x_1 x_2 \tag{11}$$

with $\mu = 2$ and $\gamma = 5$. This system is known to model the magnetic field reversal of the earth and similar to Lorenz system possesses a *strange* attractor.

Figure 9 shows the strange attractor determined by the Euler's method with an integration period $T = 10^4$. Figure 10 presents the strange attractor determined by the FCM. The region,

$$-8 \leq x_1 \leq 8,$$
$$-5 \leq x_2 \leq 5,$$
$$1 \leq x_3 \leq 10, \tag{12}$$

in the phase space was considered with $60 \times 60 \times 60 = 216000$ cells.

Note that, if an initial point is taken on the x_3-axis, the solution will go out of the region defined by (12) and fall into the sink cell. Thus, x_3-axis is not in the domain of attraction. Therefore, one should pay special attention when constructing the cell structure to avoid such points being taken as centers of the cells.

Conclusions

From the above study, one can see that the FCM method

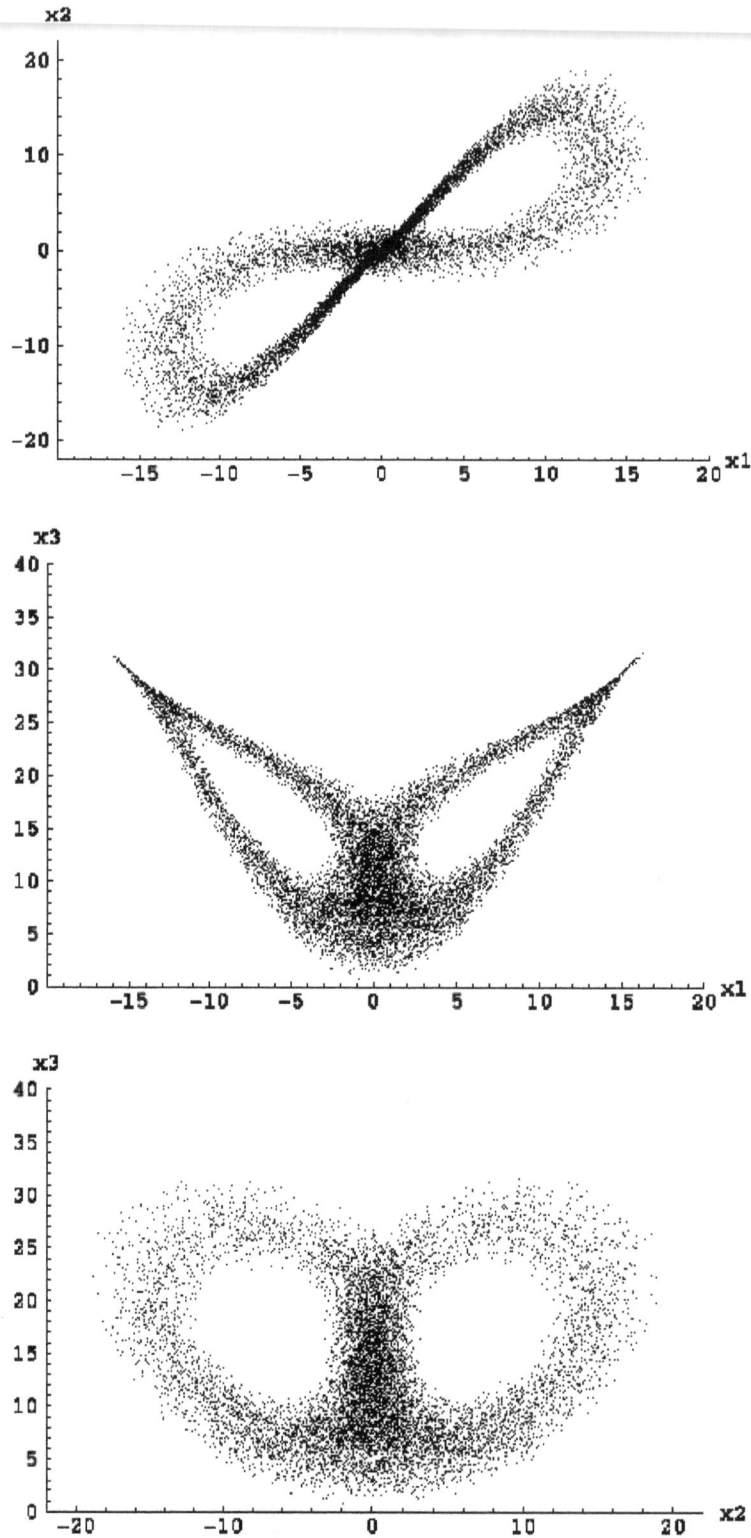

Figure 7. Strange attractor of the Lorenz equation (9) determined by the Euler's method ($T = 10^4$).

gives very good estimates on the global behavior for a variety of nonlinear dynamical systems. The results compare pretty well with that of Euler's method. We have demonstrated the versatility of FCM on autonomous

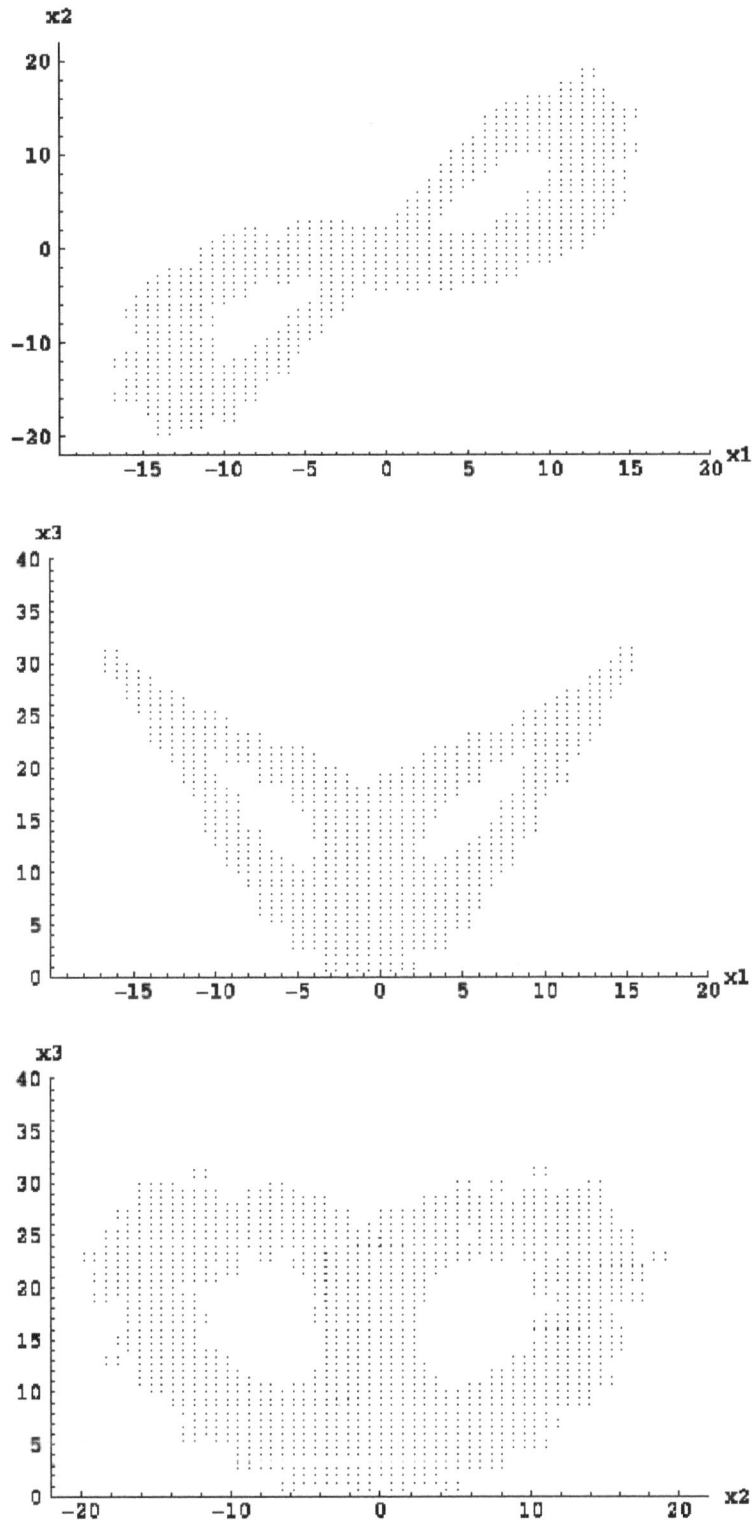

Figure 8. Strange attractor of the Lorenz equation (9) determined by the FCM $(61 \times 61 \times 61$ cells$)$.

systems with limit cycles and strange attractors. Best of all, the method is simple and straight forward. Therefore, the method is easy to understand and implement. One observation worth mentioning is that the FCM method

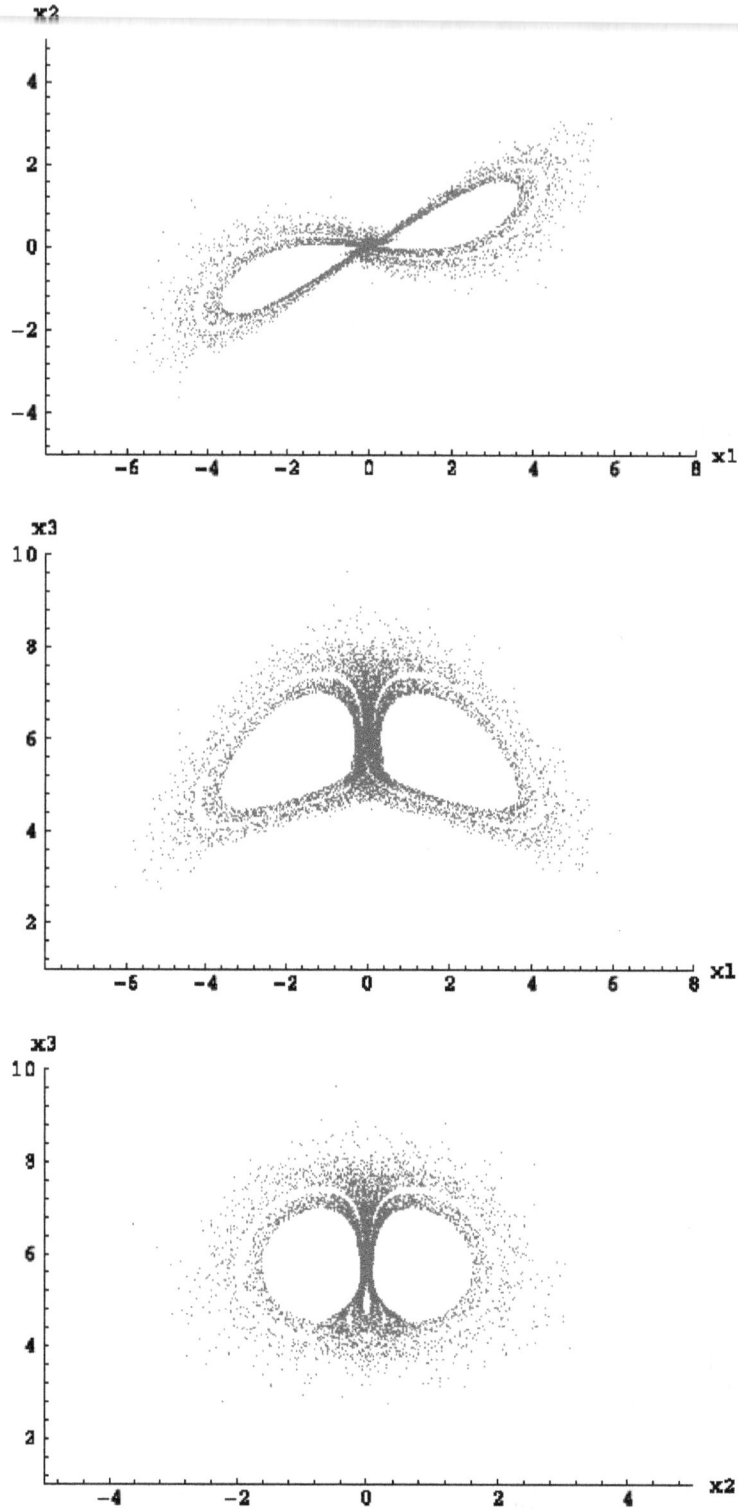

Figure 9. Strange attractor of the chaotic model (11), determined by the Euler's method ($T = 10^4$).

uses a very sparse matrix, C_F , although it is usually large. Actually only 2^N rows are nonzero in each column

in a matrix of size $m^N \times m^N$, where N is the dimension of the phase space and m is the number of cells the

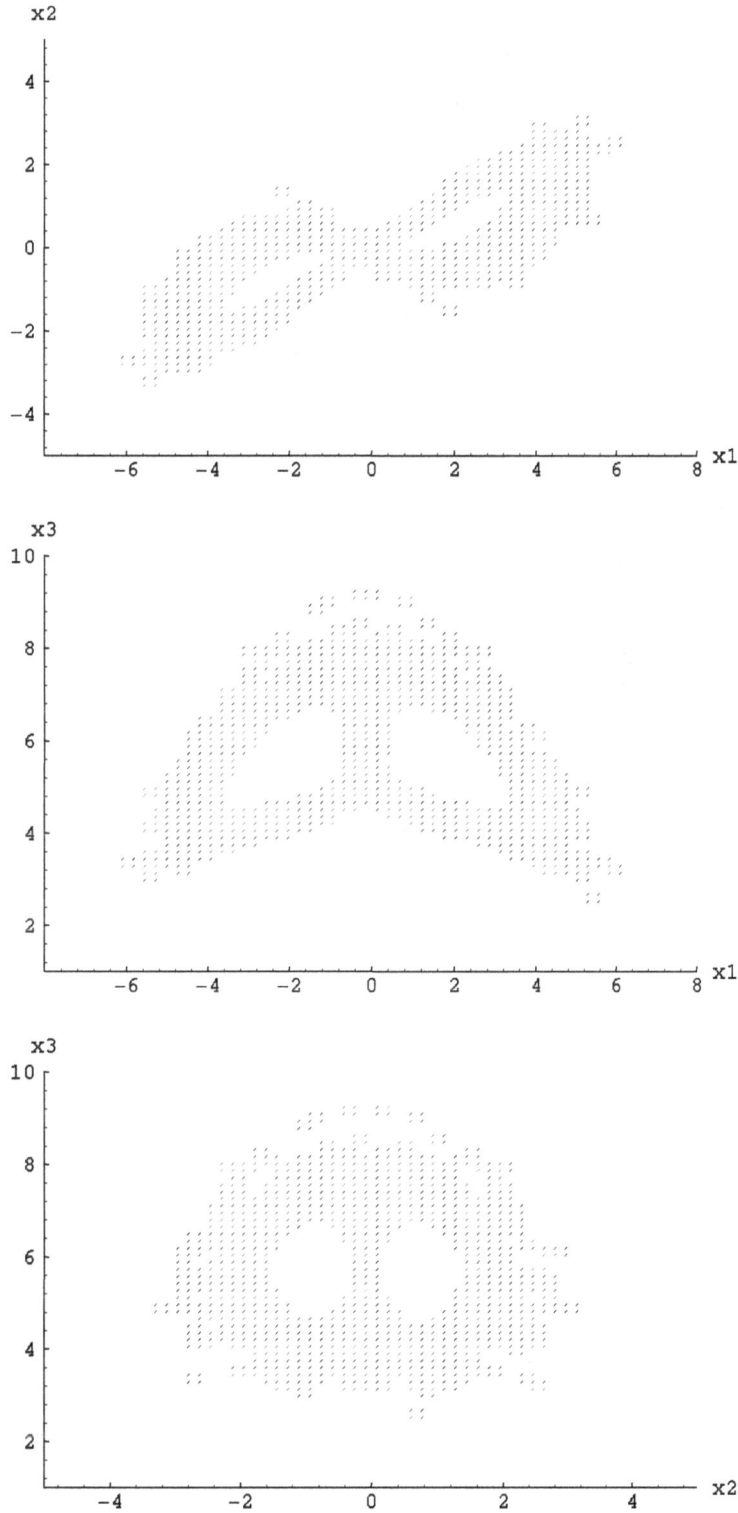

Figure 10. Strange attractor of the chaotic model (11), determined by the FCM ($60 \times 60 \times 60$ cells).

phase space is divided into along each direction. When proper storage techniques are used, one can usually deal with quite large matrices.

The key for the FCM method is the construction of the

membership functions. The reason for choosing the set of membership functions used here is that it is simple and it works. But it is by no means the only choice. The shortcoming for using such a generic membership function is that it ignores the details of the system and treats all kinds of systems the same.

One should be able to choose the membership functions based on the information from the system, such as the direction and the magnitude of the phase trajectory at each mapping location in the image cell. Tailoring the membership functions for the system should make the FCM method more efficient allowing the iterations to converge faster and currently we are exploring this approach.

The partition of the state space into cells can also be thought of as converting the nonlinear dynamical system into a symbolic dynamical system. However, not any partition results in a Markov partition. The work by Lind and Marcus (1995) presents a rigorous theory for Markov partitions and shows how topological partitions can be chosen so that a given nonlinear dynamical system can be faithfully represented by a symbolic dynamical system. Our study has demonstrated the power of fuzzification and how faithfully the fuzzy cell mapping method was able to produce the limiting behaviors of various nonlinear dynamical systems. In our future study, using the ideas presented in Lind and Marcus (1995), we intend to develop some theoretical results for the fuzzy cell mapping method.

REFERENCES

Beltrami E (1987). Mathematics for Dynamic Modeling. Academic Press: San Diego.

Edwards D, Choi HT (1997). Use of fuzzy logic to calculate the statistical properties of strange attractors in chaotic systems. Fuzzy Sets Syst. 88:205-217.

Hofbauer J, Sigmund K (1998). Evolutionary Games and Population Dynamics. Cam-bridge University Press: Cambridge.

Hsu CS (1987). Cell-to-cell mapping. Springer-Verlag: New York.

Jordan DW, Smith P (1989). Nonlinear Ordinary Differential Equations. Clarendon Press: Oxford.

Lind D, Marcus B (1995). An Introduction to Symbolic Dynamics and Coding. Cambridge University Press: Cambridge.

Lorenz EN (1963). Deterministic nonperiodic flow. J. Atmos. Sci. 20:130-141.

Manoranjan VS, Gomez MAO, Harwood RC (2008). Modelling Algae Self-Replenishment. J. Interdiscip. Math. 11:681-694.

Manoranjan VS, Rajapakse I, Krueger JM (2006). Oscillations in a Neuronal Assembly - A Phenomenological Model. Int. J. Comput. Appl. Math. 1:57-64.

Seydel R (1988). From Equilibrium to Chaos. Elsevier Science Publishing: New York.

14

Water defluoridation by bauxite-gypsum-magnesite (B-G-Mc) based filters calcined at 350 – 500°C

Thole B.[1] , Masamba W. R. L.[2] and Mtalo F. W.[3]

[1]Ngurdoto Defluoridation Research Station, P. O. Box Usa River, Arusha, Tanzania.
[2]Harry Oppenheimer Okavango Research Centre, University of Botswana, P/Bag 285, Maun, Botswana.
[3]College of Engineering and Technology, University of Dar Es Salaam, P. O. Box Dar Es Salaam, Tanzania.

Research was carried out at Ngurdoto research station in Tanzania to ascertain the potential development of a water filter made of bauxite, gypsum and magnesite in an attempt to enhance the availability of low-fluoride water. The materials were sourced within Tanzania. The X-Ray fluorescence technique showed that the major components of the materials were: bauxite: Al_2O_3 (30.33%), SiO_2 (15.0%) and Fe_2O_3 (14.3%); gypsum: CaO (28.09%), SO_3 (34.96%), and SiO_2 (9.01%); and magnesite: MgO (34.57%) and SiO_2 (19.3%). The materials were calcined at 350, 400, 450 and 500°C. The activated materials were then mixed in mass ratios of 1:2:3, 1:3:2, 2:1:3, 2:3:1, 3:1:2 and 3:2:1 (bauxite: gypsum: magnesite). One gram of each composite was employed in the batch defluoridation of 1 L of water with fluoride concentration of 12.62 mg/L. The highest defluoridation capacity, 11.89 mg F^-/g, was obtained with the 3:2:1 to 500°C composite. The quality of the treated water fell short of WHO standards on sulphates and iron but residual concentrations of Cl^-, Al^{3+}, Ca^{2+}, Mg^{2+}, Fe^{2+} were within the prescribed limits. Sorption behavior followed strongly to Langmuir isotherm, except for the 450°C calcined samples for which the Temkin isotherm behavior was pronounced. Despite the limitations of high residual sulphates and iron, a composite filter of bauxite, gypsum and magnesite was shown to be workable.

Key words: Defluoridation, bauxite, gypsum, magnesite, composite, calcine, isotherm.

INTRODUCTION

Fluoride and human health

Fluoride has been identified as a cause of dental and skeletal fluorosis world over (Maliyekkal et al., 2010; Rango et al., 2010; Peter, 2009; Onyango et al., 2009). The World Health Organisation (WHO) set a guideline value of 1.5 mg/L as an acceptable upper limit (WHO, 2006). It has been established that drinking of water having fluoride concentrations between 1.5 to 3.0 mg/L results in dental fluorosis and the browning and mottling of teeth. Fluoride levels beyond 3 mg/L in potable water result in skeletal fluorosis and when extreme concentrations above 10 mg/L are obtained, crippling fluorosis ensues (Zhu et al., 2009; Sajidu et al., 2008). It is worth noting that low concentrations in drinking water, below 1.5 mg/L, are beneficial to dental health. Dissanayake (1991) summarized the fluoride linkage to human health as presented in Table 1. Proxy indicators for high fluoride in ground water are high levels of sodium, bicarbonate and pH above 7. High-fluoride ground waters typically have relatively low calcium and magnesium concentrations with sodium and bicarbonate as the dominant dissolved constituents (WHO, 2003). There are however, some exceptions.

Table 1. Summary of linkage of fluoride ingestion through water to human health.

Level in drinking water (mg/L)	Effect on human health
<0.5	Dental caries,
0.5-1.5	Promotes dental health,
1.5-4	Dental fluorosis,
>4	Dental and skeletal fluorosis,
>10	Crippling fluorosis.

Fluoride occurrence in Tanzania and Malawi

Fluoride occurs in a number of localized areas in Malawi, among them the well known areas are Nkhotakota boma and Nathenje in the central ragion, Mtubwi, Kangankhunde, Chiradzulu, and Bangula in the southern region (Sajidu et al., 2008). In Tanzania high fluoride levels in groundwater are found at Shingida, Shinyanga, Arumeru, Arusha and other places (Mjengera, 2002; Singano, 2000). In Malawi the fluoritic areas have fluoride levels ranging from 2 to 9 mg/L whereas in Tanzania the levels are as high as 3 to 30 mg/L. Dental fluorosis is observable in the fluoritic areas in Malawi where dental, skeletal and crippling fluorosis are evident in fluoride endemic areas. The research on fluoride contamination in water has been going in both countries to address this public health challenge.

Research on defluoridation

The research carried out on defluoridation in Malawi includes synthesis and use of hydroxyapatite and clay, the optimization of gypsum and bauxite for fluoride removal (Thole, 2005; Masamba et al., 2005), dental fluorosis mapping among school children (Msonda et al., 2007), fluoride mapping in southern Malawi (Sajidu et al., 2008) and specific short term investigations at a number of other sites. These studies have identified the fluoride spread in groundwater across the country and determined the defluoridation potential and characterized locally available materials for defluoridation (Sajidu et al., 2008; Msonda et al., 2007; Thole, 2005; Masamba et al., 2005). Research in Tanzania has been ongoing and includes the WHO's Best Available Demonstrated Technology (BADT) – use of the bone char (Mjengera, 2002; Mjengera and Mkongo, 2002; Mtalo, 1997), employment of magnesite (Singano, 2000) and use of bauxitic and kaolinitic clay (Peter, 2009). Fluorosis is greatly recognized in Tanzania, probably because of its severe extent in some regions such as Shingida, Arumeru and Shinyanga. World over, the search for fluoride removal options has revealed that there are number of synthetic and naturally occurring materials that have defluoridating properties. Some of these researched materials are zeolites (Onyango et al., 2009), Mg/Al-CO_3

hydrotalcite (Hongtao et al., 2007), magnesia amended SiO_2 (Zhu et al., 2009), Mg-Al-Zn alloys (Vasudevan et al., 2009), activated carbon (Melisa, 2001; Bablia, 1996), and alum and lime (Suneetha et al., 2008). Most of the technologies reported altered pH and/or other water quality parameters beyond recommended limits. It is reported that a number of technologies change the water-pH beyond standard limits, for examples, alum and polyaluminium chloride (PAC) reduce pH (the latter reduces pH to 4.3), whereas magnesite and lime increase pH, (magnesite increases pH to 10). The WHO standard water pH is 6.5 to 8.0. The other technologies change the physicochemical water quality beyond acceptable health standards. The Nalgonda technology increases concentration of Al, gypsum introduces Ca and SO_4^{2-} in the water, whereas bauxite and clays raise the turbidity and colour of the water.

The current research investigated the defluoridation potential of a composite filter of calcined bauxite, gypsum and magnesite with the aim of identifying an optimum composite filter that would not alter the water quality beyond recommended WHO standard limits. The scope of the reported work was within calcine temperatures of 350 to 500 °C with bauxite, gypsum and magnesite obtained within Tanzania.

EXPERIMENTAL

Bauxite, gypsum and magnesite were calcined in an open air furnace at temperatures of 350, 400, 450 and 500 °C for 2 h. Finely divided powder (≤0.5 mm diameter) of each material were then mixed in ratios of 1:2:3, 1:3:2, 2:1:3, 2:3:1, 3:1:2, and 3:2:1 (bauxite: gypsum: magnesite). One gram of each composite was then placed in 1 L of water at an initial fluoride concentration of 12.62 mg/L. Fluoride concentrations, pH and the concentrations of $Cl^-, Al^{3+}, Ca^{2+}, Mg^{2+}, Fe^{2+}, SO_4^{2-}$ were measured until equilibrium fluoride concentration was obtained. This was done for each calcine-temperature and for each composition at four different temperatures and six different compositions. Each experiment was replicated 3 times. Sorption capacity was determined through a mass balance as per Equation (1):

$$q_e = \frac{(C_o - C_e)V}{m} \tag{1}$$

where q_e is the amount of adsorbed fluoride at equilibrium (mg g^{-1}); V is the volume of the solution (L); C_o and C_e are the initial fluoride

Table 2. Alkalinity, pH and ion concentrations in water before defluoridation.

Concentrations of selected ions in raw water (mg/L)						Alkalinity (mg/L)				pH
Al^{3+}	Ca^{2+}	Mg^{2+}	Fe^{2+}	SO_4^{2-}	Cl^-	Phenol	OH^-	CO_3^{2-}	HCO_3^{2-}	8.60
0.00	0.11	0.01	0.02	20	20	20	0	40	160	

WHO recommended values of the upper limits/ range of a few parameters					
Al^{3+} (mg/L)	Hardness (as CaCO₃) (mg/L)	Fe^{2+} (mg/L)	SO_4^{2-} (mg/L)	Cl^- (mg/L)	pH
0.2	500	0.3	400	250	6.5 – 8.5

concentration and the fluoride concentration at equilibrium (mgL⁻¹), respectively, and, m is the mass of adsorbent (g). Sorption isotherm was studied by fitting the data with the Langmuir, Freundlich and Temkin isotherm expressions as per Equations 2 to 4. The Langmuir isotherm assumes that the adsorption sites are energetically the same and there is a monolayer formation with no movement of particles over the surface from one site to another.

$$\frac{c_e}{q_e} = \frac{1}{Qb} + \frac{c_e}{Q} \qquad (2)$$

The term c_e is the equilibrium fluoride concentration (mg/L), q_e is the equilibrium amount on adsorbent (mg/g) Q (mg/g) and b (dm³/mg) are the Langmuir isotherm constants related to capacity and energy, respectively. A plot of $\frac{c_e}{q_e}$, c_e yields a straight line with slope and intercept $\frac{1}{Q}$, $\frac{1}{Qb}$ respectively. The Freundlich isotherm is associated with multi-layer formation on heterogeneous sites and follows the expression;

$$\ln q_e = \ln K_f + \frac{1}{n} \ln c_e \qquad (3)$$

K_f (dm³/g) and $1/n$ signify capacity and intensity, respectively. The intensity values <1 indicate favorable adsorption. A plot of $\ln q_e$ vs $\ln c_e$ should yield straight line with slope $1/n$ and intercept $\ln K_f$, if the interaction between the adsorbents and fluoride is established by the Freundlich isotherm. The terms q_e and c_e denote equilibrium concentrations in the solid phase and liquid phase, respectively. The Temkin isotherm considers chemisorption of the solute on the adsorbent media. The mathematical expression for the Temkin isotherm is given as:

$$q_e = a + b \log c_e \qquad (4)$$

The constants a and b obtainable from the plot of q_e vs Log c_e are the Temkin constants. q_e and c_e have the same meaning as earlier described.

RESULTS AND DISCUSSION

Capacity and water quality

Table 2 contains the initial alkalinity, pH and concentrations of selected ions in the water before defluoridation. The raw water quality was within the WHO recommended limits, except for pH that was 0.1 unit above the upper limit. The hardness of the water, 0.32 mg/L, also adhered well to WHO limits. The raw water was found to be very soft which confirmed the earlier findings (Mjengera, 2002; Singano, 2000). Table 3 presents the highest capacity filter (C_1), the optimum filter (C_2) (best possible filter with respect to WHO potable water standards) and the average capacity (C_μ) for all the composite filters at each temperature. pH, residual hardness and residual concentrations for the ions Fe^{2+}, Cl^-, SO_4^{2-}, and Al^{3+} are included in the table. The maximum defluoridation capacity, 11.86 mg F/g, was obtained for the 3:2:1 composite calcined at 500°C. The recommended upper limit for sulphates in water being 400 mg/L, this calcine composite had a limitation of high residual sulphates, 2200 mg/L. Water quality from all composites did not adhere to the upper sulphate limit, but the 400°C calcined composite of the ratio 2:3:1 had the lowest sulphate concentration of 900 mg/L, giving some promise of the development of a potential composite with respect to the residual sulphate. The results portray that the best and optimum composites adhered to most of the water quality parameters considered in this research, except for the residual sulphates (all composites) and iron (2:3:1 to 350°C calcined sample). When the bauxite, gypsum and magnesite filters were compared it was observed that bauxite did not adhere to the Al upper limit, whereas gypsum failed with respect to sulphate. Magnesite increased residual hardness beyond the recommended upper limit. Bauxite obtained significantly greater capacities compared to gypsum and magnesite. Bauxite, gypsum and the composite filters decreased pH of the water, whereas magnesite resulted in the increase of pH. This was attributed to the possible fluoride sorption mechanisms of the bauxite, gypsum and magnesite.

The decrease in water pH following defluoridation using bauxite and gypsum may have resulted from two attributes; (1) the amphoteric nature of oxides of aluminium, and (2) the acidic behaviour of the oxides from anionic species in water, in particular the sulphite ions. Al₂O₃ is amphoteric. Therefore, it reacts with both acids (Equation 5) and bases (Equation 6) as shown below:

Table 3. Summary of highest and optimum capacities, residual ion concentrations, hardness and pH

Parameter		Residual Ion Concentration and hardness (CaCO₃) in mg/l					pH
		Al^{3+}	Fe^{2+}	SO_4^{2-}	Cl^-	**Hardness**	
	WHO limit	0.2	0.3	400	250	500	6.5 – 8.5
350 °C	Capacity (mg F/g)	Al^{3+}	Fe^{2+}	SO_4^{2-}	Cl^-	Hardness	pH
Bauxite	7.00	8.90	0.21	100	48.1	16	8.69
Gypsum	1.88	0.00	0.27	1800	51.2	32	8.29
Magnesite	1.19	0.00	0.38	100	29.3	2220	8.93
C_1 (3:1:2)	11.61	0.16	0.28	2500	45.21	230	7.90
C_2 (2:3:1)	10.30	0.00	0.31	1800	70.12	212	8.08
C_μ	10.28	0.14	0.28	2217	65.48	221	7.94
400 °C							
Bauxite	7.41	7.70	0.19	100	33.12	54	8.49
Gypsum	0.96	0.00	0.24	1200	47.90	90	8.57
Magnesite	1.03	0.00	0.31	100	22.13	1840	9.00
C_1 (3:2:1)	11.38	0.20	0.27	1300	68.97	198	8.19
C_2 (2:3:1)	8.96	0.00	0.19	900	62.87	170	8.24
C_μ	9.61	0.09	0.25	1420	53.67	238	8.20
450 °C							
Bauxite	10.54	8.09	0.23	100	28.74	24	8.62
Gypsum	3.30	0.00	0.15	1200	42.33	78	8.50
Magnesite	2.56	0.00	0.29	100	19.87	1820	9.28
C_1 (3:1:2)	10.17	0.19	0.23	1600	41.20	188	8.13
C_2 (2:3:1)	8.29	0.00	0.11	1300	57.80	140	8.23
C_μ	8.70	0.12	0.22	2050	50.20	179	8.15
500 °C							
Bauxite	8.63	7.11	0.19	100	22.34	28	8.80
Gypsum	6.00	0.00	0.16	1400	40.17	168	8.40
Magnesite	3.99	0.00	0.21	100	20.03	2220	9.09
C_1 (3:2:1)	11.89	0.18	0.22	2200	47.20	280	8.19
C_2 (2:3:1)	9.44	0.00	0.15	1800	41.30	220	8.23
C_μ	10.36	0.21	0.15	2417	38.60	252	8.18

$$Al_2O_3(s) + 6H_3O^+(aq) + 3H_2O(l) \rightarrow 2[Al(OH_2)_6]^{3+}(aq) \quad (5)$$

$$Al_2O_3(s) + 2OH^-(aq) + 3H_2O(l) \rightarrow 2[Al(OH)_4]^-(aq) \quad (6)$$

It has also been demonstrated by the XPS analysis that the sorption of fluoride at low pH (pH <pH$_{pzc}$) for nano-AlOOH can be explained by a two step protonation / ligand exchange reaction mechanism as described by Equations (7) and (8) (Shriver et al., 1994).

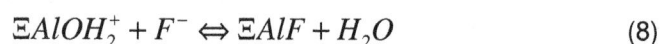

$$\Xi Al - OH + H^+ \Leftrightarrow \Xi AlOH_2^+ \quad (7)$$

$$\Xi AlOH_2^+ + F^- \Leftrightarrow \Xi AlF + H_2O \quad (8)$$

The overall reaction is represented by Equation 9

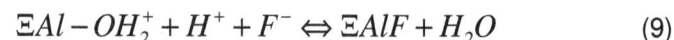

$$\Xi Al - OH_2^+ + H^+ + F^- \Leftrightarrow \Xi AlF + H_2O \quad (9)$$

The sorption model proposed by Equations (8) and (9) should result in increase of pH, because there is a net removal of H^+ ions from solution. However, when the initial pH is greater than the point zero charge (pH$_{pzc}$ = 7.8), nano-AlOOH functions as a cation exchanger and adsorbs the sodium ions present in the solution, releasing protons resulting in the decrease of pH (Wang et al., 2009; Emamjomeh and Sivakumar, 2009). The probable mechanisms for the fluoride sorption at pH>7.8 by nano-AlOOH can be explained by Equation (10).

Table 4. Percentages of sulphites and metal oxides present in the three raw materials.

Material constituent	SO_3	Al_2O_3	CaO	MgO
Bauxite	5.18	30.33	0.76	0.56
Gypsum	34.96	2.26	28.09	1.02
Magnesite	0.51	0.51	1.89	34.57

$$AlOOHxH_2O(solid) + Na^+(aq) + F^-(aq) \rightarrow AlOOH(x-1)H_2OOHNa^+F^-(solid) + H^+(aq) \quad (10)$$

In the present experiments the initial water pH was 8.6, fairly larger than the pH$_{pzc}$ of AlOOH. The X-ray fluorescence (XRF) technique showed that the bauxite employed in these experiments contained 30.33% Al_2O_3. The decrease in pH after defluoridation using the bauxite may entail the formation of AlOOH in water and subsequent protonation / ligand exchange reactions, as suggested earlier (Wang et al., 2009). The Al_2O_3 being amphoteric may have reacted as acid in the basic water medium (pH = 8.60), as illustrated earlier via Equation (6). This lowers pH because it diminishes the OH^- concentration in water and, then the following reaction may have occurred enhancing the decrease of water pH;

$$[Al(OH)_4]^-(aq) + F^- + 2H_2O \rightarrow [Al(OH)_2F] + 2H_3O^+(aq) \quad (11)$$

The bauxite and gypsum contained 5.18 and 34.96% SO_3, respectively obtained through the XRF analysis. It is established that covalent oxides are largely acidic. Such oxides on dissolution bind water molecules and release protons to the surrounding medium (Shriver et al., 1994). The behaviour of carbon dioxide in aqueous medium illustrates this point.

$$CO_2(g) + H_2O(l) \rightarrow [OC(OH)_2](aq) \quad (12)$$

$$[OC(OH)_2](aq) + H_2O(l) \Leftrightarrow [O_2C(OH)]^-(aq) + H_3O^+(aq) \quad (13)$$

From the foregoing reactions (Equations 12 and 13) it is highly plausible that the sulphite from the bauxite and gypsum, upon entering the aqueous medium, reacted similarly as depicted via Equations (14) and (15).

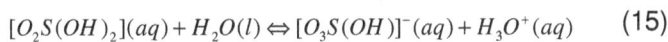

$$SO_3(aq) + H_2O(l) \rightarrow [O_2S(OH)_2](aq) \quad (14)$$

$$[O_2S(OH)_2](aq) + H_2O(l) \Leftrightarrow [O_3S(OH)]^-(aq) + H_3O^+(aq) \quad (15)$$

The reactions proposed in Equations (14) and (15) would result in decrease of pH. This entails that the presence of both Al_2O_3 and SO_3 in bauxite would function in synergy for decrease in pH during defluoridation using bauxite, in agreement with the experimental data, which shows that defluoridation using bauxite obtained lowest pH values amongst the three materials. A recap of the compositional quantities of SO_3, Al_2O, CaO, and MgO in bauxite, gypsum and magnesite depicted in Table 4 shows that bauxite and gypsum contained the sulphite in greater proportions compared to magnesite. Therefore, the effect of the sulphite reactions proposed in Equations (14) and (15) may have been negligible for magnesite. Ionic oxides being largely basic (Shriver et al., 1994) will lower pH of an aqueous medium as illustrated below.

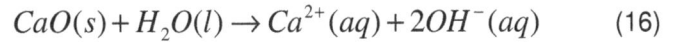

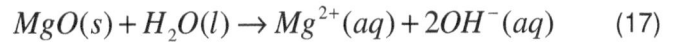

$$CaO(s) + H_2O(l) \rightarrow Ca^{2+}(aq) + 2OH^-(aq) \quad (16)$$

$$MgO(s) + H_2O(l) \rightarrow Mg^{2+}(aq) + 2OH^-(aq) \quad (17)$$

The effect of the CaO reaction may have been overshadowed by the effect of the reactions of SO_3 as proposed in Equations (16) and (17) resulting in a net decrease of pH during defluoridation with gypsum, SO_3 being in greater proportion to CaO in gypsum. On the other hand, magnesite having a very low content of SO_3, increased pH as effects of MgO reactions overshadowed the effect of SO_3 reactions because of the greater proportion of MgO compared to SO_3. From the foregoing discussions, it is plausible that defluoridation with gypsum and magnesite followed the scheme proposed in Equations (18) and (19), respectively;

$$CaO(s) + H_2O + F^-(aq) \rightarrow Ca(OH)F(aq) + OH^-(aq) \quad (18)$$

$$MgO(aq) + H_2O + F^- \rightarrow Mg(OH)F(aq) + OH^-(aq) \quad (19)$$

Sorption isotherm

Figures 1, 2 and 3 exemplify sorption isotherm plots obtained from the experimental data. Most of the data sets adhered more strongly to Langmuir sorption except for the data from the 450°C calcined samples that followed the Temkin isothermal sorption. Mixed sorption were, however, deciphered, the correlation coefficients of all three sorption isotherms being close to unity. The isotherm parameters obtained from the best fit linear equations are depicted in Table 5.

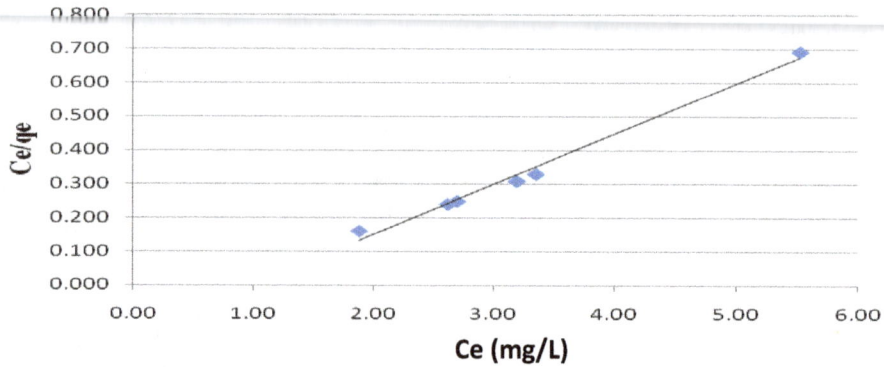

Figure 1. Langmuir isotherm for the 3:2:1 to 350°C calcined sample.

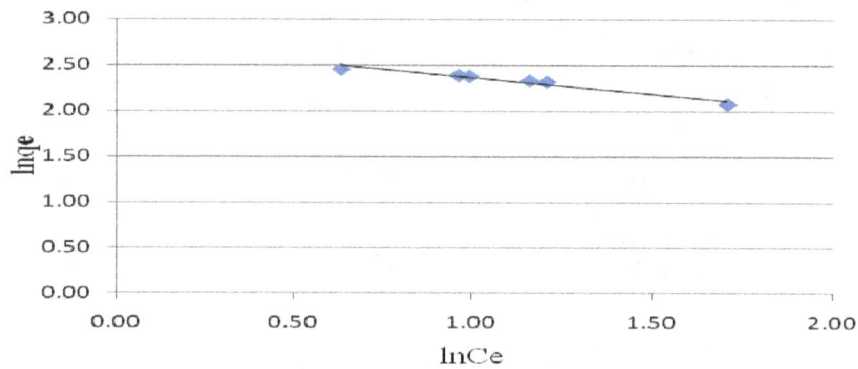

Figure 2. Freundlich isotherm for the 3:2:1 to 350°C calcined sample.

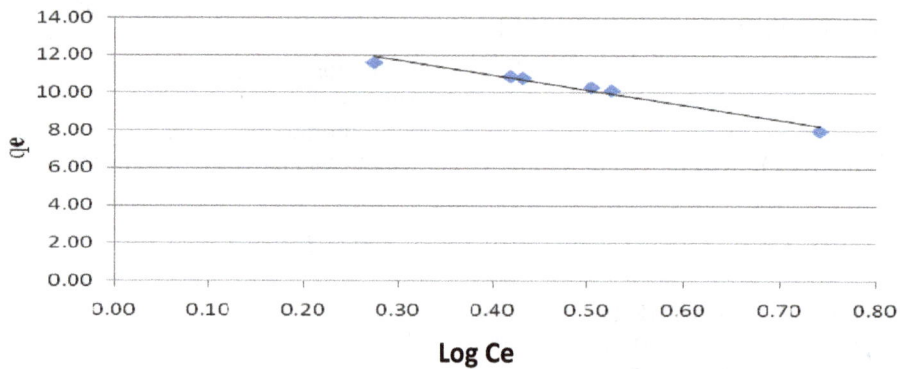

Figure 3. Temkin isotherm for the 3:2:1-350°C calcined sample.

Conclusions

The composite-filter prepared by calcination obtained higher capacities compared to the stand-alone filters of bauxite, gypsum and magnesite, depicting a synergetic relationship among the three materials. The residual sulphate is the most important limitation of defluoridation using these composites. This appears to stem from the gypsum component. The composite materials have a net effect of slightly decreasing pH that is attributable to the behaviour of Al_2O_3 and sulphite in aqueous media. Bauxite generally attained greater capacities compared to gypsum and magnesite at four temperatures. The composites with relatively larger bauxite quantities obtained the larger sorption capacities. The water quality adhered to the WHO upper limits of pH, hardness,

Table 5. Isotherm parameters obtained for Langmuir, Freundlich and Temkin isotherms.

Variable	Parameter	Temperature (°C) 350	400	450	500
Langmuir	Θ	6.70	6.20	5.31	7.59
	B	- 0.97	- 0.87	- 0.52	- 1.15
	R^2	0.989	0.982	0.986	0.992
Freundlich	k_f	15.15	15.93	22.44	15.28
	$1/n$	- 0.36	- 0.40	- 0.58	- 0.32
	r^2	0.944	0.957	0.975	0.962
Temkin	A	14.09	14.43	16.99	14.44
	B	- 7.90	- 8.66	- 11.63	- 7.60
	r^2	0.966	0.979	0.990	0.977

Fe^{2+}, Cl^-, and Al^{3+} indicating a promising scenario for developing a composite filter made of these three materials.

Further work

Defluoridation in fixed bed will be carried out to obtain the breakthrough characteristics using the composites of the three materials used in this study.

REFERENCES

Bablia K (1996). Studies of water defluoridation using activated carbons and activated carbons loaded separately with Magnesia, Alumina and Calcium. MSc Thesis, University of Dar es Salaam, Tanzania.

Dissanayake CB (1991). The fluoride problem in the groundwater of Sri Lanka – Environmental management and health. Int. J. Environ. Stud. 19:195-203.

Emamjomeh MM, Sivakumar M (2009). Fluoride removal by a continuous flow electro-coagulation reactor. J. Environ. Manage. 90:1204-1212.

Hongtao W, Jun C, Yuanfeng C, Junfeng J, Lianwen L, Henry T (2007). Defluoridation of drinking water by Mg/Al hydrotalcite-like compounds and their calcined products. Appl. Clay Sci. 35:59-66.

Maliyekkal SM, Anthony AR, Pradeep T (2010). High combustion synthesis of nanomagnesia and its application for fluoride removal. Sci. Total Environ. 408:2273-2282.

Masamba WRL, Sajidu SMI, Thole B, Mwatseteza JF (2005). Water defluoridation using Malawi's locally sourced gypsum. Phys. Chem. Earth, A/B/C, 30(11–16):846–849.

Melisa J (2001). Defluoridation of drinking water by adsorption of fishbone. MSc Thesis, University of Dar es Salaaam, Tanzania.

Mjengera H (2002). Optimisation of bone char filter column for defluoridating drinking water at household level in Tanzania, PhD thesis, University of Dar es Salaam.

Mjengera H, Mkongo G (2002). Appropriate technology for use in fluoritic areas in Tanzania. 3rd Waternet/WARFSA Symposium on water demand management for sustainable use of water resources, University of Dar Es Salaam.

Msonda KWM, Masamba WRL, Fabiano E (2007). A study of fluoride ground water occurrence in Nathenje, Lilongwe, Malawi. Phys. Chem. Earth, A/B/C, 32(15–18):1178–1184.

Mtalo FW (1997). Fluoride occurence in water in Arusha region. Tanzania Eng. J. 6(2):50 – 57.

Onyango MS, Leswifi Y, Ochieng A, Kuchar D, Otieno FO, Matsuda H (2009). Breakthrough analysis for water defluoridation using surface-tailored zeolite in a fixed bed column. Ind. Eng. Chem. Res. 48(2):931-937.

Peter KH (2009). Defluoridation of high fluoride waters from natural water sources by using soils rich in bauxite and kaolinite. J. Eng. Appl. Sci. 4(4):240 – 246.

Rango T, Bianchini G, Beccaluva L, Tassinari R (2010). Geochemistry and water quality assessment of central main Ethiopian rift natural waters with emphasis on source and occurrence of fluoride and arsenic. J. Afr. Earth Sci. 57(5):479 – 491.

Sajidu SMI, Masamba WRL, Thole B, Mwatseteza JF (2008). Ground water fluoride levels in villages of Southern Malawi and removal studies using bauxite. Int. J. Phys. Sci. 3(1):001 – 011.

Shriver DF, Atkins PW, Langford CH (1994). Inorganic chemistry. 2nd Ed., ELBS, Oxford University Press, pp. 196–198.

Singano JJ (2000). Investigation of the mechanisms of defluoridation of drinking water by using locally available magnesite, PhD thesis, University of Dar es Salaam.

Suneetha N, Rupa PK, Sabitha V, Kumar KK, Mohanty S, Kanagasabapathy, AS, Rao P (2008). Defluoridation of water by a one step modification of the Nalgonda technique. Ann. Trop. Med. Public Health 1(2):56–58. Http//www.atmph.org.

Thole B (2005). Water defluoridation with Malawi bauxite, gypsum and synthetic hydroxyapatite, bone and clay: Effects of pH, temperature, sulphate, chloride, phosphate, nitrate, carbonate, sodium potassium and calcium ions. MSc Thesis, Zomba, University of Malawi.

Vasudevan S, Lakshmi J, Sozhan G (2009). Studies on Mg-Al-Zn as anode for the Removal of Fluoride from Drinking Water in an Electrocoagualtion Process. Clean. 37(4-5):372-378.

Wang SG, Ma Y, Shi YJ, Gong WX (2009). Defluoridation performance and mechanisms of nano-scale aluminium oxide hydroxide in aqueous solution. J. Chem. Technol. Biotechnol. 83:1043–1050.

WHO (2006). Guidelines for drinking-water quality. 1st Addendum to volume 1. Recommendations, 3rd ed., Geneva: World Health Organisation.P. 595.

WHO (2003). Removal of Excessive Fluoride; a World Health Organisation paper: www.who.int/environmental-information/informationresources/htmdocs/fluoride/fluor.

Zhu P, Wang H, Sun B, Deng P, Hou S, Yu Y (2009). Adsorption of fluoride from aqueous solution by magnesia-amended silicon dioxide granules. Wiley Interscience: www.interscience.wiley.com. DOI 10.1002/jctb.2197.

15

A tuneable metamaterial design using microelectromechanical system (MEMS) based split ring resonator (SRR)

Tanuj Kumar Garg[1] , S. C. Gupta[2], S. S. Patnaik[3] and Vipul Sharma[1]

[1]Deptartment of Electronics and Communication Engineering, Gurukul Kangri University, Haridwar, India.
[2]Deptartment of Electronics and Communication Engineering, DIT, Dehradun, India.
[3]Deptartment of ETV, NITTTR, Sector-26, Chandigarh, India.

In this paper, we present a study of tuneable equilateral triangular shaped spilt ring resonator (ETSRR). In this ETSRR we rotate the inner and outer rings by varying the position of the splits in rings. For this we used radio frequency microelectromechanical system (RF MEMS) switches. By making MEMS switches ON/OFF, the positions of splits in the rings were varied which can be considered as rotation of rings. As we rotate the inner and outer rings (by varying the position of splits), the configuration is tuned to different frequency from its basic configuration, thus we get tunability.

Key words: Split ring resonator (SRR), metamaterials, equilateral triangle, radio frequency microelectromechanical system (RF MEMS) switch.

INTRODUCTION

Nowadays, metamaterial becomes most popular among the researchers because it shows simultaneously negative values of effective permittivity (ε_{eff}) and effective permeability (μ_{eff}) over a common frequency band. Metamaterials are also regarded as left handed materials (LHMs) or negative refractive index materials (NIMs) because these materials exhibits the properties like backward propagation, reverse Doppler effect and reverse Vavilov - Cerenkov effect which are not possessed by natural material (Ziolkowski, 2003). The negative values of effective permittivity (ε_{eff}) can be obtained by using metal rod and effective permeability (μ_{eff}) can be obtained by Split ring (Huang et al., 2010). The design of metamaterial based on shape and geometry is most popular work among the others. Various types of ring type structures like circular, square,

U-shaped, S- shaped, Ω- shaped, elliptical shaped, phi-shaped (Sharma et al., 2011) have been proposed till now. The split ring resonator (SRR) structures which are most famous, circular or rectangular. The triangular shaped metamaterial resonator was first studied by Sabah and Uckun (2008) although now few studies are there in literature (Zhu et al., 2009; Jalali et al., 2009; Sabah, 2010). Metamaterials can also be used in antenna designing to enhance the gain and directivity of the antennas (Wu et al., 2005; Lee and Hao, 2008; Gil et al., 2006; Qureshi et al., 2005).

Compared to PIN diodes and FET transistors, RF MEMS switches have better performance in terms of isolation, insertion loss, power consumption and linearity (Muldavin et al., 2000a, 2000b).

Wang et al. (2008) purposed a theory about SRR with

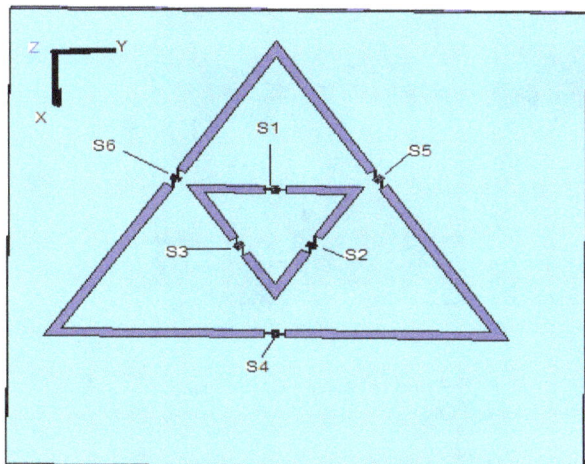

Figure 1. Equilateral triangular spilt ring resonator with splits and RF MEMS switches in each arm.

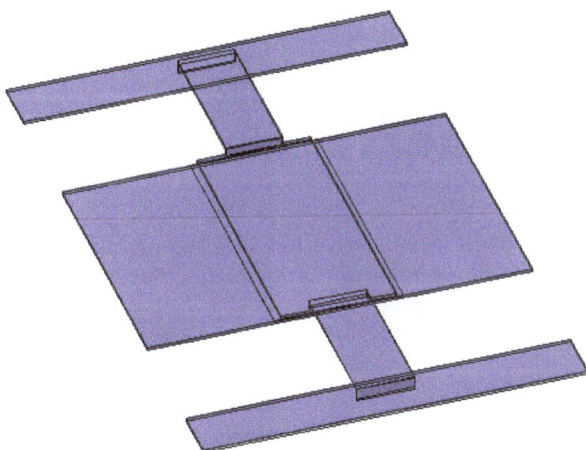

Figure 2. The structure of RF MEMS shunt switch.

rotated inner ring to analyze the controllability of its magnetic resonant frequency. The inner ring was rotated by means of control bars and by rotation of inner ring as the angle between the two splits decreases magnetic resonant frequency increases.

Sabah (2010) proposed tuneable metamaterial (MTM) structure composed of triangular spilt ring resonator and wire strip. These MTMs are formed from FR4 and RT/duriod 5880 substrates show tunability in terms of substrate thickness. The results shown are very promising.

The rotation of rings can also be achieved by putting splits in each arm of rings and then the position of splits can be made ON/OFF by using MEMS switches. By this, the magnetic resonant frequency gets shifted and thus getting tunability.

In present paper, authors have obtained frequency tuneable MTM triangular SRR (Sabah, 2010) by rotating

the inner (Wang et al., 2008) and outer ring. The rotation of rings is implemented by change in position of splits in each arm by using MEMS switches. Excellent performance is achieved.

DESIGN

In this design, RogersRT /duriod5880 (relative permittivity = 2.2) is used as a substrate with a thickness of 0.8 mm .The length and width of the substrate is 28 and 30 mm, respectively. The dimensions of outer ETSRR base length is 22.52 mm and height is 19.5 mm; 8.66 and 7.52 mm for inner ETSRR.

The separation between outer and inner ETSRR is 9.5 mm from vertex of outer ETSRR to base of inner ETSRR. The width of each strip is 0.5 mm. The split gap in each ETSRR is 1.0 mm. Splits are made at each arm of inner and outer ETSRR along with RF MEMS switches placed in each split. Switches S1, S2, S3 are placed in inner ETSRR and switches S4, S5, S6 are placed in outer ETSRR. The proposed design is shown in Figure 1.

The structure of RF MEMS shunt switch (Figure 2) consist of thin metal (gold in this case) membrane bridge that is suspended over the central conductor of coplanar waveguide (CPW) and fixed on the ground conductor. The dimensions of shunt switch are: length of the bridge = 200 μm, width of the bridge = 90 μm, thickness of the bridge = 2 μm, silicon nitrate (relative permittivity = 7) is used as the dielectric having a thickness of 0.2 μm, air gap between lower conductor and upper conductor is 0.9 μm.

When a switch is in ON position in a particular arm, that means there is no split in that particular arm; whereas, when the switch is in OFF position, then it means there is presents of a split in that arm. The whole structure is designed and placed in two port waveguide formed by a pair of both perfect magnetic conductor (PMC) walls in z-direction and perfect electric conductor (PEC) walls in y-direction. The whole structure is excited by an electromagnetic wave with propagation vector in x-direction. The structure is designed and simulated by using Ansoft HFSS simulator, finite element based electro-magnetic mode solver.

To show the physical properties of the designed structure, S parameters are calculated and effective permeability is extracted by using effective parameter retrieval method (Smith et al., 2005).

ANALYSIS AND DISCUSSION

Metamaterials type structures can be considered as LC resonant circuit whose resonance frequency can be determined by $\omega = 1/\sqrt{LC}$. When the switch S4 was OFF in outer ETSRR and switch S1 was OFF in inner ETSRR, while rest of switches were ON (Figure 3a); the

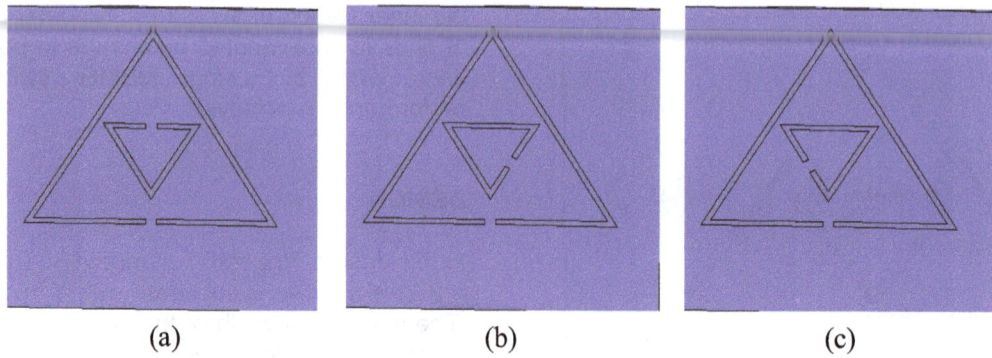

(a)   (b)   (c)

Figure 3. ETSRR Configuration: When switch S4 in outer ETSRR is OFF and (a) switch S1, (b) switch S3, (c) switch S2 in inner ETSRR are OFF; and rest of the switches are ON.

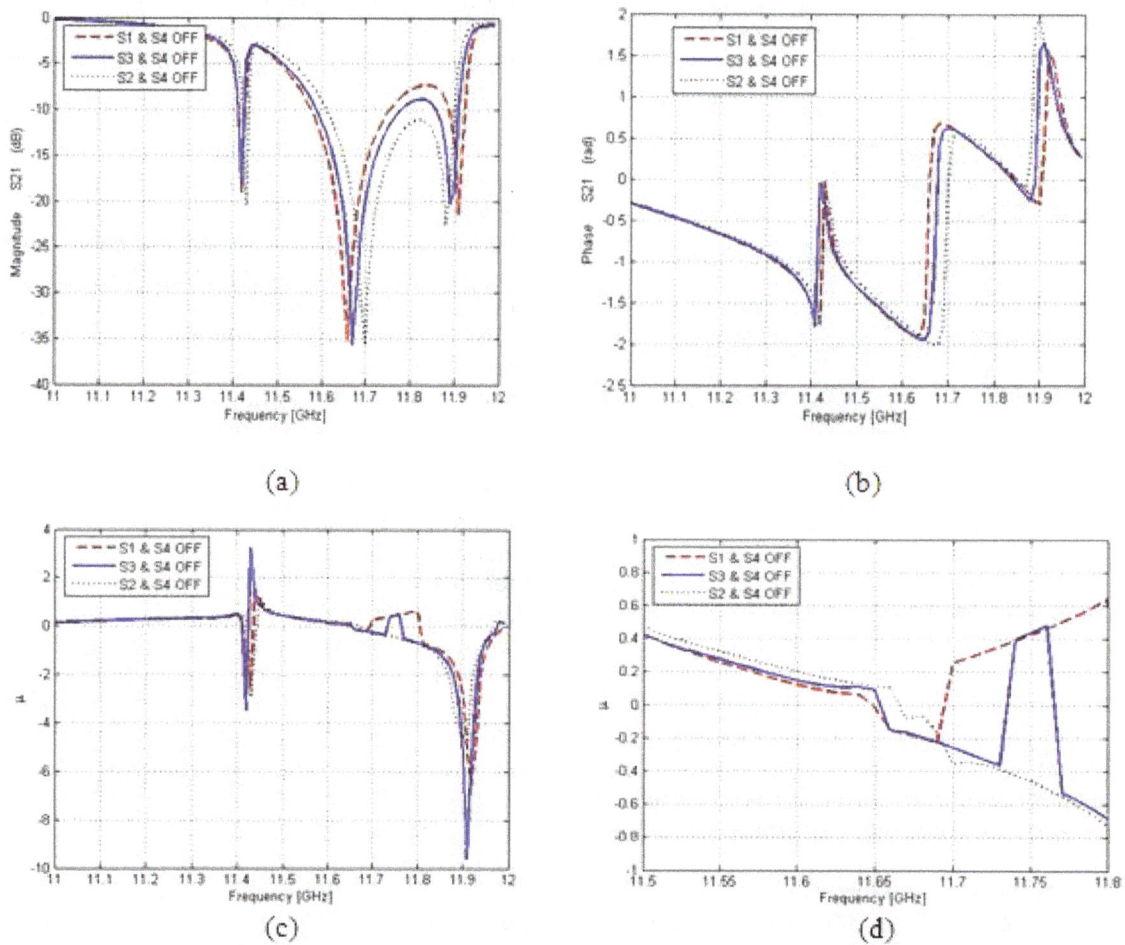

(a)   (b)

(c)   (d)

Figure 4. (a) minimum of transmission (S_{21}) in dB, (b) dip in phase of S_{21} (rad), (c) effective permeability, (d) Zoom of effective permeability.

minimum of transmission (S_{21}) was observed at 11.66 GHz (Figure 4a, red curve), the dip in phase of S_{21} was observed at 11.63 GHz, (Figure 4b), red curve), negative permeability occurred in frequency regime 11.65 GHz ~ 11.69 GHz (Figure 4c), red curve), but they overlap each other at frequency 11.66 GHz. Because of negative permeability, EM waves cannot transmit through the structure which result in dip in transmission spectrum. In the original ETSRR the angle between two splits is π.

Now, when we rotated the inner ring by making the

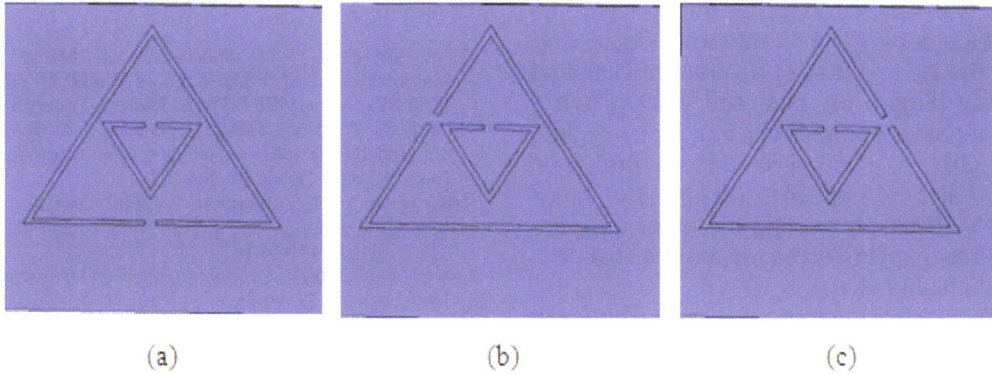

Figure 5. ETSRR Configuration: When switch S1 in inner ETSRR is OFF and (a) switch S4, (b) switch S6, (c) switch S5 in outer ETSRR are OFF; and rest of the switches are ON.

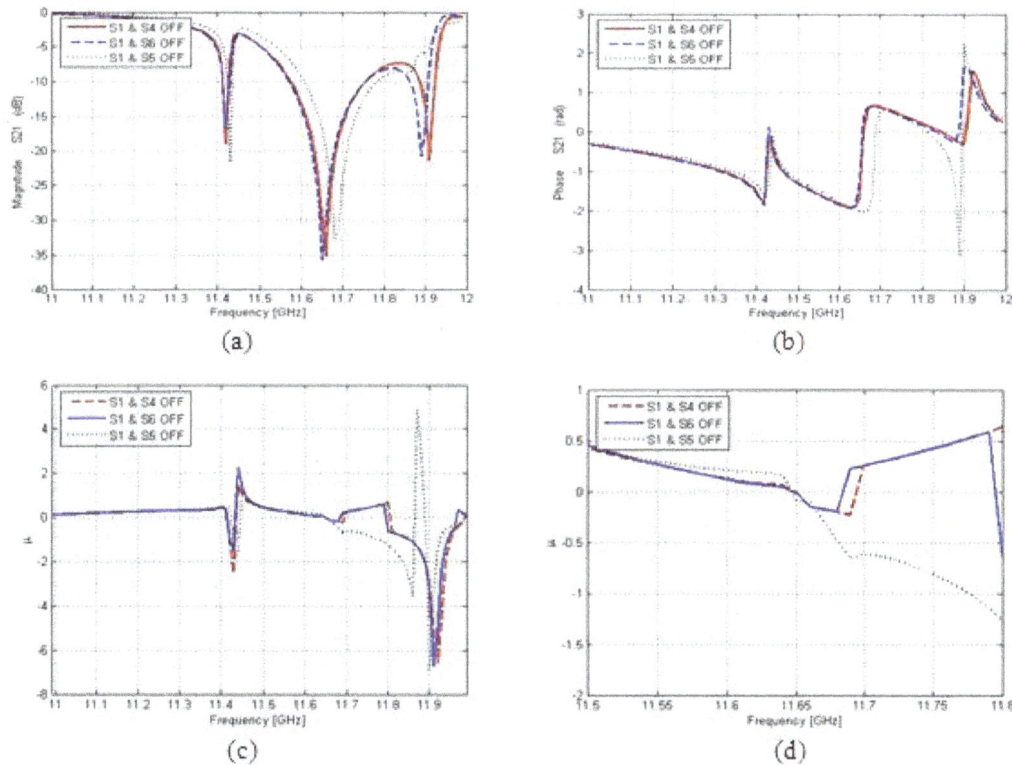

Figure 6. (a) minimum of transmission (S_{21}) in dB, (b) dip in phase of S_{21} (rad), (c) effective permeability, (d) Zoom of effective permeability.

switch S1 transit from OFF to ON position and S3 or S2 were transited from ON to OFF position in inner ETSRR (S4 in outer ETSRR remained OFF) while rest of the switches remained as they were (Figure 3b, c); the minimum of transmission (S_{21}) were shifted to higher frequency 11.67 GHz (Figure 4a) blue curve) and 11.70 GHz (Figure 4a) black curve), and the dip in phase of S_{21} were also shifted to higher frequency (Figure 4b) blue and black curve); accordingly, the negative permeability frequency regime shifted to 11.66 GHz ~ 11.73 GHz

(Figure 4c, blue curve) and 11.67 GHz ~11.96 GHz (Figure 4c, black curve), respectively. Thus, by rotating the inner ring, the angle between the two splits decreases, so the resonant frequency shifted to higher side.

When we rotated the outer ring by making the switch S4 transit from OFF to ON position and S6 or S5 were transit from ON to OFF position in outer ETSRR (S1 in inner ETSRR remained OFF) while rest of the switches remained as they were (Figure 5b, c); the minimum of

transmission (S$_{21}$) were shifted to higher frequency 11.65
GHz (Figure 6a, blue curve) and 11.68 GHz (Figure 6(a)
black curve), and the dip in phase of S$_{21}$ were also shifted
to higher frequency (Figure 6b, blue and black curve);
accordingly, the negative permeability frequency regime
shifted to 11.65 GHz ~ 11.68 GHz (Figure 4(c) blue
curve) and 11.65 GHz ~11.86 GHz (Figure 4(c) black
curve), respectively. Thus, by rotating the outer ring, the
angle between the two splits decreases, so the resonant
frequency shifted to higher side.

Thus, we can control the magnetic resonant frequency
by rotating the inner and outer ETSRR. In each case, the
configuration is tuned to different frequency. If we
compare this with purposed technique in Sabah (2010) in
which he got tunability by varying the substrate thickness;
the technique presented in this paper is easy to get
tunability because in this we rotate the rings by means of
MEMS switches (making then ON/OFF to make splits
present or not); as well as from technique presented in
Wang et al. (2008) in which they rotated the inner ring by
using control bars.

Conclusion

In this paper we design an equilateral triangle shaped
split ring resonator with their basic configuration. Then we
varied the position of splits in inner ring by using RF
MEMS switches (making them ON/OFF) that was
considered as the inner ETSRR was rotated. So by
rotation of inner ring, the configuration was tuned to
different frequency. Similarly, when we rotated the outer
ring by varied the position of splits in outer ring using
MEMS switches; the configuration was again tuned to
different frequency. So, by rotation of inner or outer ring,
we can control the magnetic resonant frequency and thus
we get tunability. This ETSRR can be used in antenna
design to obtain high directivity and high gain of the
antennas.

REFERENCES

Gil I, Bonache J, García-García J, Martín F (2006). Tunable Metamaterial Transmission Lines Based on Varactor-Loaded Split-Ring Resonators. IEEE Trans. Microw. Theory Technol. 54:6.

Huang C, Zhao Z, Feng Q, Cui J, Luo X (2010). Metamaterial composed of wire pairs exhibiting dual band negative refraction. Appl. Phys. B Laser Optics 98:365-370.

Jalali M, Sedgji T, Zehforoosh Y (2009). Miniaturization of waveguides dual band antenna using TSRR-WS metamaterials. Int. J. Comput. Electr. Eng. 1:1793-8163.

Lee Y, Hao Y (2008). Characterization Of Microstrip Patch Antennas On Metamaterial Substrates Loaded With Complementary Split-Ring Resonators. Microw. Optical Techn. Lett. 50:8.

Muldavin B, Rebeiz GM (2000a). High-Isolation CPW MEMS Shunt Switches - Part 1: Modeling. IEEE Trans. Microw. Theory Techn. 48:1045-1052.

Muldavin JB, Rebeiz GM (2000b). High-Isolation CPW MEMS Shunt Switches - Part 2: Design. IEEE Trans. Microw. Theory Techn. 48(6):1053-1056.

Qureshi F, Antoniades MA, Eleftheriades GV (2005). A Compact and Low profile Metamaterial Ring Antenna with Vertical Polarization. IEEE Antennas Wireless Propag. Lett. Vol. 4.

Sabah C (2010). Tunable Metamaterial Design Composed of Triangular Split Ring Resonator and Wire Strip for S- and C- Microwave Bands. Prog. Electrom. Res. B, 22:341-357.

Sabah C, Uckun S (2008). Triangular split ring resonator and wire strip to form new metamaterial, Proc. Of XXIX General Assembly of Int. Union of Radio Science.

Sharma V, Pattnaik SS, Garg T, Devi S (2011). A metamaterials inspired miniaturized phi-shaped antenna. Int. J. Phys. Sci. 6(18):4378-4381.

Smith DR, Vier DC, Koschny T, Soukoulis CM (2005). Electromagnetic parameter retrieval from inhomogeneous metamaterials, Phy. Rev. E, 71.

Wang J, Qu A, Xu Z, Ma H, Yang Y, Gu C (2008). A Controllable Magnetic Metamaterial: Split Ring Resonator with Rotated Inner Ring, IEEE Trans. Antennas Propag. 56:7.

Wu BI, Wang W, Pacheco J, Chen X, Grzegorczyk T, Kong JA (2005). A Study Of Using Metamaterials As Antenna Substrate To Enhance Gain. Prog. Electromag. Res., PIER 51:295-328.

Zhu C, Ma JJ, Chen L, Liang CH (2009). Negative index metamaterials composed of triangular open loop resonator and wire structures. Microw. Optical Tech. Lett. 51:2022-2025.

Ziolkowski RW (2003). Design, fabrication, and testing of double negative metamaterials. IEEE Trans. Antennas Propag. 51:1516-152.

16

Applying the fuzzy analytic network process to the selection of an advanced integrated circuit (IC) packaging process development project

Yung-Hsiang Hung*, Mei-Ling Huang and Kun-Liang Fanchiang

Department of Industrial Engineering and Management, National Chin-Yi University of Technology, 35, Lane215, Section 1, Chung-Shan Road, TaiPing, TaiChung, 411, Taiwan, R.O.C.

In the modern business environment, process technology evaluation and selection (PTES) is a crucial component of innovation in new product development (NPD). The most difficult task for project managers in PTES is to make the optimal technology choice for a Research and Development (R&D) project, and there are many attendant uncertainties and risks in process technology R&D projects for NPD. Recently, Integrated Circuit (IC) Packaging has become an equal part of the cost-performance equation in the silicon world, and packaging foundries have responded quicker than many other semiconductor companies to the rapidly changing requirements of chip-scale packaging. This facilitates the transfer of new technology from the assemblers to the chip suppliers. This study applies the fuzzy analytic network process (FANP) model to evaluate the strategic impact of new IC manufacturing technologies in firms within Taiwan's IC packaging industry. Our study will determine the key decision-making factors affecting R&D project selection using FANP, and additionally we develop an optimal manufacturing process. As a case study, the ongoing "Controller IC packaging R&D project A" was chosen to minimize warpage of controller IC.

Key words: Integrated Circuit (IC) packaging technology, controller integrated circuit (IC), new product development, Research and Development (R&D) project selection, fuzzy analytic network process (FANP).

INTRODUCTION

In today's fiercely competitive global economy, new product development (NPD) is widely considered as an essential activity contributing towards the success, survival and renewal of organizations (Brown and Eisenhardt, 1995). A key NPD activity that firms use to reduce the risks and uncertainties associated with new products is the careful selection of new potentially successful product ideas and technological process innovations. As process technology evaluation and selection (PTES) is vital to technological process innovation in NPD, technology is seen as a driving force of innovation. In particular, PTES involves determining the process technical requirements of a new product and assessing how well they match the firm's technical capabilities. This can assist firms in addressing technical and manufacturing problems early in the NPD process, and permits the rapid introduction of new products onto the market. However, in the PTES production process, the most difficult task for project managers is making the correct technology Research and Development (R&D) project choice, as a poor R&D selection commonly leads to either failure of that product in the market place or an extended product development time. Understanding customer needs and making the correct R&D project choice will lead to the development of successful products and a reduced development time.

R&D project selection is an indispensable resource for progressive hi-tech companies in the semiconductor industries, especially those that depend on innovation such

*Corresponding author. E-mail: hung.yunghsiang@gmail.com.

as Integrated Circuit (IC) Packaging companies.

The key to continued competitiveness lies in their ability to develop and implement new product and process technologies. Project selection is the process of evaluating individual R&D projects, that is, to choose the right project based analysis, with the aim of achieving the company's objectives. It involves a thorough analysis, including the important time-to-market aspect to determine the optimum project from the alternatives. Unfortunately, there are attendant uncertainties and risks associated with R&D projects for NPD. Uncertainty arises from multiple sources, including technical, management and commercial issues, and may be both internal and external to the project (Feyzioğlu and Büyüközkan, 2006). However, R&D project managers are faced with complex decision-making environments and project problems. Recent literature on R&D project evaluation and selection features of many decision models uses a wide range of mathematically-based approaches. Criticisms of these techniques include their inability to consider strategic factors and their mathematical complexity (Albala, 1975; Fahrni and Spätig, 1990; Lockett et al., 1986; Meade and Presley, 2002). A multi-attribute decision-making approach is commonly used in the assessment and selection of alternative projects. The analytic hierarchy process (AHP) is perhaps the most widely used decision-making approach (Sevkli et al., 2008). It is a decision-aiding method developed by Saaty (2008), and its validity is based on the many hundreds (and now thousands) of actual applications in which AHP results were accepted and used by cognizant decision-makers (DMs). AHP is mainly used to solve the decision-making problems of multi-factor assessment with uncertainties (Chan et al., 2004). However, in the problem-solving process, the hierarchical analysis conducts an assessment by systematizing complex problems while assuming the factors of each hierarchy are independent. Many decision-making problems cannot be structured hierarchically; otherwise, strong interactions and dependencies would exist between inter-level and intra-level elements.

To solve the aforementioned problems, Saaty (2001) proposed the analytic network process (ANP) to take into consideration the inter-hierarchy relations, that is, the relations and interactions of factors of both the high-hierarchy and low-hierarchy. The hierarchical structure is linear in the hierarchical analysis, and has a non-linear network structure in network hierarchal analysis. The network hierarchical analysis features interdependence and feedback characteristics, and a super-matrix is applied to the weight calculation. This process has a higher level strategic hierarchy that controls all the benefit, cost, risk and opportunity sub networks required by the specific R&D project management problem. Despite their popularity, AHP and ANP have been frequently criticized for their inability to adequately handle the inherent uncertainty and imprecision associated with the mapping of a DM's perception to group members.

Because conflicts always occur in group decision making, and members may not always be in agreement initially, ANP has an inherent weakness in capturing the vagueness, uncertainty and imprecision of judgments given by different members of an expert group. This may be caused by their varying levels of experience, a lack of experimental data and other unknown factors. In real world decision-making problems, decision-making with fuzzy set theory enabled one to reach the aim in a quicker, easier and more sensitive way (Nataraja et al., 2006).

Due to the trend of developing thinner consumer electronics products, the R&D of thinner IC packaging technology has become a major focus of technological innovation for IC packaging factories around the world. IC manufacturing technologies have continued to evolve from their original prototypes. Thinner IC packaging makes the quality problem of warpage in the IC packaging process a serious issue. The semiconductor foundry industry, whose core business is IC manufacturing, has been greatly influenced and shaped by the flow of these newly developed technologies. As the R&D of IC packaging manufacturing technology requires considerable time and cost to be expended, the product R&D and marketing period can be shortened if the opinions of experts in marketing, R&D and manufacturing can be effectively and quickly integrated. The problems discussed in the present study are common and widespread in the management of R&D in today's major packaging factories. In fact, the R&D of thinner IC manufacturing process technologies involves the expertise of many different areas. The fuzzy ANP (FANP) project selection method applied in the present study can systematically and rapidly integrate the expertise of various areas, saving a significant amount of time and enabling firms to rapidly carry out the R&D of the product manufacturing process technologies. This method allows group members to express fuzzy preferences for alternatives and individual judgments for solution selection criteria. It has been proven to overcome the limitations of the compensatory approach and the inability of ANP in handling proper linguistic variables.

LITERATURE REVIEW

In today's rapidly changing business environment, innovative R&D projects for new products are growing quickly in terms of employment and profitability. New IC product design problems can be divided into four groups: strategic design, innovative design, variant design and repeat order electronics engineering product development (Culverhouse, 1993). In general, innovative design changes the product by 20 to 50% and requires the considerable input of either product or manufacturing technologies. It consists of four basic components: concept selection, component selection, material selection and process technology selection. In particular, process technology selection signifies the determination of the best pathway from which a specified product or service can be provided, through selection from a number of competing alternative processes. Accordingly, the first step in any such R&D activity is to understand the critical success factors of R&D competence which make the difference between success and failure at NPD.

Cooper and Kleinschmidt (1995) pointed out three factors critical to the success of a project: 1) the nature of the product, as a uniquely high quality product will yield better than expected economic returns for the customers; 2) the nature of the market, measured by market demand intensity, market growth rate and market scale; and 3) technical implementation and synergy of new products and existing products. In recent years, many studies have investigated a wide variety of factors affecting a new product's viability, both technologically as well as commercially. Meade and Presley (2002) reviewed literature and classified current R&D project selection methods into three major themes, that is, the need to relate selection criteria to corporate strategies, the need to consider the qualitative benefits and risks of candidate projects, and the need to reconcile and integrate the needs and desires of different stakeholders. In spite of this, different views exist regarding the relevance of success factors. On the whole, there is a consensus of opinions, as researchers seem to agree that the market, technology, environment and organization classes of variables are important (Lilien and Yoon, 1989).

R&D is seen as a driving force of innovation. However, given the magnitude of variables and stakeholders involved, R&D managers face a difficult challenge determining the measures that are useful for measuring product development success. R&D project selection involves uncertainty and a high level of risk (Mohamed and McCowan, 2001). Therefore, the decision-making aspect of R&D project selection requires the cooperation of business organizations at different levels. R&D decisions which are necessary at early stages of development contain a considerable amount of elements which cause uncertainty, potentially confusing the DM's efforts to achieve the target performance. For the above reasons, the project selection process and result are a key step in the success of R&D for IC packaging technology development (Martino, 2004).

Considerable efforts have been made in the past four decades to assist organizations in making better decisions in R&D project evaluation and selection (Martino, 2004; Henriksen and Traynor, 1999; Ringuest et al., 2004). Some models are strictly empirical and are based on statistical analysis of the correlation between project characteristics and project success (Cooper and Kleinschmidt, 1995). AHP is the most widely used decision-making approach in the world today (Sevkli et al.,

2008). AHP combines qualitative analysis with quantitative analysis, and uses a fundamental scale of absolute numbers that has been proven in practice and validated by physical and decision problem experiments. A number of scholars have studied AHP to improve the quality of PTES decision-making (Gerdsri, 2005; Bhattacharya et al., 2004; Ong et al., 2003; Chen et al., 2005) applied AHP to evaluate the strategic impact of new IC manufacturing technologies in the semiconductor foundry industry in Taiwan (Chen et al., 2005). The results show the relative importance of competitive goals in the semiconductor foundry industry. Each competitive goal is aligned to technology strategies as well as to emerging technologies in the prioritized order. However, in a multi-project environment, the success of an R&D project is not solely dependent on the project management team, as no functional managers can be omitted from the decision-making process (Mohanty et al., 2005). Thus, to analyze project alternatives, a feedback loop is necessary for each of these functional organizations at each level of maturity of the project. As such, ANP has been chosen as a decision-making tool to aid such analysis. The most important advantage of ANP over AHP is that ANP is a holistic approach in which all criteria and alternatives involved are connected in a network system that accepts various dependencies (Saaty, 2008). ANP enables users to take into consideration the degree of interdependences between the judgments of DMs and the processes' technical requirements by means of AHP. The key to success in an R&D project is to establish consistent group judgment (Schmidt and Freeland, 1992; Åstebro, 2004). Nevertheless, conflict always occurs in group decision-making since members in a group generally do not reach a unanimous decision. Previous studies have already considered the fuzzy set theory for prioritizing R&D project decisions (Saaty, 2008). For example, some researchers used the concept of fuzzy theory combined with AHP to address the uncertainty of human thinking (Kahraman et al., 2006; Wu et al., 2004; Wang et al., 2005; Mikhailov and Singh, 2003). Meade and Presley (2002) applied the ANP proposed by Saaty (2001) to the selection of the R&D project while developing a complete decision-making support system. The R&D project selection related literature is summarized in Table 1.

FUZZY ANALYTIC NETWORK PROCESS (FANP)

Review of the analytic network process (ANP)

The ANP is the most comprehensive framework for the analysis of societal, governmental, and corporate decisions available to the modern DM. The key concept of the ANP is that influence does not necessarily have to flow only downwards, as is the case with the hierarchy in the AHP. Influence can flow between any factors in the

Table 1. Summary of the R&D project selection related literature.

Author	Method
Chan et al., (2004) and Mohamed and McCowan (2001).	Using AHP to solve the decision-making problems of multi-factor assessment with uncertainties.
Saaty (2001)	Proposed the ANP to take into consideration the inter-hierarchy relations, to solve the aforementioned problems.
Cooper and Kleinschmidt (1995)	Pointed out three factors critical to the success of a project: 1) the nature of the product, 2) the nature of the market, and 3) technical implementation and synergy of new products and existing products.
Martino (2004), Henriksen and Traynor (1999) and Ringuest et al. (2004)	To assist organizations in making better decisions in R&D project evaluation and selection.
Chen et al. (2005)	Applied AHP to evaluate the strategic impact of new IC manufacturing technologies in the semiconductor foundry industry in Taiwan.
Schmidt and Freeland (1992) and Åstebro (2004)	The key to success in an R&D project is to establish consistent group judgment.
Kahraman et al. (2006), Wu et al. (2004), Wang et al. (2005) and Mikhailov and Singh (2003)	Using the concept of fuzzy theory combined with AHP to address the uncertainty of human thinking
Meade and Presley (2002)	Applied the ANP proposed by Saaty (2001) to the selection of the R&D project while developing a complete decision-making support system.

network, causing non-linear results for the priorities of scenario choices. In general, ANP models have two parts. The first is a control hierarchy or network of objectives and criteria that control the interactions in the system under study. The second part of the ANP model is the many sub-networks of influences among the elements and criteria of the problem, with one for each control criterion. An outline of the steps of the ANP is as follows:

Step 1: Determine the control hierarchies, including their criteria for comparing the components of the system and their sub-criteria for comparing the elements of the system.

Step 2: Computed derived weights are used later to weigh the elements of the corresponding column criteria of the super-matrix corresponding to the control criterion. Firstly, a pair-wise comparison matrix is set up. Secondly, the super-matrix limit (eigenvector) is computed. Thirdly, consistency analysis is performed. Lastly, the limiting priorities using each super matrix are computed.

Step 3: Synthesize the limiting priorities by weighing each limiting super-matrix with the weight of its control criterion and adding the resulting super-matrices.

The ANP allows one to include all factors and criteria, both tangible and intangible, which relate to making the best decisions. It allows both interaction and feedback within clusters of elements (inner dependence) and between clusters (outer dependence). Such feedback best captures the complex effects of interplay in human society, especially when risk and uncertainty are involved (Sevkli et al., 2008; Saaty, 2008).

A so-called super-matrix, describing the interaction between the components of the system, is constructed from the priority vectors (eigenvectors). It can be used to assess the results of feedback. Each of the columns is an eigenvector that represents the impact of all the elements in the ith component on each of the elements in the jth component. Interaction in the super-matrix is measured according to several possible criteria, in which priorities and relations are represented in a control hierarchy as shown in Figure 1. Sub-matrix X represents a pair-wise comparison matrix of cluster A under hierarchy C, sub-matrix D represents a pair-wise comparison matrix of cluster C under hierarchy A and sub-matrix E represents a pair-wise comparison matrix of cluster A with a dependency relationship.

There is no pair-wise comparison matrix in hierarchy C, due to the absence of a dependency relationship. Hence, the super-matrix is shown as follows:

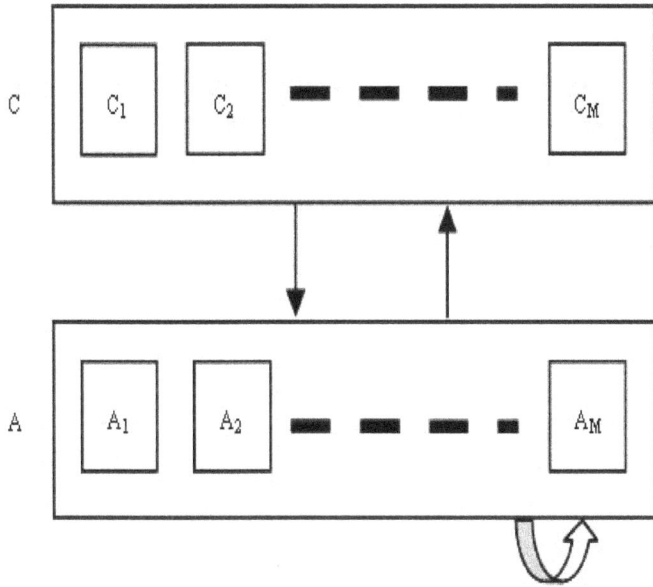

Figure 1. The structure of the ANP.

$$M' = \begin{array}{c} A \\ C \end{array} \begin{array}{c} A \quad C \\ \begin{bmatrix} E & X \\ D & 0 \end{bmatrix} \end{array} \qquad (1)$$

M' is an unweighted super-matrix. If the matrix does not conform to the column-stochastic rule (sum of column value =1), the weighted super-matrix M can be obtained through the weight conversion procedure until the sum of each column is 1. The weighted super-matrix is then limited, namely, M is multiplied by M to the power of $2K+1$ to converge the dependency relationship. The relative weight of all elements is computed (Saaty, 2008). Finally, the desirability index (DI) in Equation 2 is used to judge an optimal solution.

$$DI_i = \sum_{j=1}^{r} S_{ij} = \sum_{j=1}^{r} R_j W_{ij} \qquad (2)$$

DI_i, Desirability index of ith feasible solution; S_{ij}, weight of ith feasible solution under j^{th} element; R_j, relative weight of jth element; W_{ij}, relative weight of ith solution under j^{th} element.

A feasible solution with the highest DI is optimal solution A^*, as shown in Equation 3:

$$A^* = \left\{ A_i \mid DI_i = \max_{k=1,2,...,n} imum\left(DI_k\right) \right\} \qquad (3)$$

Fuzzy pair-wise comparison matrix

A set of ANP pair-wise comparison matrices was constructed for each of the lower levels with one matrix for each element in Equation 3, using the relative scale measurement as shown in Table 2. If each entry in E is denoted by e_{ij}, then $e_{ij} = 1/e_{ij}$ (the reciprocal property) holds, and so does $e_{jk} = e_{ik}/e_{ij}$ (the consistency property). By definition, $e_{ii} = e_{jj} = 1$ (when comparing two elements which are the same). Given the subjectivity, uncertainty, and fuzziness of experts' evaluation on an R&D project, the results evaluated from the ANP may differ from the actual situation. Some researchers represented uncertain judgments as fuzzy sets of fuzzy numbers e_{ij}. The fuzzy scale relating relative preferences, measuring the relative weights is given in Table 2.

Laarhoven and Pedrycz (1983) and Buckley et al. (2001) modified the AHP for a fuzzy hierarchical analysis using comparison matrices with triangular fuzzy numbers. They obtained fuzzy priorities $\tilde{a}_{ij}$, $i,j =1, 2, ..., n$ by applying a fuzzy version of the Logarithmic least squares method. If the pair-wise comparison matrix of the kth expert amongst N experts is transformed into a fuzzy pair-wise comparison matrix $\tilde{A}^K\left(a_{ij}\right)$, the result is as shown in Equation 4.

$$\tilde{A} = \begin{bmatrix} 1 & \tilde{a}_{12}(L,M,R) & \cdots & \tilde{a}_{1n}(L,M,R) \\ \dfrac{1}{\tilde{a}_{12}(L,M,R)} & 1 & \cdots & \tilde{a}_{2n}(L,M,R) \\ \vdots & \vdots & \ddots & \vdots \\ \dfrac{1}{\tilde{a}_{1n}(L,M,R)} & \dfrac{1}{\tilde{a}_{2n}(L,M,R)} & \cdots & 1 \end{bmatrix}$$

$$(4)$$

After the fuzzy pair-wise comparison matrices of all experts are established, the weights of all experts are collectively computed by the geometric average (shown in Equation 5) recommended by Buckley et al. (1999).

$$\widetilde{W} = \left[\prod_{k=1}^{N} \tilde{A}^k \right]^{1/N} \qquad (5)$$

There are a number of defuzzification methods in fuzzy set theory, of which the commonly used methods include center of gravity, area, height, midpoint of maximum and the weighted average method. In this paper, defuzzification is affected with center of gravity of Equation 6 proposed by Hsieh et al. (2004).

$$DF = \left[(R-L) + (M-L) \right] / 3 + L \qquad (6)$$

Table 2. Pair-wise comparison scales for AHP and triangular fuzzy scales.

Verbal judgments of preferences	AHP numerical scale	Triangular fuzzy scale
Extremely preferred	9	(8, 9, 9)
Very strongly preferred	7	(6, 7, 8)
Strongly preferred	5	(4, 5, 6)
Moderately preferred	3	(2, 3,4)
Equally preferred	1	(1, 1, 2)

CASE STUDY

Semiconductor manufacturing has many distinct segments, including design, marketing, masking, manufacturing, testing, and packaging. Packaging protects the IC and provides its interconnections, power, and cooling. As packaging sizes continue to decrease, the level of integration of semiconductor devices continues to increase in complexity as well as the number of components. Competitive IC manufacturing must accommodate this trend without increasing the cost of manufacture. In order to develop competitive IC packaging products with a high performance/cost ratio, design schemes should be optimized in terms of technical capacity, economic benefit, product performance, risk management, and so on (Balachandra and Friar, 1997). Most semiconductor IC packaging foundries have developed fine-pitch ball grid array (FPBGA) packages using laminate and polyimide interposers. Dozens of new packaging styles have been developed for specific applications, and much of this work has occurred at the assemblers. Therefore, IC packaging must also evolve to accommodate the changing trends in IC technology (such as BGA, Chip Scale Package, Multi Chip Modules and Flip-Chip etc) (Tummala, 2001). However, R&D personnel are technically trained perfectionists who believe that cost and time are relatively unimportant when it comes to improving a technology. Very few people in an organization truly understand the R&D environment and the problems faced by R&D managers. In fact, technology is the strategic problem of the IC packaging process requiring a model that evaluates several criteria in different dimensions. These dimensions are required for ranking them according to the likelihood of their being a goal, according to Cheng and Wu (2004), Huang et al. (2008) and Taiwan's IC packaging experts. This study employed four dimensions, that is, technological merit (TM), potential benefits (PB), availability of resources (AOR), and R&D risks (RDR), to evaluate the micro HDD controller IC's packaging project in a central Taiwan IC packaging company. The experts surveyed for the case study include a project manager, risk analysis and assessment engineer, material science engineer, structure engineer (model construction and simulation analysis), equipment engineer, manufacturing process engineer, debugging engineer and product engineer. The

standard network and a description of the criteria are listed and shown in Figure 2.

The controller IC packaging process of the R&D network is constructed based on the dimensions in Figure 2. However, project selection in the IC packaging process involves many uncertainties, such as different expert opinions and project team members have varying attitudes to the project. Ambiguity and uncertainty inevitably produce risks which may hamper project objectiveness when an assessment is carried out. Ambiguity may compensate for decision-making behavior in the traditional hierarchical analysis approaches that do not take uncertainties, ambiguities, and lack of information into consideration, making it possible to reflect the environment of the real world. In the subsequent analysis, FANP was used to compute the relative weight of items at every hierarchy, whereby the priority of every IC packaging R&D solution was evaluated as the basis for selection.

Application of fuzzy pair-wise comparison matrix

As mentioned earlier, defuzzification aims to address the uncertainty of the experts' subjective suggestions using the following steps: 1) transform the pair-wise comparison matrix of each expert into a fuzzy pair-wise comparison matrix; 2) combine these fuzzy pair-wise comparison matrices into a single fuzzy pair-wise comparison matrix; and 3) defuzzify to obtain the relative weight of individual criteria and sub-criteria. As an example, consider the TM dimension in the second hierarchy of Figure 2. Assuming there are 2 experts, the pair-wise comparison matrices of the 1st and 2nd experts are separately transformed into corresponding fuzzy pair-wise comparison matrices, as shown in Tables 3 and 4. After computing the fuzzy pair-wise comparison matrix of individual expert suggestion (Tables 3 and 4), combine the two fuzzy pair-wise comparison matrices into the fuzzy pair-wise comparison matrix of the expert group using the geometric average method (Table 5). Then, defuzzify the weight (the approximate value of the eigenvector) of the calculating level using the center of gravity equation (Hsieh et al., 2004), with the computed weight of the two experts' suggestions listed in Table 6. Finally, substitute the approximate value of the eigenvector

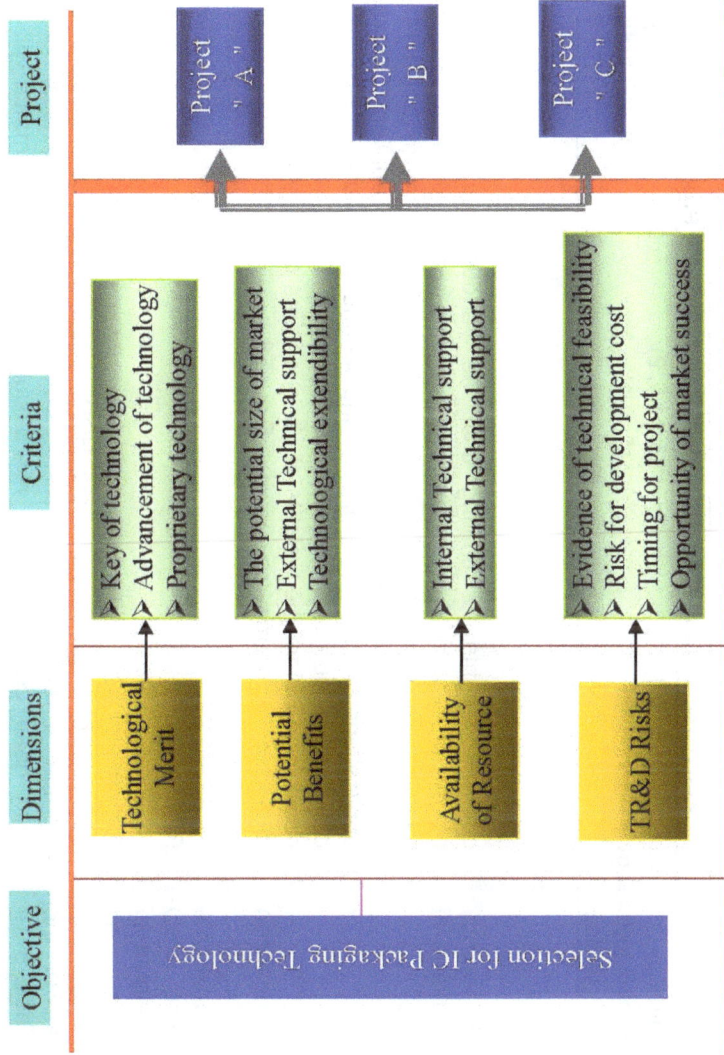

Figure 2. The ANP structure of control for IC packaging R&D.

Objective	Dimensions	Criteria	Project
Selection for IC Packaging Technology	Technological Merit	Key of technology / Advancement of technology / Proprietary technology	Project "A"
	Potential Benefits	The potential size of market / External Technical support / Technological extendibility	Project "B"
	Availability of Resource	Internal Technical support / External Technical support	Project "C"
	TR&D Risks	Evidence of technical feasibility / Risk for development cost / Timing for project / Opportunity of market success	

Table 3. The first expert's fuzzy pair-wise comparison matrix transformation.

	Key of technology	Advancement of technology	Proprietary technology		Key of technology	Advancement of technology	Proprietary technology
Key of technology	1	1/2	1/4		(1, 1, 1)	(1/3, 1/2, 1)	(1/5, 1/4, 1/3)
Advancement of technology	2	1	1/3	→	(1, 2, 3)	(1, 1, 1)	(1/4, 1/3, 1/2)
Proprietary technology	4	3	1		(3, 4, 5)	(2, 3, 4)	(1, 1, 1)

λ_{max}, 3.0183; C.I, 0.0092; C.R, 0.016.

Table 4. The second expert's fuzzy pair-wise comparison matrix transformation.

	Key of technology	Advancement of technology	Proprietary technology		Key of technology	Advancement of technology	Proprietary technology
Key of technology	1	3	3		(1, 1, 1)	(2, 3, 4)	(2, 3, 4)
Advancement of technology	1/3	1	2	→	(1/4, 1/3, 1/2)	(1, 1, 1)	(1, 2, 3)
Proprietary technology	1/3	1/2	1		(1/4, 1/3, 1/2)	(1/3, 1/2, 1)	(1, 1, 1)

Table 5. Combined fuzzy pair-wise comparison matrix for two experts.

	Key of technology	Advancement of technology	Proprietary technology
Key of technology	(1, 1, 1)	(4/5, 11/9, 2)	(5/8, 6/7, 7/6)
Advancement of technology	(1/2, 9/11, 5/4)	(1, 1, 1)	(1/2, 4/5, 11/9)
Proprietary technology	(6/7, 7/6, 8/5)	(9/11, 5/4, 2)	(1, 1, 1)

in Table 6 into the super-matrix of ANP, thereby obtaining the weight of dimension and criteria at the final level.

Operation of the super-matrix

Firstly, construct the fuzzy pair-wise comparison matrices of the 44 experts according to the steps specified earlier, and compute the approximate eigenvector of controller IC packaging R&D hierarchy. As shown in Table 7, the weights of various hierarchies are combined into an unweighted super-matrix, which is used in super-matrix operation to obtain the weight of the controller IC packaging R&D project (Table 8). If the weighted super-matrix cannot meet the column-stochastic requirement in the calculating process, the super-matrix in Table 8 is given

different weights for limiting until the various dimensions and criteria are converged, that is, the sum of each column is 1.

Table 9 shows the super-matrix after 27 applications of limiting (M^{27}). After the weights of the limiting super-matrix are normalized, it is possible to obtain the weights of various items in the criterion hierarchy as shown in Figure 3.

RESULTS

A controller IC is the key electronic component of our object of study, the micro HDD with 20GB storage. It reduces mechanical shock, provides read head control, and is responsible for the performance and effectiveness of the micro HDD. Control in the vertical dimension (Z-axis) of the controller IC is crucial, owing to the limited space

in a micro HDD. For the controller IC packaging, the ratio of chip area to package area is close to more than 1.14 when chip scale package (CSP) is applied. This ratio approaches the ideal value of unity. However, this challenge will result in warpage of the components or destruction of the shape of the components and finally deteriorate the quality of the controller IC. In fact, the warpage of a micro HDD affects the manufacturing process yield of the controller IC. When the total height of the package is limited to 0.65 mm (Figure 4) and the warpage limit is less than 3 mil (Figure 5). With the selection of different combinations of substrate thickness and mold thickness, coefficient of thermal expansion (CTE) mismatch will occur in thermal processes during assembly. These include molding, post-mold curing and re-flow processes. In addition, different combinations can result in large variations in the

Table 6. Defuzzificated fuzzy pair-wise comparison matrix.

	Key of technology	Advancement of technology	Proprietary technology	Approximate value of eigenvector
Key of technology	1	4/3	8/9	0.3508
Advancement of technology	3/4	1	6/7	0.2862
Proprietary technology	9/8	7/6	1	0.3629

λ_{max} 3.007; C.I, 0.0035; C.R, 0.006.

Table 7. Unweighted super-matrix.

		Technological merit	Availability of resource	Potential benefits	TR&D risks	Internal technical support	External technical support	Proprietary technology	Opportunity of market success	Risk for development cost	The potential size of market	Technological extendibility	Advancement of technology	Evidence of technical feasibility	Effects of patents technology	Timing for project	Key of technology	Selection for IC packaging technology
Dimensions	Technological merit	0.262	0.228	0.212	0.202	0	0	0	0	0	0	0	0	0	0	0	0	0.154
	Availability of resource	0.202	0.3	0.281	0.443	0	0	0	0	0	0	0	0	0	0	0	0	0.237
	Potential benefits	0.286	0.134	0.16	0.159	0	0	0	0	0	0	0	0	0	0	0	0	0.337
	TR&D risks	0.251	0.339	0.348	0.196	0	0	0	0	0	0	0	0	0	0	0	0	0.272
Criteria	Internal technical support	0	0.799	0	0	0.125	0.139	0.08	0.076	0.076	0.083	0.009	0.079	0.074	0.077	0	0.083	0
	External technical support	0	0.201	0	0	0.061	0.057	0.08	0.076	0.076	0.083	0.009	0.079	0.074	0.077	0	0.083	0
	Proprietary technology	0.44	0	0	0	0.101	0.08	0.101	0.072	0.124	0.083	0.009	0.151	0.105	0.077	0	0.082	0
	Opportunity of market success	0	0	0.326	0	0.082	0.08	0.08	0.076	0.076	0.083	0.009	0.079	0.08	0.077	0.2	0.083	0
	Risk for development cost	0	0	0.439	0	0.082	0.08	0.08	0.085	0.085	0.085	0.009	0.079	0.08	0.077	0.049	0.083	0
	The potential size of market	0	0	0.235	0	0.082	0.08	0.08	0.076	0.076	0.083	0.302	0.079	0.074	0.077	0	0.085	0
	Technological extendibility	0	0	0	0.414	0.082	0.08	0.063	0.076	0.076	0.083	0.314	0.079	0.08	0.158	0	0.082	0
	Advancement of technology	0.24	0	0	0	0.082	0.08	0.08	0.074	0.119	0.083	0.009	0.079	0.165	0.077	0.204	0.078	0
	Evidence of technical feasibility	0	0	0	0.155	0.082	0.08	0.08	0.076	0.076	0.083	0.009	0.079	0.074	0.077	0	0.083	0
	Effects of patents technology	0	0	0	0.223	0.082	0.08	0.08	0.076	0.066	0.083	0.3	0.079	0.057	0.077	0	0.083	0
	Timing for project	0	0	0	0.209	0.082	0.08	0.118	0.162	0.076	0.083	0.009	0.061	0.074	0.077	0.105	0.093	0
	Key of technology	0.32	0	0	0	0.082	0.08	0.082	0.076	0.076	0.083	0.009	0.079	0.074	0.077	0	0.083	0
Objective	Selection for IC packaging technology	0	0	0	0	0	0	0	0	0	0	0	0	0	0	0	0	0
Total		2	2	2	2	1	1	1	1	1	1	1	1	1	1	1	1	1

Table 8. Weighted super-matrix.

Dimension / Criteria / Objective		Technological merit	Availability of resource	Potential benefits	TR&D risks	Internal technical support	External technical support	Proprietary technology	Opportunity of market success	Risk for development cost	The potential size of market	Technological extendibility	Advancement of technology	Evidence of technical feasibility	Effects of patents technology	Timing for project	Key of technology	Selection for IC packaging technology
Dimensions	Technological merit	0.131	0.114	0.106	0.101	0	0	0	0	0	0	0	0	0	0	0	0	0.154
	Availability of resource	0.101	0.15	0.14	0.222	0	0	0	0	0	0	0	0	0	0	0	0	0.237
	Potential benefits	0.143	0.067	0.08	0.079	0	0	0	0	0	0	0	0	0	0	0	0	0.337
	TR&D risks	0.125	0.17	0.174	0.098	0	0	0	0	0	0	0	0	0	0	0	0	0.272
Criteria	Internal technical support	0	0.4	0	0	0.125	0.139	0.08	0.076	0.076	0.083	0.009	0.079	0.074	0.077	0	0.083	0
	External technical support	0	0.101	0	0	0.061	0.057	0.08	0.076	0.076	0.083	0.009	0.079	0.074	0.077	0	0.083	0
	Proprietary technology	0.22	0	0	0	0.082	0.08	0.101	0.076	0.076	0.083	0.009	0.151	0.074	0.077	0	0.082	0
	Opportunity of market success	0	0	0	0.111	0.082	0.08	0.08	0.072	0.124	0.083	0.009	0.105	0.105	0.077	0.2	0.083	0
	Risk for development cost	0	0	0	0.077	0.082	0.08	0.08	0.085	0.085	0.085	0.009	0.079	0.08	0.077	0.049	0.083	0
	The potential size of market	0	0	0.163	0	0.082	0.08	0.08	0.083	0.083	0.085	0.009	0.079	0.074	0.077	0.2	0.083	0
	Technological extendibility	0	0	0	0	0.082	0.08	0.08	0.083	0.083	0.085	0.314	0.079	0.074	0.077	0	0.083	0
	Advancement of technology	0.12	0	0.22	0	0.082	0.08	0.063	0.076	0.076	0.083	0.302	0.079	0.074	0.158	0	0.078	0
	Evidence of technical feasibility	0	0	0	0.207	0.082	0.08	0.08	0.074	0.119	0.083	0.009	0.079	0.165	0.077	0.204	0.083	0
	Effects of patents technology	0	0	0	0	0.082	0.08	0.08	0.162	0.066	0.083	0.3	0.079	0.074	0.077	0	0.083	0
	Timing for project	0	0	0	0.104	0.082	0.08	0.08	0.076	0.076	0.083	0.009	0.079	0.057	0.073	0.105	0.083	0
	Key of technology	0.16	0	0	0	0.082	0.08	0.118	0.076	0.076	0.083	0.009	0.061	0.074	0.077	0	0.093	0
Objective	Selection for IC packaging technology	0	0	0	0	0	0	0	0	0	0	0	0	0	0	0	0	0
Total		1	1	1	1	1	1	1	1	1	1	1	1	1	1	1	1	1

warpage level. The product development and key technologies of the HDD controller IC involve the substrate and lens thickness manufacturing technologies. For example, using the lower cost substrate of relatively higher thickness (0.136 mm) as the base would result in a more favorable hardness. However, given the product's height limit of 0.65 mm, thinner lenses of 1.5 mil need to be fabricated. Warpage is relatively low at 1.56 mil as evidenced by the simulation software. However, the manufacturing process and quality control becomes increasingly difficult and

Table 9. Limiting super-matrix.

		Dimension				Criteria												Objective
		Technological Merit	Availability of Resource	Potential Benefits	TR&D Risks	Internal Technical support	External Technical support	Proprietary technology	Opportunity of market success	Risk for development cost	The potential size of market	Technological extendibility	Advancement of technology	Evidence of technical feasibility	Effects of patents technology	Timing for project	Key of technology	Selection for IC Packaging Technology
Dimensions	Technological merit	0	0	0	0	0	0	0	0	0	0	0	0	0	0	0	0	0
	Availability of resource	0	0	0	0	0	0	0	0	0	0	0	0	0	0	0	0	0
	Potential benefits	0	0	0	0	0	0	0	0	0	0	0	0	0	0	0	0	0
	TR&D risks	0	0	0	0	0	0	0	0	0	0	0	0	0	0	0	0	0
Criteria	Internal technical support	0.072	0.072	0.072	0.072	0.072	0.072	0.072	0.072	0.072	0.072	0.072	0.072	0.072	0.072	0.072	0.072	0.072
	External technical support	0.062	0.062	0.062	0.062	0.062	0.062	0.062	0.062	0.062	0.062	0.062	0.062	0.062	0.062	0.062	0.062	0.062
	Proprietary technology	0.071	0.071	0.071	0.071	0.071	0.071	0.071	0.071	0.071	0.071	0.071	0.071	0.071	0.071	0.071	0.071	0.071
	Opportunity of market success	0.088	0.088	0.088	0.088	0.088	0.088	0.088	0.088	0.088	0.088	0.088	0.088	0.088	0.088	0.088	0.088	0.088
	Risk for development cost	0.106	0.106	0.106	0.106	0.106	0.106	0.106	0.106	0.106	0.106	0.106	0.106	0.106	0.106	0.106	0.106	0.106
	The potential size of market	0.096	0.096	0.096	0.096	0.096	0.096	0.096	0.096	0.096	0.096	0.096	0.096	0.096	0.096	0.096	0.096	0.096
	Technological extendibility	0.105	0.105	0.105	0.105	0.105	0.105	0.105	0.105	0.105	0.105	0.105	0.105	0.105	0.105	0.105	0.105	0.105
	Advancement of technology	0.063	0.063	0.063	0.063	0.063	0.063	0.063	0.063	0.063	0.063	0.063	0.063	0.063	0.063	0.063	0.063	0.063
	Evidence of technical feasibility	0.094	0.094	0.094	0.094	0.094	0.094	0.094	0.094	0.094	0.094	0.094	0.094	0.094	0.094	0.094	0.094	0.094
	Effects of patents technology	0.095	0.095	0.095	0.095	0.095	0.095	0.095	0.095	0.095	0.095	0.095	0.095	0.095	0.095	0.095	0.095	0.095
	Timing for project	0.078	0.078	0.078	0.078	0.078	0.078	0.078	0.078	0.078	0.078	0.078	0.078	0.078	0.078	0.078	0.078	0.078
	Key of technology	0.067	0.067	0.067	0.067	0.067	0.067	0.067	0.067	0.067	0.067	0.067	0.067	0.067	0.067	0.067	0.067	0.067
Objective	Selection for IC packaging technology	0	0	0	0	0	0	0	0	0	0	0	0	0	0	0	0	0

complicated as the lens becomes thinner. This leads to additional manufacturing time and an increased risk in the development process. In contrast, if a relatively thicker lens (for example, 3.5 mil) is used, the development process can be completed in less time. However, as the thickness of the substrate is only 0.2 mm, the control of warpage will become more difficult due to insufficient substrate thickness and hardness, as well as the influence of the molding process, leading to damage to the lens properties. To summarize, lens thickness not only affects the development of lens manufacturing technologies, but also directly influences the sizes of the mold compound. Additionally, different combinations may differ greatly in terms of their effect on the warpage of the HDD controller IC, as well as directly affecting the marketing time and cost of the product. This study analyzed the R&D project selection of a controller IC for a micro HDD packaging process, and designed three possible manufacturing processes (Table 10) using finite element method (FEM) and Taguchi methods. Using the desirability index, it is hoped that these three manufacturing processes could be integrated into projects with network hierarchies, thereby enabling executives to determine the optimal packaging project for R&D under the existing state of operation. In this paper, the four dimensions and twelve criteria in Figure 2 were

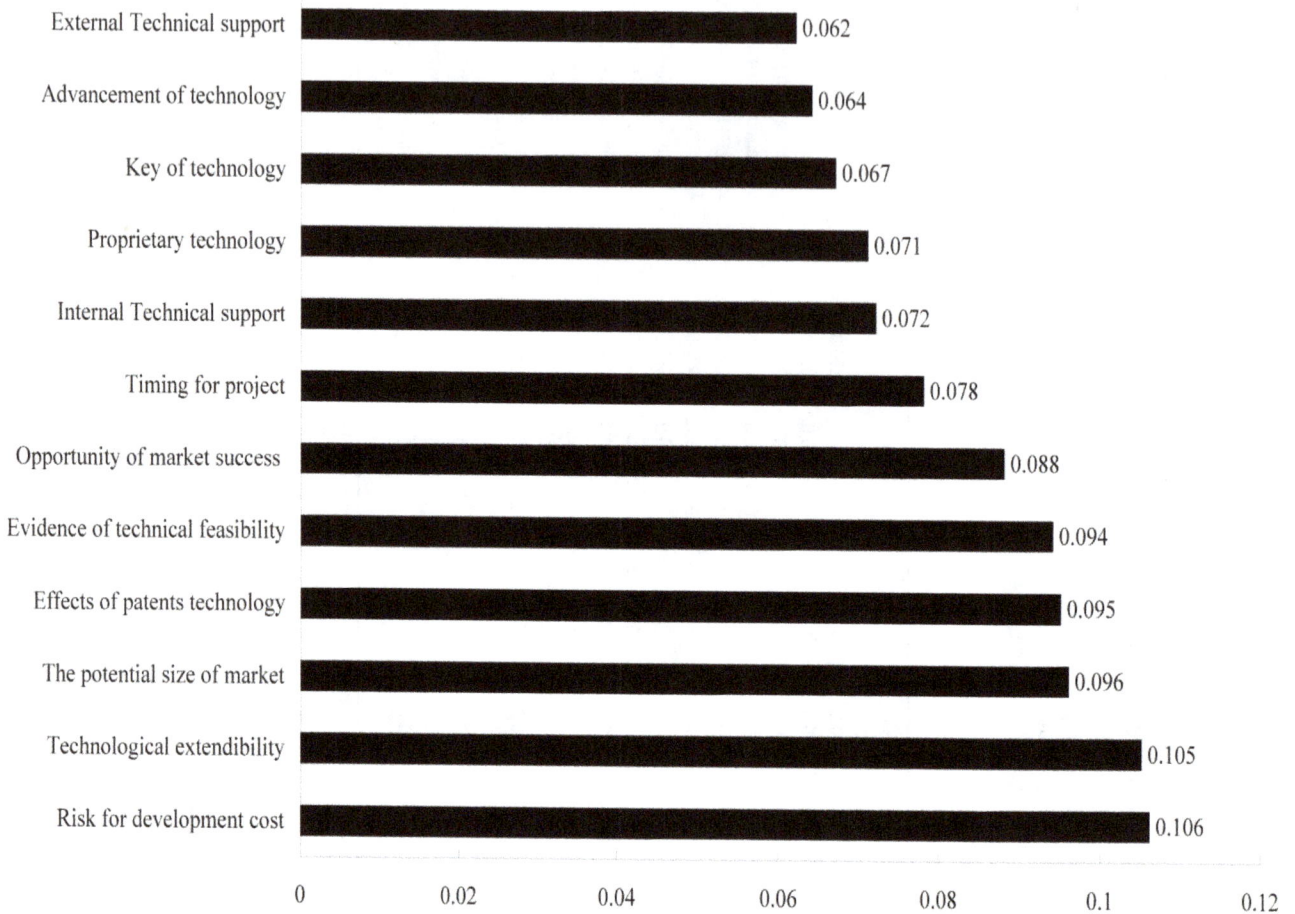

Figure 3. Criteria priority for controller IC packaging process R&D.

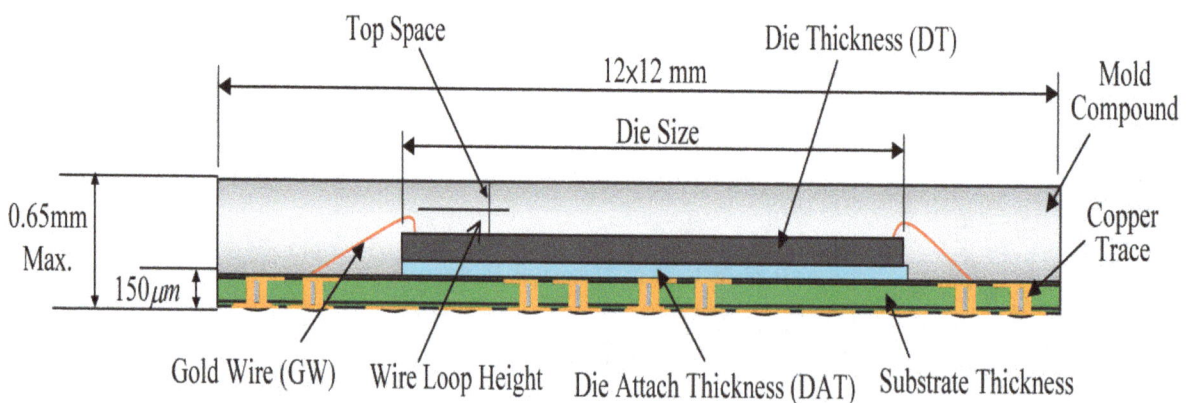

Figure 4. Controller IC package for micro HDD.

taken as the evaluation benchmark, and the three packaging projects in Table 10 were compared. The process, as described earlier is to construct a fuzzy pair-wise comparison matrix according to fuzzy theory, combine the questionnaire for expert groups with Equation 4, and then obtain the fuzzy pair-wise comparison matrix of expert groups. Following this, defuzzify the fuzzy pair-wise comparison matrix using the center of gravity Equation 5, computing the weights of the three scenarios (Table 11) and testing the consistency of λ_{max}, C.I and

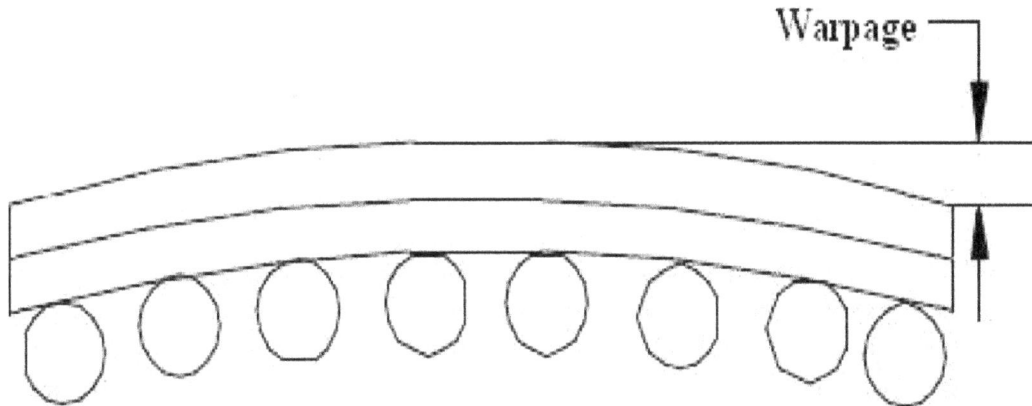

Figure 5. Convex warpage, bent downward.

Table 10. Projects for controller IC packaging R&D.

	Project "A"	Project "B"	Project "C"
Pkg.	UTLGA 12×12/172	XTLGA 12×12/172	XTLGA 12×12/172
Substrate	BT, 0.136mm	BT, 0.106mm	BT, 0.09mm
Mold thickness	0.3mm	0.25mm	0.2mm
Die size	5.2×5.2mm	5.2×5.2mm	5.2×5.2mm
Die thickness	1.5mil	2.5mil	3.5mil
Film	EM-500 M3, 25um thick	EM-500 M3, 25um thick	EM-500 M3, 25um thick
Compound	Kyocera KE-1150 UM	Kyocera KE-1150 UM	Kyocera KE-1150 UM
warpage predicts	1.56mil	2.78mil	3mil

C.R. According to the computed results in Table 11, the weight of Projects "A", "B", and "C" are substituted into Equation 3, and the relative weight and DI of three manufacturing packages are computed as shown in Table 12. Among the expected value of these manufacturing processes, Project "A" (DI$_A$ = 0.025+ 0.012+ 0.019+ 0.039+ 0.021+ 0.044+ 0.027+ 0.02+ 0.043+ 0.048+ 0.023+ 0.03= 0.351) has the highest weight, followed by Project "B"(0.329) and Project "C"(0.32). The analytical results show that the priority of consideration should be given to scenario A during the controller IC packaging R&D project selection.

CONCLUSION AND DISCUSSION

This study is the first to apply FANP model to evaluate the strategic impact of new IC manufacturing technologies. Although, there are some related works dealing with similar topic, which we have mentioned in literature section, the methodologies applied in the mentioned research are different from ours. As a result of the FANP analysis, the following priorities are given by DI, as for the alternative: Project "A" (0.351) > Project "B" (0.329) > Project "C" (0.32). The results indicate that the selection for the R&D project should be 1) lower risk for development cost (Substrate: 0.136 mm and mold thickness: 0.3 mm), 2) technological extendibility and potential size of the market (Die thickness: 1.5 mil and film: 25 um). Since there is a large variation of CTE characteristics among different mold thickness and mold compound types, as well as substrate thickness and mold thickness, selecting the proper combination of the aforementioned four variables is essential to control the warpage level (warpage predicts: 1.56 mil). Lens thickness not only affects the development of lens manufacturing technologies, but also directly influences their time-to-market and relative cost. Our case study summarized the four dimensions and twelve criteria of the HDD controller IC project through interviews with experts, and constructed an ANP network using the dependency relationship of dimensions and criteria. The study determined the key decision-making factors for R&D project selection using FANP, proposed three packaging projects in cooperation with the sample company, and combined them into an optimal manufacturing process by calculating the desirability index.

During controller IC packaging R&D selection, five leading items associated with packaging R&D were obtained from FANP, namely: 1) risks for development cost

Table 11. Pair-wise comparison matrix for controller IC packaging R&D projects.

Criteria	Project	Weight	λ_{max}	C.I	C.R
Key of technology	A	0.368			
	B	0.326	3.015	0.007	0.013
	C	0.306			
Advancement of technology	A	0.187			
	B	0.356	3.019	0.01	0.016
	C	0.456			
Proprietary technology	A	0.273			
	B	0.386	3.002	0.001	0.002
	C	0.341			
The potential size of market	A	0.402			
	B	0.222	3.050	0.025	0.043
	C	0.376			
Effects of patents technology	A	0.218			
	B	0.347	3.005	0.003	0.004
	C	0.435			
Technological extendibility	A	0.42			
	B	0.344	3.087	0.044	0.075
	C	0.235			
Internal technical support	A	0.377			
	B	0.378	3.061	0.030	0.052
	C	0.246			
External technical support	A	0.32			
	B	0.351	3.006	0.003	0.006
	C	0.328			
Evidence of technical feasibility	A	0.455			
	B	0.304	3.115	0.057	0.099
	C	0.241			
Risk for development cost	A	0.449			
	B	0.299	3.113	0.057	0.097
	C	0.252			
Timing for project	A	0.297			
	B	0.289	3.003	0.002	0.003
	C	0.415			
Opportunity of market success	A	0.339			
	B	0.384	3.079	0.039	0.068
	C	0.277			

(0.106); 2) technological extendibility (0.105); 3) the potential size of the market (0.096); 4) effects of patents technology (0.095); and 5) evidence of technical feasibility (0.094). In other words, the company should attach great importance to possible risks on packaging R&D, such as a shortage of resources and labor force, time wasting, and so on. The company should then consider the potential benefits of a R&D project, such as the possible profitability of technological extendibility, potential market scale and manufacturability.

Finally, the company should take into consideration the project's technical feasibility. We conclude that project

Table 12. Desirability calculation for controller IC packaging R&D project scenarios.

Criteria	Project	Weight	Criteria	Scenario	Weight
Key of technology (0.067)	A (0.368)	0.025	Internal technical support (0.072)	A(0.377)	0.027
	B (0.326)	0.022		B(0.378)	0.027
	C (0.306)	0.021		C(0.246)	0.018
Advancement of technology (0.064)	A (0.187)	0.012	External technical support (0.062)	A(0.32)	0.02
	B (0.356)	0.023		B(0.351)	0.022
	C (0.456)	0.029		C(0.328)	0.02
Proprietary technology (0.071)	A (0.273)	0.019	Evidence of technical feasibility (0.094)	A(0.455)	0.043
	B (0.386)	0.027		B(0.304)	0.029
	C (0.341)	0.024		C(0.241)	0.023
The potential size of market (0.096)	A (0.402)	0.039	Risk for development cost (0.106)	A(0.449)	0.048
	B (0.222)	0.021		B(0.299)	0.032
	C (0.376)	0.036		C(0.252)	0.027
Effects of patents technology (0.095)	A (0.218)	0.021	Timing for project (0.078)	A(0.297)	0.023
	B (0.347)	0.033		B(0.289)	0.023
	C (0.435)	0.041		C(0.415)	0.032
Technological extendibility (0.105)	A (0.42)	0.044	Opportunity of market success (0.088)	A(0.339)	0.03
	B (0.344)	0.036		B(0.384)	0.034
	C (0.235)	0.025		C(0.277)	0.024

"A" is an optimum solution in terms of implementation feasibility.

ACKNOWLEDGEMENTS

The authors thank the anonymous referees for their careful reading of the paper and for making several suggestions that improved it. The authors also thank the National Science Council of the Republic of China for financially supporting this research under Contract No. 98-2221-E-167 -010 -MY2.

REFERENCES

Albala A (1975). "Stage approach for the evaluation and selection of R&D projects." IEEE Trans. Eng. Manage., EM-22: 153–164.

Chan AHS, Kwok WY, Duffy VG (2004). "Using AHP for determining priority in a safety management system." Ind. Manage. Data Syst., 104: 430–445.

Åstebro T (2004). "Key success factors for technological entrepreneurs' R&D projects." IEEE Trans. Eng. Manage., 51: 314–321.

Balachandra R, Friar JH (1997). "Factors for success in R&D projects and new product innovation: A Contextual Framework." IEEE Trans. Eng. Manage., 44: 276–287.

Brown SL, Eisenhardt KM (1995). Product development: Past Research, Present Findings, and Future Directions, Acad. Manage, pp. 343–378.

Bhattacharya S, Sarkar B, Mukherjee RN (2004). "Evaluation of a convey or belt material based on multi-criteria decision making."

Indian J. eng. Mater. Sci., 11: 401–405.

Buckley JJ, Feuring T, Hayashi Y (2001). "Fuzzy hierarchical analysis revisited." Eur. J. Oper. Res., 129: 48–64. DOI: 10.1016/S0377-2217(99)00405-1.

Buckley JJ, Feuring T, Hayashi Y (1999). "Fuzzy hierarchical analysis." IEEE International Fuzzy System Conference Proceedings, Seoul Korea, pp. 1009–1013.

Culverhouse PF (1993). "Four design routes in electronics engineering product development." Int. J. Design Manufact,, 3: 147–158.

Cooper RG, Kleinschmidt EJ (1995). "Benchmarking the firm's critical success factors in new project development." J. Prod. Innov. Manage., 12: 374–391.

Cheng CY, Wu CY (2004). The survey of the R&D performance metrics—A case study in the IC packaging and testing industry, Thesis, Dept. of Industrial Engineering and Management, Chaoyang University of Technology, Taiwan, pp. 54-55.

Huang CC, Chu PY, Chiang YH (2008). "A fuzzy AHP application in government-sponsored R&D project selection." Omega-Int. J. Manage. Sci., 36 : 1038–1052. DOI:10.1016/j.omega.2006.05.003.

Feyzioğlu O, Büyüközkan G (2006). "Evaluation of new product development projects using artificial intelligence and fuzzy logic." Proceedings of World Academy of Science, Engineering and Technology 11: 183–189.

Fahrni P, Spätig M (1990). An application oriented guide to R&D selection and evaluation methods, R&D Management, 20: 155–171.

Wu FG, Lee YJ, Lin MC (2004). "Using the fuzzy analytic hierarchy process on optimum spatial allocation." Int. J. Ind. Ergon., 33: 553–569.

Gerdsri N (2005), An Analytical Approach to building a Technology Development Envelope (TDE) for Roadmapping of Emerging Technologies, PICMET '05. Portland International Conference on Technology Management: A Unifying Discipline for Melting the Boundaries, pp. 123–135.

Henriksen A, Traynor A (1999). "A practical R&D Project-Selection scoring tool." IEEE Trans. Eng. Manage., 46: 158–170.

Hsieh TY, Lu ST, Tzeng GH (2004). "Fuzzy MCDM approach for planning and design tenders selection in public office buildings." Int. J. Project Manage., 22: 573–584. DOI:10.1016/j.ijproman.2004.01.002.

Chen H, Kocaoglu DF, J Ho (2005), Applying sensitivity analysis to the strategic evaluation of emerging technologies in Taiwan semiconductor foundry industry, PICMET '05. Portland International Conference on Technology Management: A Unifying Discipline for Melting the Boundaries, pp. 166–173.

Kahraman C, Ertay T, Büyüközkan G (2006). "A fuzzy optimization model for QFD planning process using analytic network approach." Eur. J. Oper. Res., 171: 390–411.

Wang K, Wang CK, Hu C (2005). "Analytic Hierarchy Process with fuzzy scoring in evaluating multidisciplinary R&D projects in China." IEEE Trans. Eng. Manage., 52: 119–129.

Lockett G, Hetherington B, Yallup P, Stratford M, Cox B (1986). Modeling a research portfolio using AHP: A group decision process, R&D Management, 16: 151–160.

Lilien GL, Yoon E (1989). "Determinants of new industrial product performance: A strategic reexamination of the empirical literature." IEEE Trans. Eng. Manage., 36: 3–10.

Laarhoven PJM, Pedrycz W (1983). A fuzzy extension of Saaty's priority theory, Fuzzy Sets Syst., 11: 229–241. doi:10.1016/S0165-0114(83)80083-9.

Meade LM, Presley A (2002). "R&D Project Selection using the Analytic Network Process." IEEE Trans. Eng. Manage., 49: 59–66.

Mohamed S, McCowan AK (2001). "Modelling project investment decisions under uncertainty using possibility theory." Int. J. Project Manage., 19: 231–241.

Martino JP (2004). Research and Development Project Selection, Wiley and Sons, New York, pp. 151-155.

Mohanty RP, Agarwal R, Choudhury AK, Tiwari MK (2005). "A fuzzy ANP based approach to R&D project selection: a case study." Int. J. Prod. Res., 43: 5199–5216.

Nataraja MC, Jayaram MA, Ravikumar CN (2006) "Prediction of Early Strength of Concrete: A Fuzzy Inference System Model." Int. J. Phys. Sci., 1(2): 047-056.

Ong SK, Sun MJ, Nee AYC (2003). "A fuzzy set AHP-based DFM tool for rotational parts." J. Mater. Process. Technol., 138: 223–230.

Ringuest JL, Graves SB, Case RH (2004). "Mean-Gini analysis in R&D portfolio selection."Eur. J. Oper. Res., 154: 157–169.

Sevkli M, Lenny KSC, Zaim S, Demirbag M, Tatoglu E (2008). "Hybrid analytical hierarchy process model for supplier selection." Ind. Manage. Data Syst., 108: 122–142.

Saaty TL, (2008). Decision making the Analytic Hierarchy Process. Int. J. Services Sci., 1: 83–98.

Saaty TL (2001). The Analytic Network Process: Decision Making With Dependence and Feedback, USA, RWS, pp. 224-226.

Schmidt RL, Freeland JR (1992). "Recent process in modeling R&D project selection processes." IEEE Trans. Eng. Manage., 39: 189–201.

Tummala R (2001). Fundamentals of Microsystems Packaging. New York, McGraw-Hill, pp. 73-74.

17

Simulation of 4π HPGe Compton-Suppression spectrometer

M. E. Medhat[1,2] and Yifang Wang[2]

[1]Experimental Nuclear Physics Department, Nuclear Research Centre, P. O. 13759, Cairo, Egypt.
[2]Institute of High Energy Physics, CAS, Beijing 100049, China.

Compton-suppression spectrometer is well suited to the analysis of low levels of radioactive nuclides. Monte Carlo simulations can be a powerful tool in calibrating these types of detector systems, provided enough physical information on the system is known. A simplified Compton-suppression spectrometer model using the Geant–4 simulation toolkit was discussed. The spectrometer model was tested to evaluate photo peak efficiency of detecting point and disk sources. The efficiency calibration was calculated for incident gamma energy from 200 to 3000 keV in both the suppressed and unsuppressed mode of operation. The applicability of the efficiency transfer method in various measurement geometries was tested successfully. It can save time and avoid tedious experimental calibration for different samples geometries.

Key words: Compton-suppression spectrometer, photopeak efficiency, Geant–4 Monte Carlo simulation.

INTRODUCTION

Gamma ray spectrometry based on high pure germanium (HPGe) detectors is an important tool in the field of radioactivity measurements. The reason is due to the excellent energy resolution of HPGe detectors that permits the analyses of various radionuclides in composite samples selectively as well as the high efficiency of recently produced HPGe detectors (L'Annunziata, 2012).

Improved detection system for experiments need low background environment and is strongly needed especially in the field of high energy physics; dark matter, low-energy solar neutrino experiments. Compton-suppression spectrometer is considered a powerful technique to reduce the contribution from Compton scattered photons in a measured sample. Generally, it consists of primary detector surrounded by secondary detectors, and the pulse produced by the primary detector is accepted by analyzer only when the

secondary detectors do not produce a pulse within a time period. The most wide Compton-suppression spectrometer consists of high pure germanium (HPGe) detector surrounded by scintillator crystals such as CsI (Tl), NaI(Tl) or BGO in 4π geometry coupled by photo-multipliers (PMTs) as in Exogam, Miniball, Gamma-Sphere, Euroball, GASP. By operating spectrometer in a fast anticoincidence regime (tens of nanoseconds time resolution) can be obtained as a significant reduction of the Compton background with high resolved spectra (Fan et al., 2013; Breier and Povinec, 2009 and detectors websites).

Samples suitable to be measured are generally disk sources (various geometries and matrices) which would require certain measurements. The spectrometer accuracy is essential for the absolute measurements of radioactive materials. Experimental measurements can be applied to limited different geometries, compositions

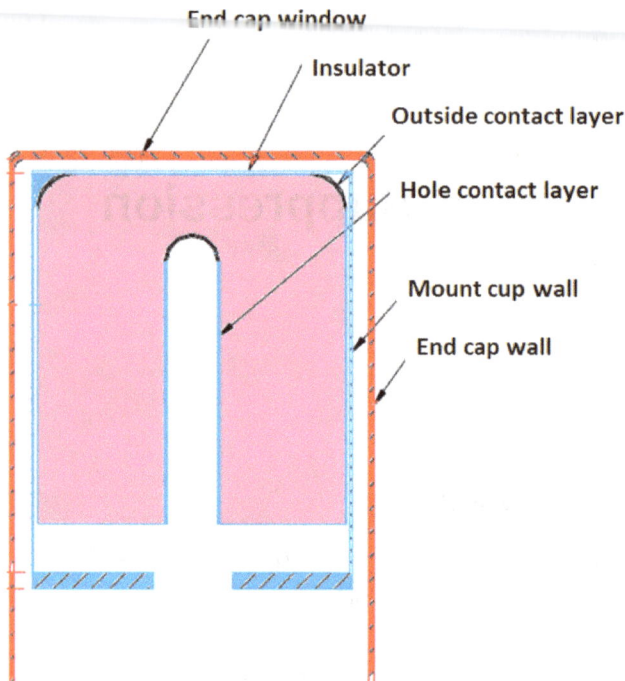

Figure 1. Schematic illustration of the HPGe detector.

and densities of sources and cannot be applied directly to all configurations. This can be a time consuming process and in some cases impossible to replicate. As an alternative approach, Monte Carlo simulations are recommended to do this task and then to continue with experimental measurements. Monte Carlo simulation is a powerful and flexible tool for simulating various physical phenomena. These types of simulations can be powerful complement to experimental installations. Modeling the geometry in computer environment gives flexibility and ease of use instead of performing experiment in different geometries (Britton, 2012; McNamara et al., 2012; Rehman et al., 2011).

One of the main problems faced in any detection system is evaluation of detection efficiency. So, instead of performing efficiency calibration for each sample, which of course is impractical, the model can be used to calculate the efficiency of the spectrometer for that particular geometry. This work aims to present a preliminary evaluation of detection efficiency of Compton-suppression spectrometer in both of the suppressed and unsuppressed mode using Geant–4Monte Carlo simulation based on object-oriented methodology and C++ language (Agostinelli et al., 2003).

SPECTROMETER SIMULATION

Geant–4Toolkit

Geant–4 is a simulation tool kit that can simulate accurately the detectors and interactions of photon and particles through matters.

The code was written in C++ and developed by CERN (CERN, 2007). It simulates accurately the passage of particles through matter. It offers the possibility to include a complete description of an experiment and extract information that might be useful in many different fields, such as nuclear and high energy physics, medical physics and astrophysics.

Geant–4 toolkit contains a complete range of functionalities including tracking, geometry; physics models and hits. It is controlled through the instantiation of the appropriate Geant4 classes to define the geometry, applicable physics, and particle source and to control the execution. The key class for all Geant4 applications is the G4RunManager which controls the initializations of geometry, physics list and primary particle generation. The user has full freedom to develop an own simulation program. The user must implement several mandatory classes to describe the detector geometry, the primary particle generator and a class to describe the relevant particles and physics processes. Other non-mandatory classes must be created to resolve proper objectives.

Simulation of spectrometer geometry

Modeling the spectrometer geometry in computer environment gives flexibility and ease of use, instead of performing an experimental determination of detection efficiency for different geometries. For this reason, the model of spectrometer system would be useful for further experiments when the geometry is changed. Then, instead of performing efficiency calibration for each sample, which may sometimes be impractical, the model can be used to calculate the efficiency of the system for that particular geometry (Chirosca et al., 2013; Baccouche et al., 2012).

The supposed Compton-suppression spectrometer geometry HPGe detectors surrounded by a CsI(Tl) cylindrical scintillator crystal with photomultipliers attached. The simulated detector is p coaxial type detector. A sketch of the HPGe detector, adapted from the manufacturer manual, is shown in Figure 1 and the provided dimensions of the HPGe detector are further summarized in Table 1. The HPGe cylinder is oriented inside a scintillator cylinder with diameter 150 mm and length 250 mm. Sample with diameter 5 mm and length 15 mm is placed at distance 10 mm from the top of the HPGe crystal. The detection system of germanium crystal works in anticoincidence with photomultipliers, trying to eliminate a part of Compton electrons generated by gamma rays which are not absorbed by photoelectric effect in HPGe crystal. The proposed Geant–4code was written using ten user classes: three user mandatory classes and some user action classes. The geometry of the Compton-suppression spectrometer was coded in the mandatory class (Detector Construction). For this study, physics process was defined in the other mandatory class (Physics List). For gamma rays, Compton scattering, photo electric absorption, pair production and Rayleigh (coherent) scattering processes are defined with valid energy range down to 250 eV. The definition is also disk to include electrons and positron multiple scattering, ionization and bremmsstrahlung processes. Atomic effects after photoelectric effect, as X-rays emission and Auger effect are included. Two Sensitive Detector classes, one for HPGe detector and one for CsI(Tl) scintillator were built. Registration of interesting processes is made in Stepping Action class. Algorithm for generating computing efficiency was implemented in (Run Action) class.

RESULTS AND DISCUSSION

The full model of efficiency calculation and efficiency transfer has been written in ten user classes: three user mandatory classes and some user action classes are

Table 1. Summary of the components and dimensions of the HPGe detector provided by the manufacturer.

Component	Dimension (mm)
Ge crystal diameter	69.8
Ge crystal length	89.5
Core hole diameter	11.6
Core hole depth	80.8
Outer electrode thickness	0.5
Inner electrode thickness	0.003
Window electrode thickness	0.003
Endcap window thickness	0.76
Aluminium endcap thickness	1.5

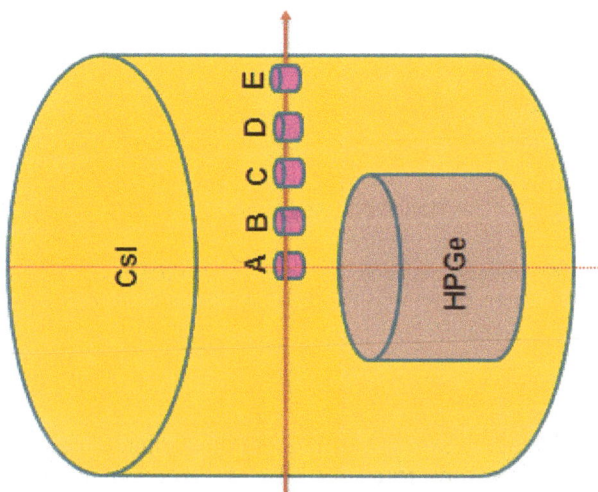

Figure 2. Detected source at different positions.

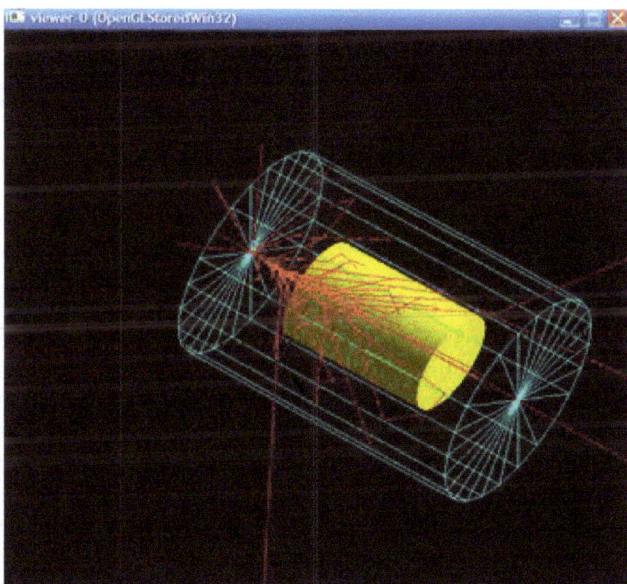

Figure 3. Visualization model of the detection system.

written. Registration of interesting processes is made in Stepping Action class. In Run Action class the simplified algorithm for computing efficiency was implemented. The detector system was modelled in both the suppressed and unsuppressed modes. In the unsuppressed mode, the coincidence registering (sensitive detector) of the surrounding CsI is turned off. While for the suppressed mode, a photon hit in the HPGe detector is triggered if a hit in the NaI detector does not occur within the same event. Both the HPGe and CsI detectors were set as sensitive detectors in the model and photon hits in each were collected using the G4HCofThisEvent class.

The principle of calculating full-energy peak efficiency is started by initialization of the decay process; then the deposited energy of each photon in the detector is summed after completing the photon tracking. The tracking of a single photon is stopped when it leaves the volume of interest or when the energy of photon becomes lower than a specified threshold value called cut-off energy. Consequently, a realistic spectrum of energy deposited in the detectors is obtained through simulation and the experimental efficiency can be compared to the simulated one so-called apparent efficiency obtained by calculating the peak area, after correcting for the continuum to match the measured one.

The efficiency of the investigated Compton suppression γ-spectrometer was generated using Geant-4 model for both isotropical point source and volume source, placed at five different positions (A, B, C, D, E) of the spectrometer as shown in Figure 2. The visualization model of the investigated Compton suppression γ-spectrometer generated by Geant-4 is shown in Figure 3. Total of 10^6 events for each energy for each different source positions, from 200 to 3000 keV, were utilized to determine system efficiency. The summation effects are also considered a source of errors in the simulated efficiency values. This effect appears at ^{60}Co (1173.2, 13323.5 keV). Geant 4 code has been applied to correct these summation and to improve their values of peak efficiencies.

The first round of simulations was performed using a

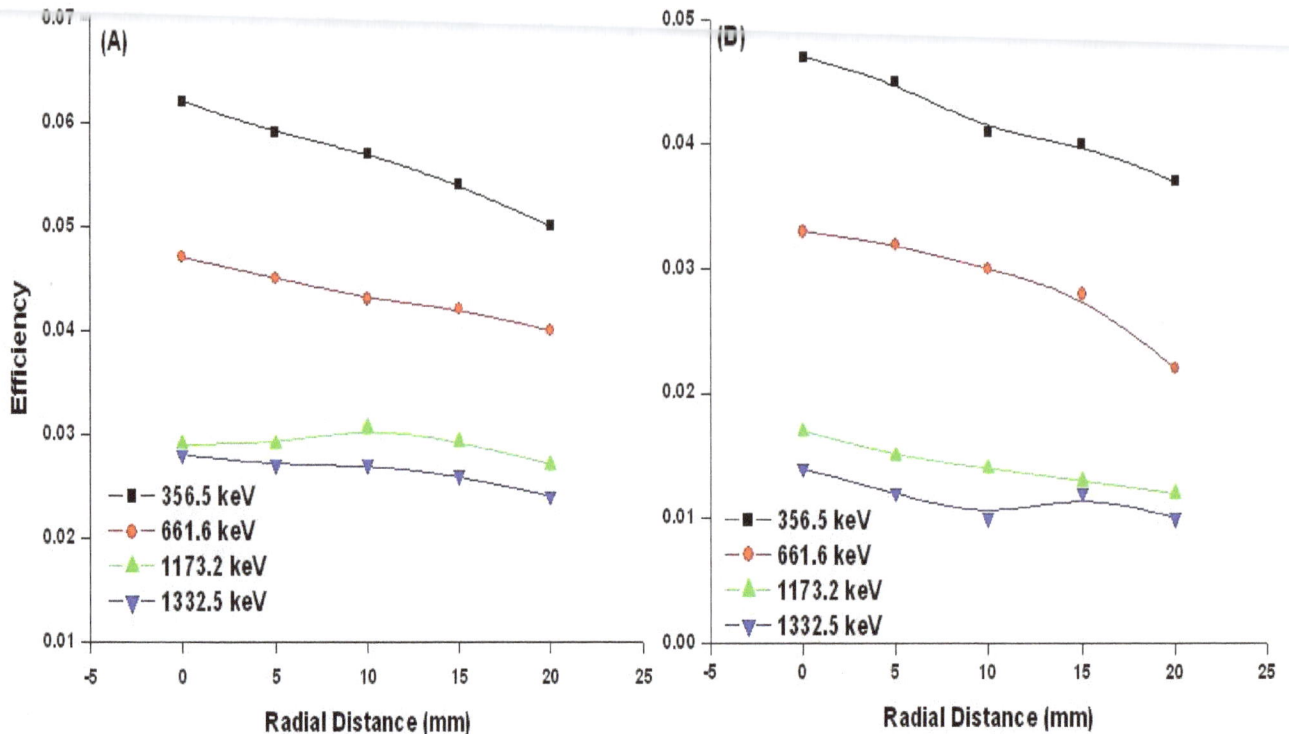

Figure 4. Variation of simulated photopeak efficiency for a point source at various positions for Compton suppression γ-spectrometer:- A) without CsI(Tl) anticoincidence scintillator (anti-suppressed mode); B) with CsI(Tl) anticoincidence scintillator (suppressed mode).

point source. Variation of photo-peak efficiency in cases of including CsI (Tl) scintillator as a suppressed mode to the detection system at different positions is shown in Figure 4. The trend is same for various gamma ray energies (356.5; 661.6; 1173.2; 1332.5) but is more pronounced for lower gamma ray energies than higher energies as presented in Table 2.

The second round of simulations was performed using a disk source. The variation of simulated efficiency in the two spectrometer cases is shown in Figure 5 and presented in Table 3. It is clear from Tables 2 and 3 that the efficiency in the suppressed mode is lower. This is largely due to the coincidence detection of two different photopeak photons from the same source, in a time frame shorter than the timing resolution of the detector system.

The applicability of the efficiency transfer in various measurement geometries on the basis of the simulated efficiency for reference point source geometry can be applied successfully. The detector efficiency was calculated for the same locations of the point sources and also for the disk sources. Figure 6 shows relation between point and disk source. It is clear that there is an increase at 1000 to 14500 keV in comparison with the fitting. It returns to the coincidence summing effects of [60]Co (1173.2, 13323.5 keV) for the transformation between point and bulk samples. The transfer efficiency

is computed for discreet values of the fitted efficiency data of point source to derive new efficiencies values for a disk source which can be applied successfully to transfer efficiency data between the two geometries as presented in Table 4.

Conclusion

Monte Carlo simulation is a powerful tool in designing and calibrating different types of detection system and spectrometers. The applicability of this method is greatly dependent on the accuracy of detector geometry model. Additionally, the geometry, composition and density distribution of the sample matrix may be particularly important in models used for low level background. The Geant-4simulation toolkit was used to model a simplified Compton-suppression spectrometer operating in both the suppressed and unsuppressed mode to determine the photo peak efficiencies incident gamma energy in a range 200 to 3000 keV. The simulations show that the efficiency in the suppressed detector mode is lower than anti-suppression mode. This is largely due to the coincidence detection of two different photons from the same source which is not within the analyzer time period. The simulated efficiency values for point as well as for disk sources at different position can be mathematically

Table 2. Simulated photopeak efficiency for a point source at various positions.

Energy (keV)	Without CsI(Tl)	With CsI(Tl)
Source position: 0 mm (A)		
356.5	0.062	0.047
661.6	0.047	0.033
1173.2	0.029	0.017
1332.5	0.028	0.014
Source position: 5 mm (B)		
356.5	0.059	0.045
661.6	0.045	0.032
1173.2	0.029	0.015
1332.5	0.027	0.012
Source position: 10 mm (C)		
356.5	0.057	0.041
661.6	0.043	0.030
1173.2	0.032	0.014
1332.5	0.027	0.010
Source position: 15 mm (D)		
356.5	0.054	0.040
661.6	0.042	0.028
1173.2	0.031	0.013
1332.5	0.026	0.012
Source position: 20 mm (E)		
356.5	0.050	0.037
661.6	0.040	0.022
1173.2	0.030	0.012
1332.5	0.024	0.010

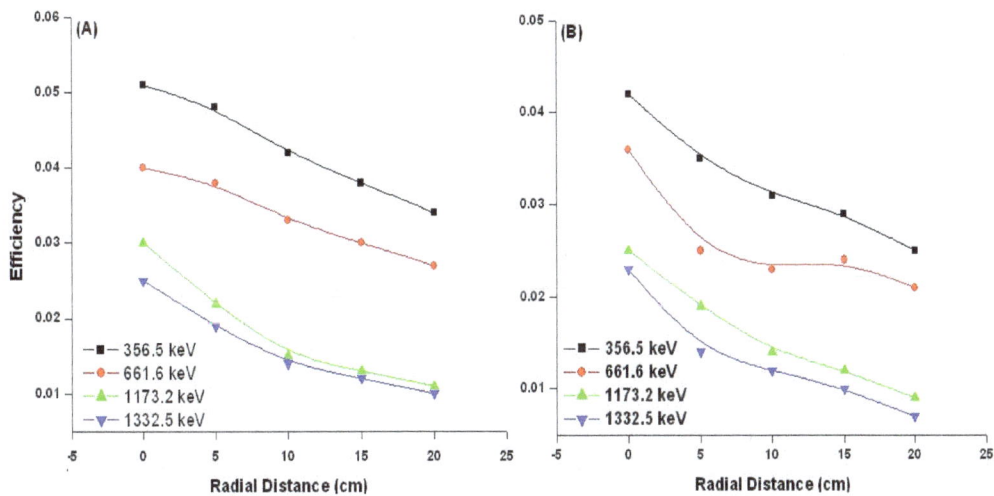

Figure 5. Variation of simulated photopeak efficiency for a disk source at various positions for Compton suppression γ-spectrometer:- A) without CsI(Tl) anticoincidence scintillator (anti-suppressed mode)., B) with CsI(Tl) anticoincidence scintillator (suppressed mode).

Table 3. Simulated photopeak efficiency for disk source at various positions.

Energy (keV)	Without CsI(Tl)	With CsI(Tl)
Source position: 0 mm (A)		
356.5	0.051	0.042
661.6	0.040	0.036
1173.2	0.030	0.025
1332.5	0.025	0.023
Source position: 5 mm (B)		
356.5	0.048	0.035
661.6	0.038	0.025
1173.2	0.022	0.019
1332.5	0.019	0.014
Source position: 10 mm (C)		
356.5	0.042	0.031
661.6	0.033	0.023
1173.2	0.015	0.014
1332.5	0.014	0.012
Source position: 15 mm (D)		
356.5	0.038	0.029
661.6	0.030	0.024
1173.2	0.013	0.012
1332.5	0.012	0.010
Source position: 20 mm(E)		
356.5	0.034	0.025
661.6	0.027	0.021
1173.2	0.011	0.009
1332.5	0.010	0.007

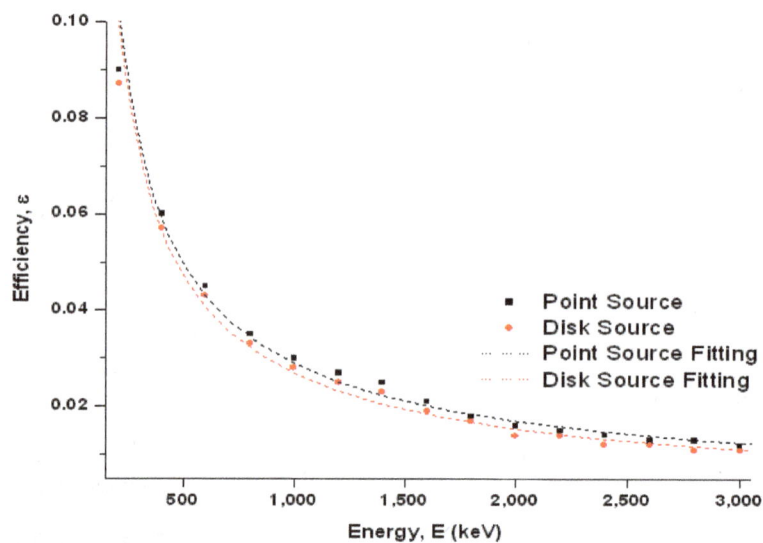

Figure 6. Mathematical efficiency transfer relation between point and disk sources.

Table 4. Photopeak efficiency transfer between point and disk sources.

Energy (keV)	Point →Disk		Disk → Point	
	Obs.	Calc.	Obs.	Calc.
356.5	0.061	0.058	0.062	0.064
661.6	0.045	0.043	0.047	0.048
1173.2	0.027	0.026	0.029	0.029
1332.5	0.025	0.025	0.028	0.027

related for the same locations to save time and avoid experimental calibration for different samples geometries.

ACKNOWLEDGMENTS

The first author would like to thank all members in the Experimental Center of the Institute of High Energy Physics (IHEP) during hosting of his postdoctoral fellowship funded by Chinese Academy of Science (CAS), China. The authors also appreciate the reviewers due to their useful comments and suggestions on the structure of manuscript.

REFERENCES

Agostinelli S J, Allisonas J, Amako K (2003). G4—a simulation toolkit. NuclInstrum Methods Phys Res A. 506:250–303.

Baccouche S, Al-Azmi D, Karunakara A,Trabelsi A (2012). Application of the Monte Carlo method for the efficiency calibration of CsI and NaI detectors for gamma-ray measurements from terrestrial samples. Appl. Radiat. Isot. 70:227–232.

Breier R, Povinec PP (2009) Monte Carlo simulation of background characteristics of low-level gamma-spectrometers. J Radioanal. Nucl. Chem. 282:799–804.

Britton R (2012) Compton suppression systems for environmentalradiological analysis.J. Radioanal. Nucl. Chem. 292:33–39.

CERN (2007). Geant4 collaboration physics http://geant4.cern.ch.

Chirosca A, Suvaila R, Sima O (2013) Monte Carlo simulation by GEANT 4 and GESPECOR of in situ gamma-ray spectrometry measurements.ApplRadiatIsot.http://dx.doi.org/10.1016/j.apradiso.2013.03.015.

Fan Y, Wang S, Li Q, Zhao Y, Zhang X, Jia H (2013). The performance determination of a Compton-suppression spectrometer and the measurement of the low level radioactive samples. ApplRadiatIsot. http://dx.doi.org/10.1016/j.apradiso.2013.03.045. 4π geometry spectrometers, http://pro.ganilpiral2.eu/laboratory/detectors/exogam;http://www.phy.anl.gov/gammasphere; http://nnsa.dl.ac.uk/euroballhome; http://npgroup.pd.infn.it/GASP

L'Annunziata FM (2012). Handbook of Radioactivity Analysis, 3rd edition, Academic Press.

McNamara AL, Heijnis H, Fierro D, Reinhard MI (2012). The determination of the efficiency of a Compton suppressed HPGe detector using Monte Carlo simulations. J. Environ. Radioact. 106:1–7.

Rehman SU, Mirza SM, Mirza NM, Siddique MT (2011). GEANT4 simulation of photo-peak efficiency of small high purity germanium detectors for nuclear power plant applications. Ann. Nucl. Energy 38:112–117.

18

A novel method to develop an automobile assembly line system

Ali A. J. Adham and Ahmad N. N. Kamar

Faculty of Industrial Management, University Malaysia Pahang, Gambang, Pahang, 26300, Malaysia.

The assembly line is an important component of the automobile production process. The function of the assembly line is to produce different models of vehicles with minimum work in the process. For better performance, activities on the assembly line should be performed to minimise the process steps and achieve other objectives. This study develops a new dynamic sequencing method to improve activities on the assembly line and also an automated sequence-control system. Three methods, namely the Multi-Objectives Model, the Genetic Algorithm System and the Simulation Model, are integrated to enhance the efficiency of the assembly line by controlling the processing time within the workstations. The results show that the method was able to improve the working time performance and also increase throughputs.

Key words: Processing time, assembly line, mathematical method, genetic algorithm system, simulation model.

INTRODUCTION

Body Shop (BS), Paint Shop (PS), Assembly Shop (AS) and Test Shop (TS) and other sub-assemblies, also called stations, have the function of feeding the main assembly. The production of automobiles in the AS is a typical example of the mixed-model production system (Wonjoon and Hyunoh, 1997). Figure 1 shows the assembly system of the automobile production system. In the figure, all the assembly plants have their own stations (namely S_1, S_2....S_n). The sub-assembly stations are also shown in the figure.

In the automobile industry system, one of the areas under consideration is the Assembly Line Balancing Problem (ALBP) which distributes the total workload among manufacturing stages (Adham, 2012; Ali and Razman, 2011; Toshio et al., 1996). There were many researchers who studied the issues related to the ALBP and the Production Line System (PLS) in order to obtain the best solution (Razam and Ali, 2012; Minh and Soemon, 2008; Williams, 2007).

The Hybrid Model (HM), combining the Multi-Objectives Model (MOM), the Genetic Algorithm System (GAS) and the Simulation Model (SM), is presented in this study. It is a new technique and one of the most powerful methods to obtain the best balance of the cycle. Many real-world

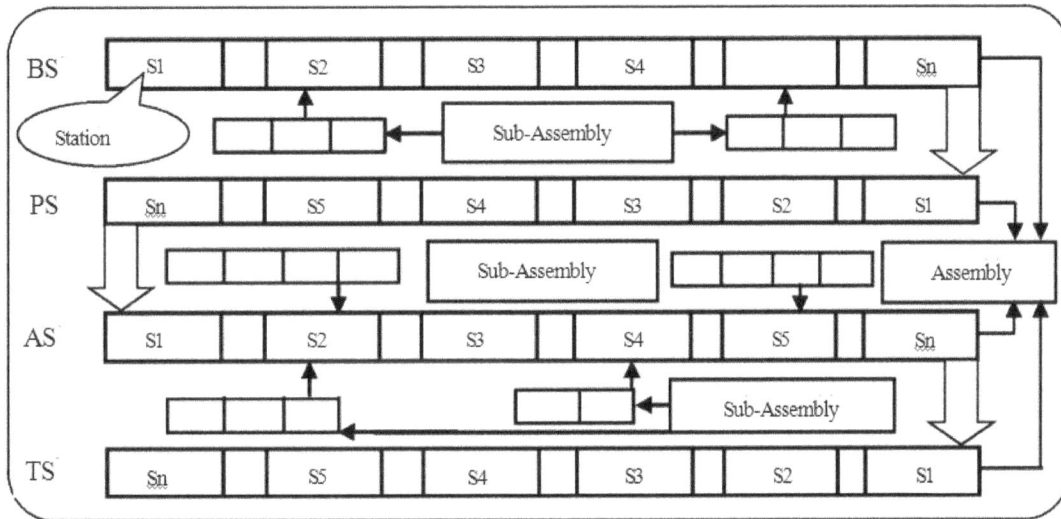

Figure 1. Assembly line system (ALS).

problems require an optimal solution that could be obtained by adopting the HM approach. The HM developed in this study is able to assist managers to have an optimal cycle time (CT), balancing the ALS and managing the production plan to know the capacity of the assembly line after solving the ALBP (Ali and Razman, 2012; Amir and Farhad, 2006; Anand et al., 2012).

The contribution of this study is to approach an integrated model (including the MOM and the GAS) to solve the queuing problem within the stations on the assembly line, with the SM to manage the capacity of the assembly line. Additionally, the integrated model can combine the unbalanced assembly line problem and the ratio of the production plan. As a result, the method will achieve the target by minimising unbalanced CT and maximising workload to achieve the production plan. This study focuses on the main problem of the production line which is balancing CT within the stations. The unbalancing problem occurs when not all stations are able to complete all tasks at the same time (Christian and Armin, 2009). As a result, it causes a congestion problem on the production line and the resources are underutilised. Figure 2 presents the ALBP which is unbalanced CT within the stations on the assembly line.

METHODOLOGY

Multi-objectives model (MOM)

The MOM is formulated to create a balanced time within stations through to obtaining optimal balance within the stations. There are two goals of the MOM: (1) to minimise the queuing time within the stations; (2) to minimise the idle time within the stations.

$$\text{Min Q} = \max \sum_{i-1}^{p} \sum_{i-1}^{sq} |QU_{ij}| \, X_{ij} \qquad (1) \; (1^{st} \text{ goal})$$

$$\text{Min Id} = \max \sum_{i=1}^{p} \sum_{j=1}^{sd} DT_{ij} \, X_{ij} \qquad (2) \; (2^{nd} \text{ goal})$$

Where: **Q**: total queuing time within the stations, **QU**: queuing between the stations, **Id**: total idle time within the stations, **DT**: idle time between the stations.

The MOM aids management to achieve either the optimum solution (Razman and Ali, 2011; Razman and Ali, 2010). Figure 3 shows the implementation of the MOM for the ALS. The MOM will reduce the queuing and the idle time to obtain the best balance, then the best solution for the PLS.

GAS

The GAS is formulated to create an advanced balance time within stations through shuffling of the tasks in order to obtain an optimum balance. The GAS will select the task that should be moved within stations according to the objectives (1) and (2). Figure 4 presents the model of moving tasks among the stations. In the figure, there are two categories of task movement: the first category is a movement (1) from station (1) towards the station (n) passing through all stations respectively. The second category is movement (2) from station (n) towards station (1) passing through all stations respectively. The aim is to find the best solution for these objects using the GAS. As seen, the solution should allow all points to be passed by choosing the closest path among them in one go. The task movement occurs after selection of the first task with a high CT from any station, which should be moved towards the next station which has a low CT and function as a final task. Otherwise, the GAS will select the final task with high CT from any station and move it towards the previous station with low CT and have it function as a first task. This formulation of GAS will be more realistic, that is, create an optimum balance of CT within stations.

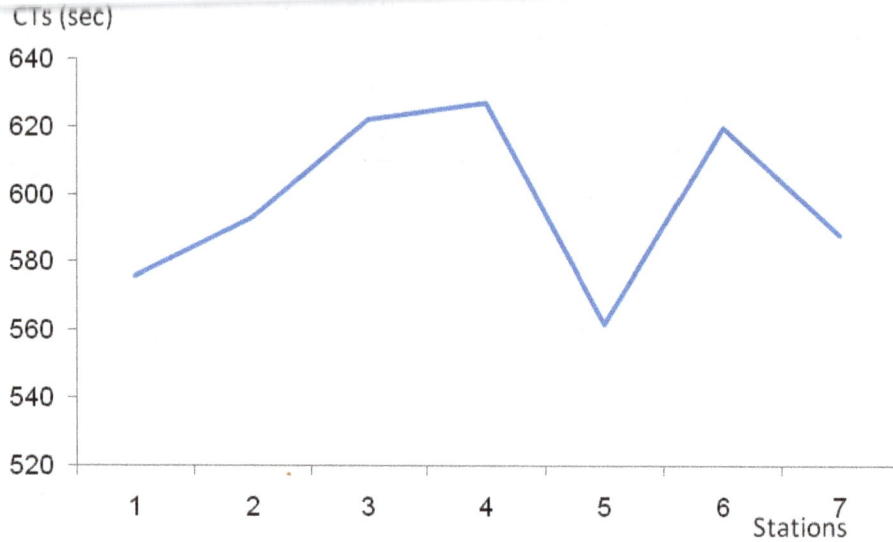

Figure 2. Unbalanced CTs.

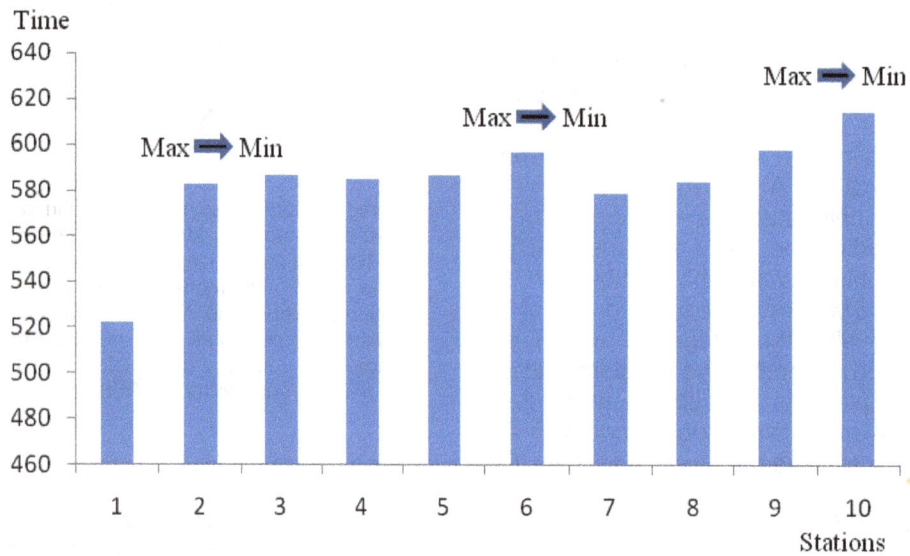

Figure 3. Unbalancing on the assembly line.

Genetic algorithm objectives

The GAS approach in this study aims to achieve two goals: rebalancing CT within the stations for each shop, through task movement; and also redistributing the jobs among the workers to obtain the optimum solution. The GAS obtains the optimum balance (optimum solution) of the ALS with two objectives which are:

(i) First goal: Moving the tasks among the stations.
(ii) Second goal: After applying the MOM, if the ALS still has queuing issues, the GAS will redistribute the jobs to the workers in order to achieve the optimum balance.

The formulas of the GAS are presented in Equations (3) and (4). These explain how the GAS achieves a time balance on the assembly line.

$$Goal1 = CTS_1 \approx CTS_2 \approx CTS_3 \approx CTS_4 \approx,,,,\approx CTS_n \quad (3) \text{ (1}^{st}\text{ goal)}$$

$$Goal2 = RJS_1 \approx RJS_2 \approx RJS_3 \approx RJS_4 \approx,,,, \quad \approx RJS_n \quad (4) \text{ (2}^{nd}\text{ goal)}$$

Where: CTS_i =CT for each station, RJS_i = ratio of the jobs of each station.

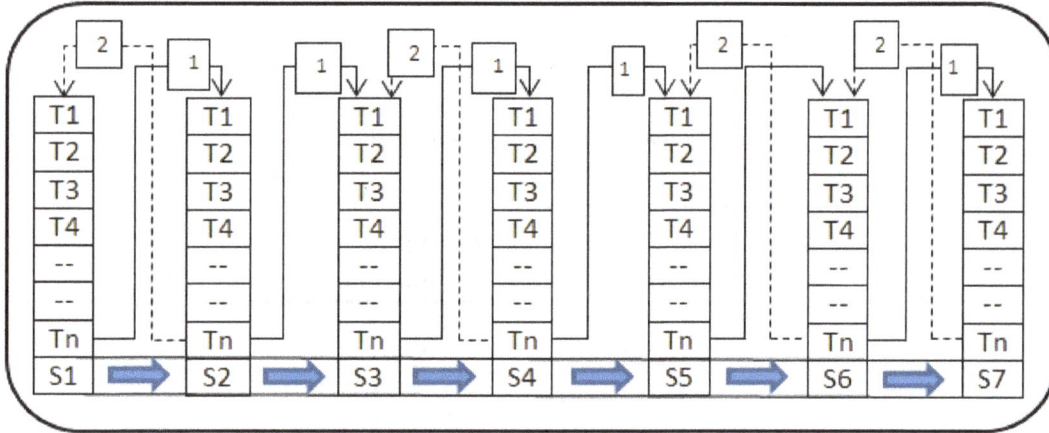

Figure 4. Task movement among the stations.

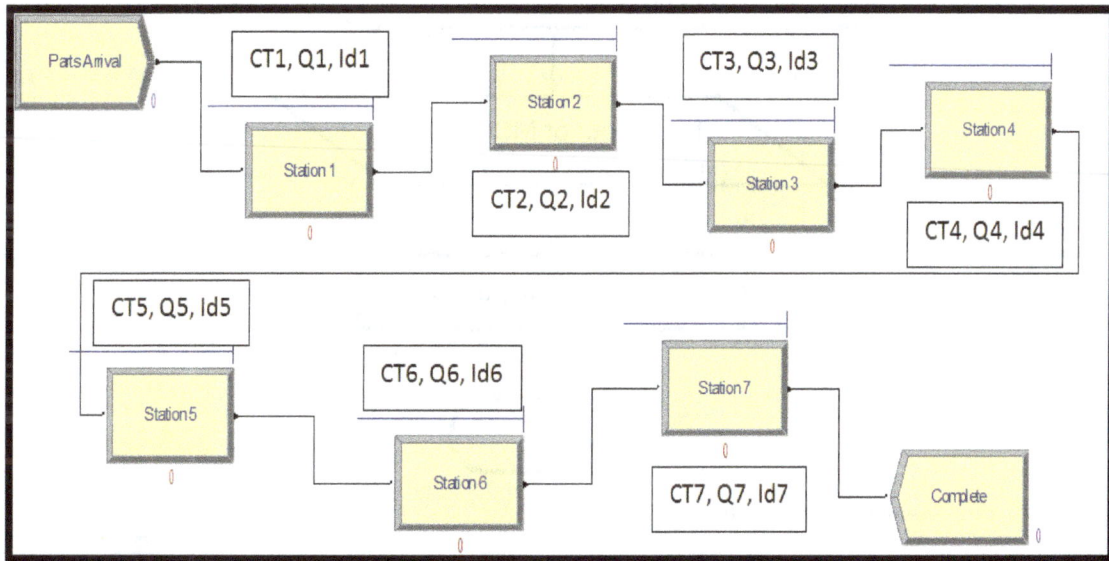

Figure 5. SM of ALS.

To achieve goal 1, the GAS should move the tasks among stations until it gets the best CT balance. This goal corresponds to the first and second objectives of MOM (1, 2), that is, to obtain the optimum balance.

Simulation model (SM)

Simulation is a technique with which a real-world problem can be mimicked and modelled with the aid of computers. The SM provides analysis and allows users to perform 'what-if' analysis where users can test different strategies or policies and observe how the model behaves before implementing it in the real world. Besides that, the simulation also serves as a training and educational tool (Holst and Bolmsjo, 2001).

In this study, the SM is developed using the ARENA simulation package. Figure 5 shows the SM of ALS. The chassis section in the ALS is modelled and inputs such as arrival time and processing time are incorporated into the model.

Hybrid model (HM)

HM flowchart

The HM, applied to solve both problems, which are queued and idle time, also manages a new plan depending on the available total working time to obtain the best balancing by applying the MOM. The SM will create new plans depending on the efficiency of the cycle time. Figure 6 shows the flowchart of the HM.

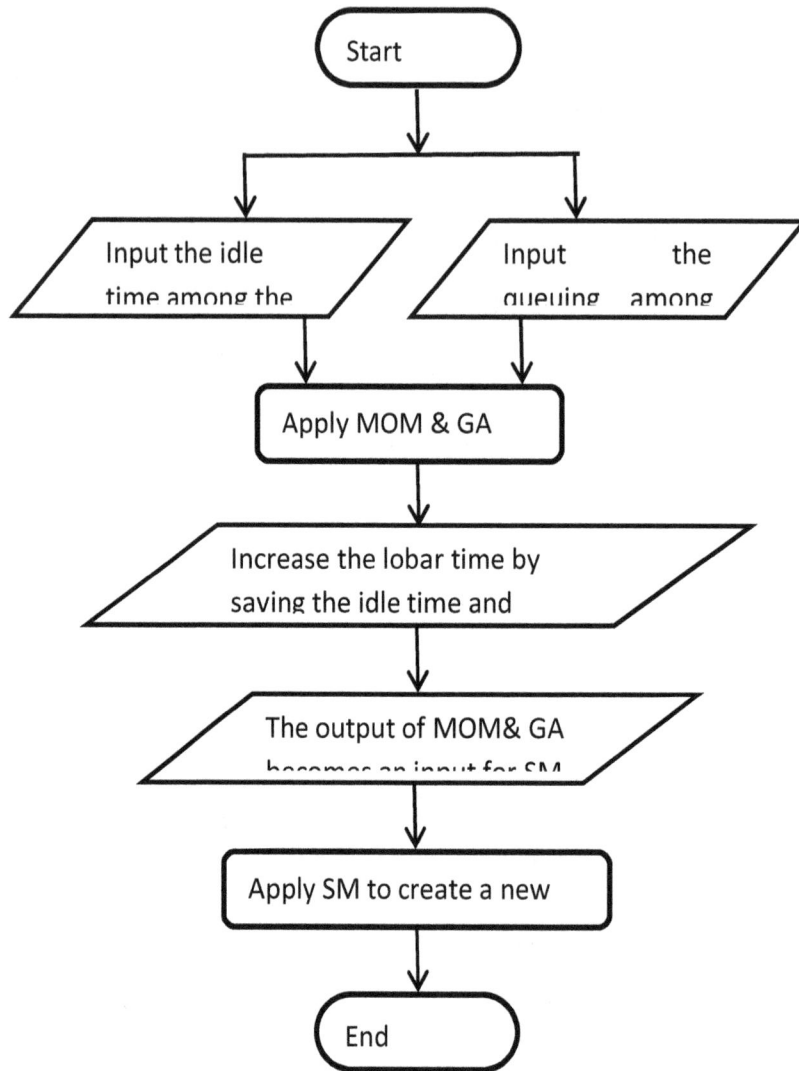

Figure 6. HM flowchart.

Procedure of HM

Computer software helps to optimise the ALS by applying the MOM and GAS. MATLAB software is used to solve the system issues to obtain the optimum balanced CT of the ALS. The procedure is as follows:

Procedure: solve the unbalanced CT of the assembly line.
Input data: CT, task number, number of workers for each station, queuing and idle time.
Output data: the optimum balancing of the ALS.
Begin
{
        Calculate the CT on the production line for each station
                While i < total number of stations
                        Balance process time tasks
                Move the tasks among the stations
        i=i+1

End
Print the optimum balance
} End
Once the optimum balancing of ALS is calculated using the MOM, a SM is developed. The SM is constructed to test the maximum number of cars that can be produced if the ALS CT is balanced in order to optimise the capacity of the production line.

MODEL RESULTS

Balancing problems of the ALS

CT is the time taken to complete all tasks at the stations of the shops. For a highly efficient ALS, the CT should be equal among the stations (Nai-Chieh and I-Ming, 2011).

Table 1. Variables of the chassis section.

Stations	CT seconds (L3)	No. tasks (L2)	Workers (L3)
1	576	122	2
2	593	116	2
3	622	130	1
4	627	125	2
5	562	118	1
6	620	127	1
7	588	120	1
Total	4,188	858	10

Table 2. Queuing and idle time at the CS.

No. station	CT before applying the model	CT after applying the model	Queuing before	Idle time before	Queuing after	Idle time after
CS_1	576	593	17	0	6	0
CS_2	593	599	29	0	0	0
CS_3	622	599	5	0	4	0
CS_4	627	603	0	65	0	1
CS_5	562	602	58	0	0	6
CS_6	620	596	0	32	0	0
CS_7	588	596	0	0	0	0
	4,188	4,188	109	97	10	7

Normally, it is a very challenging target to reach a balance of CT within the stations. An unbalanced CT is caused by the queuing problem on the assembly line. This study examined the chassis section which is one part of the assembly line.

Chassis section (CS)

The CS has seven stations. Its function is to assemble an engine, bellows, axle and other mechanical works. The total processing time in this section is 4,188 s. Table 1 describes the operational aspects of this data. It shows that station 1 has a processing time of 576 s with 122 tasks. Only two workers are involved at this station. Table 2 shows the queuing time and the idle time before and after applying the MOM. The queuing and the idle time before applying the model were 109 and 97 s, respectively. After applying the HM, the queuing and idle time become 10 and 7 s, respectively. As a result, the model reduced time arising from the queuing problem by around 99 s (1.65 min), and 90 s (1.5 min) due to idle time. The total time saved is 189 s (3.15 min) in preparation of only one car.

Figure 7 presents the CT station before and after applying the MOM and the GAS to the ALS. In the figure,

L1 represents the working CT before applying the model; L2 represents the best balancing within stations after applying the model.

Simulation results

The current assembly line operates for 7.5 h per day and produces 28 cars daily. The SM is used to test the maximum number of cars that can be produced if the assembly line operates according to the optimum CT calculated using the MOM. Table 3 shows the number of cars that can be produced daily. The simulation results reveal that by adopting the new balanced CT, the ALS can produce an additional four cars daily, given that the maximum number of operating hours of the ALS is 7.5 h. Producing more than 32 cars will require additional working hours. This finding serves as a guideline on how many additional cars can be produced daily without exceeding the current capacity and maximum duration.

DISCUSSION

The new method combines the MOM and the GAS with the SM to solve the unbalancing and planning

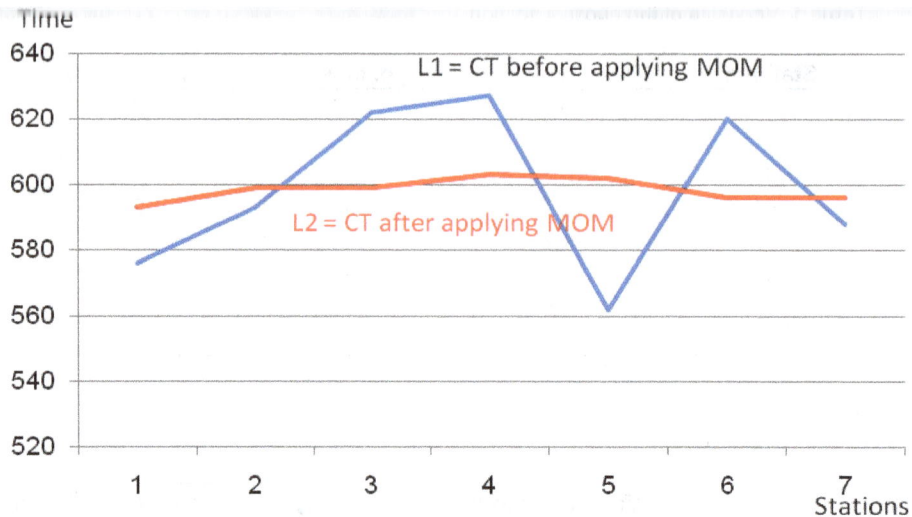

Figure 7. CTs before applying, after applying the MOM and the GAS.

Table 3. Number of cars produced and hours needed.

Number of cars produced	Hours
20	5.69
24	6.02
28	6.36
30	6.69
32	7.03
34	7.36
36	7.53
37	7.70

problems in the ALS. This method was applied to the chassis section of the ALS to obtain the optimum balance and plan. The MOM the GAS saved 189 s in the chassis section. The total queuing before applying the MOM and the GAS was 109 s (1.81 min); it became 10 s (0.16 min) after applying the MOM and GAS which saved 99 s (1.65 min) for preparation of one car. Also, the model reduced the idle time within the stations as well. It was 97 s (1.61 min) before applying the model and became 7 s (0.11 min) for preparation of one car. Therefore, the total time saved by using the MOM and the GAS is 189 s (3.15 min) for preparation of one car in respect to both issues of the queuing and the idle time. Besides minimising queuing and idle time, this study further enhanced the results by developing a SM to test the maximum number of cars that can be produced daily by the ALS if the balanced CTs are adopted. The results show that from the current number of 28 cars produced, the balanced ALS can produce a maximum of 32 cars daily. The new technique is beneficial to all workshops and sections of the production line as it increases the capacity of the production line in automobile manufacture.

Conclusion

The ALS is very important for the automobile industry. The unbalancing variations within stations are difficult problems which affect efficiency of the assembly line. This study proposed a new technique to solve the problems, such as queuing and idle time within stations. The HM combines the MOM, the GAS and the SM to obtain the optimum solution and plan. As a result, the new technique is very important for enhancement of the efficiency of the assembly line. Moreover, the HM reduces the unbalanced time within the stations and increases production by four cars per day.

Conflict of Interest

The authors have not declared any conflict of interest.

ACKNOWLEDGMENTS

This paper was supported by the University of Malaysia Pahang, under Grant No. RDU120380. The authors would like to thank the university and all staff that assisted in the success of this research. Also, thanks to anonymous referees whose insights improved the content of this manuscript.

REFERENCES

Adham AAJ (2012). An integrated model for production line balancing planning, PhD dissertation, University Malaysia Pahang, Pahang, Malaysia.

Ali AJA, Razman M (2012). Enhancing efficiency of automobile assembly line using the fuzzy logical and multi-objective genetic algorithm. WCCI 2012 IEEE World Congress on Computational Intelligence June, 10-15, Brisbane, Australia.

Ali AJA, Razman M (2011). Process queuing in the automobile manufacturing body shop: Using a multi-objective model. Int. J. Phys. Sci. 6(30):6928-6933.

Amir A, Farhad KA (2006). Multi-objective lead time control problem in multistage assembly systems using an interactive method. J. Appl. Math. Computation. 176:609-620.

Williams A (2007). Product service systems in the automobile industry: contribution to system innovation. J. Cleaner Production. 15:1093-1103.

Christian B, Armin S (2009). Balancing assembly lines with variable parallel workplaces: problem definition and effective solution procedure. Eur. J. Operational Res. 199:359-374.

Holst L, Bolmsjo G (2001). Simulation integration in manufacturing system development: A study of Japanese industry. Ind. Manage. Data Syst. 101(7):339-35.

Anand MSA, Sarkar S, Rajendra S (2012). Application of distributed control system in automation of process industries. Int. J. Emerging Technol. Adv. Eng. Website 6(2):2250-2459.

Minh DN, Soemon T (2008). Emergence of simulations for manufacturing line designs in Japanese automobile manufacturing plants. Proceeding of the Winter Simulation Conference IEEE. Japan.

Nai-Chieh W, I-Ming C (2011). A solution procedure for type e simple assembly line balancing problem. J. Comput. Industrial Eng. 61:824-830.

Razman M, Ali AAJ (2011). Optimum efficiency of production line in automobile manufacture. Afr. J. Bus. Manage. 6(20):6266-6275.

Razman MT, Ali AJ (2010). Design and analysis of automobiles manufacturing system based on simulation. Canadian Center Sci. Edu. J. Model Appl. Sci. 4(7):130-134.

Toshio S, Atsushi N, Kouske S, Takashi M, Nobuyoshi H (1996). Development of the new human-conscious automobile assembly plant. Annals of the CIRP. 46(1):381-384.

Wonjoon C, Hyunoh S (1997). A real-time sequence control system for the level production of the automobile assembly line. Comput. Ind. Eng. 33(3-4):769-772.

19

Detection of energetic particles from plasma focus using Faraday cup and SSNTD (LR-115A)

G. M. El-Aragi

Plasma Physics and Nuclear Fusion Department, Nuclear Research Center, AEA, P. O. Box 13759 Cairo, Egypt.

A Mather-type plasma focus device is used in this work which is prefilled with helium at 0.8 Torr. The total ion current density of the plasma stream is measured to be 750 mA/cm^2. From time-resolved measurements (Faraday cup), the ion beam energy was distributed with energy ranging from 0.3 to 540 keV. The ion flux density of ion beam is estimated to be 4.47 × 10^{11} ion/steradian using track etching technique (LR-115A).

Key words: Energetic ions, plasma focus, charging voltage.

INTRODUCTION

Ion diagnostics of high-temperature plasma objects are considered to be very important, because they provide essential data about plasma parameters. Ion beams are emitted from the high-temperature plasma objects are an abundant source of valuable information about fusion reaction yields, plasma ion temperatures, as well as a spatial distribution of fusion reaction sources.

In several theoretical papers (Bernstein, 1970; Jager and Herold, 1987; Pasternak and Sadowski, 1998) very simple configurations have been adopted to explain ion behavior inside of the plasma column. Experimental studies of Dense Plasma Focus (DPF) facilities proved the occurrence of high-energy ion beams, generated within the plasma focus pinch column or its vicinity. High energetic ions are considered to play an important role in the production of the intense neutron flux in the plasma focus device when using deuterium gas (Zakaullah et. al., 1999). Studies of high-energy ions emitted from plasma focus devices provide information on the ion acceleration mechanisms and are also important for various plasma focus technology. Gerdin et al. (1981) employed a Faraday cup in a time-of-flight technique to measure the ion spectrum and a careful study of the ion-neutral interactions allowed the observation of deuteron energies down to ~ 25 keV. Different kinds of acceleration mechanisms for charge particle were identified (Mather, 1971; Deutsch and Kies, 1988). General characteristics of the ion beams were studied in different laboratories (Gerdin et al., 1981; Kelly and Marquez, 1996).

Track etching technique has been successfully employed in many insulating materials for detection and identification of charged particles, e.g. in the study of heavy primary cosmic rays, the search for super heavy elements and innumerable applications in radiation dosimeters.

Early in the past decade cellulose nitrate as recognized as the most sensitive of all track detectors, and so has been used as a detector to record protons.

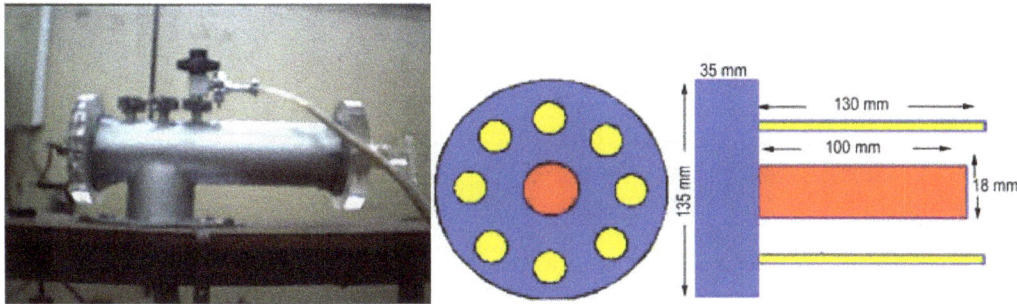

Figure 1. Schematic of experimental setup.

EXPERIMENTAL SETUP

Mather-type 112.5 J plasma focus device consists of an outer electrode, which is formed of eight copper rods, each of 130 mm length and 10 mm diameters as shown in Figure 1. The outer diameter of center electrode is 18 mm and inner diameter of squirrel cage (outer electrode) is 55 mm. A hole of 5 mm diameter and 8 mm depth was drilled in the front of the inner electrode in which different metals were filled. The cylindrical insulator ring is of 130 mm diameter and 35 mm thickness. The electrode system is enclosed in a vacuum chamber made of stainless steel tank of 350 mm length and 100 mm diameter. There are several ports in the vacuum chamber for diagnostic purposes. The condenser bank of plasma focus device consists of one condenser of 25 kV and 1 μF low inductance condenser. A capacitor bank charged at 15 kV (112.5 J), giving peak discharge current of about 5 kA, powered the focus device. The inner electrode is connected to the positive connection of high voltage supply via a triggertron-type vacuum tube (CX1159) served as a switch, whereas the outer electrode is grounded.

The vacuum chamber was evacuated up to 10^{-2} mbar pressure by a rotary pump (Edwards single stage model 1 Sc.-150B) before filling gas (helium). To avoid vapor from back streaming, the vacuum chamber is washed by gas after evacuation by rotary pump. The gas was fed into the system via flow meter (OMEGA model). The external inductance of the system including the capacitor, thyratron switch, connecting cables and coaxial electrodes (cathode and anode) is measured to about 6.5 μH.

In order to register the particle radiation within the pinhole camera, the use of nuclear track detectors of the LR-115A type was made. After irradiation those detectors were etched under standard conditions (in a 6.25-N solution of NaOH, at a temperature of 70°C) for a period ranging from one hour to several hours. To perform time-resolved measurements Faraday-type collectors were used: a single cup (FC) and a so-called double-cup system (DFC), which used two ring-shaped collectors placed at a given distance (time-of-flight basis), but adjusted along the common z-axis.

The applied voltage and the discharge current through the discharge chamber were measured using a voltage divider (Home made), which was connected between the two electrodes, and a current monitor, which can be located upon returning to the ground. The signals from the voltage divider and the current monitor were recorded in a digitizing oscilloscope (Lecroy, USA) with a 200-MHz bandwidth.

The peak value of the discharge current was measured approximately 5 kA during the pulse. Figure 2 shows the current and voltage waveforms that characterized the pulsed low energy plasma focus device. Current and voltage were measured as a function of time at an input energy of 112.5 J (maximum applied voltage 15 kV).

RESULTS AND DISCUSSION

The Plasma Focus (PF) is a device consisting in two coaxial electrodes in vacuum connected to a fast high voltage capacitor bank and separated by an insulator. When the high voltage (HV) is applied to the electrodes through a thyratron switch, an electrical breakdown develops on the surface of the insulator and the discharge is driven by the Lorentz force to run along the gap between the electrodes at a speed of some 10^6 cm/s and when reaches the end of the inner electrode collapses on the axis of the device focusing in a hot blob of plasma (Plasma Focus). In this phase strong electric fields are generated that produce intense ion beams. The duration of each discharge is of the order of 8 μs.

Preliminary time-resolved measurements of the ion pulses were carried out by means of an ion collector, which was placed at a distance of 20 cm from the top of the anode. The ion collector consists of a copper disc of 2.0 cm diameter which is connected to the ground through resistor (R =75 mΩ). The voltage develop across the resistor is fed it digital storage oscilloscope (Lecroy) to record the ion current signals. The collector plate was polarized negatively, and the whole measuring circuit was shielded against electromagnetic noise. Time-resolved studies of ions were performed by means of a double-collector of the Faraday type, which was designed especially for time of flight (TOF) measurements of pulsed charged-particle streams. That detector was equipped with two separate collectors (collector and grid) adjusted along the same Z-axis. The first ring-shaped collector (grid) was placed at a distance of 20 cm from the focus pinch, and the second one was situated about 2 cm behind the first collector. Some examples of the registered traces of the collector signals are presented in Figure 3. The velocity, energy and density of helium ions are estimated using TOF technique (Gerdin et al., 1981; Wong et al., 2002). The ion velocity is estimated by taking the ratio of the distance to the flight time of ions from source to detector (Lee and Saw, 2013). The variation of ion flux with filling helium gas pressure is shown in Figure 4. It is note that the maximum ion flux reaches at maximum value at pressure 0.8 Torr,

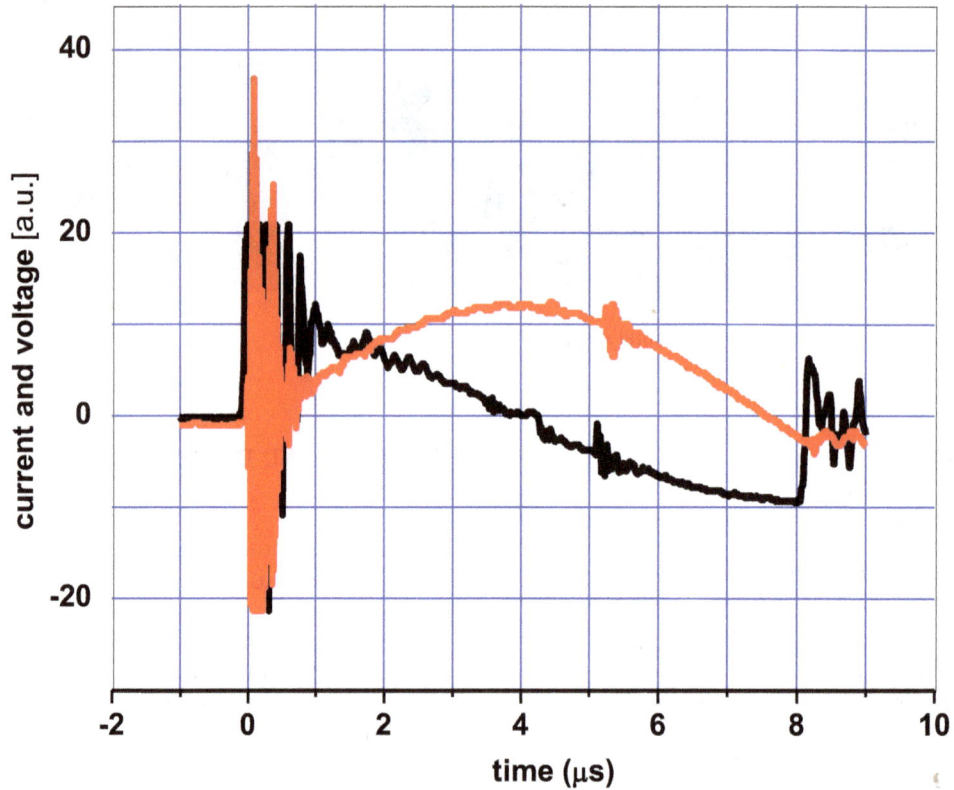

Figure 2. Discharge current (red) and voltage (black) signals from the plasma device.

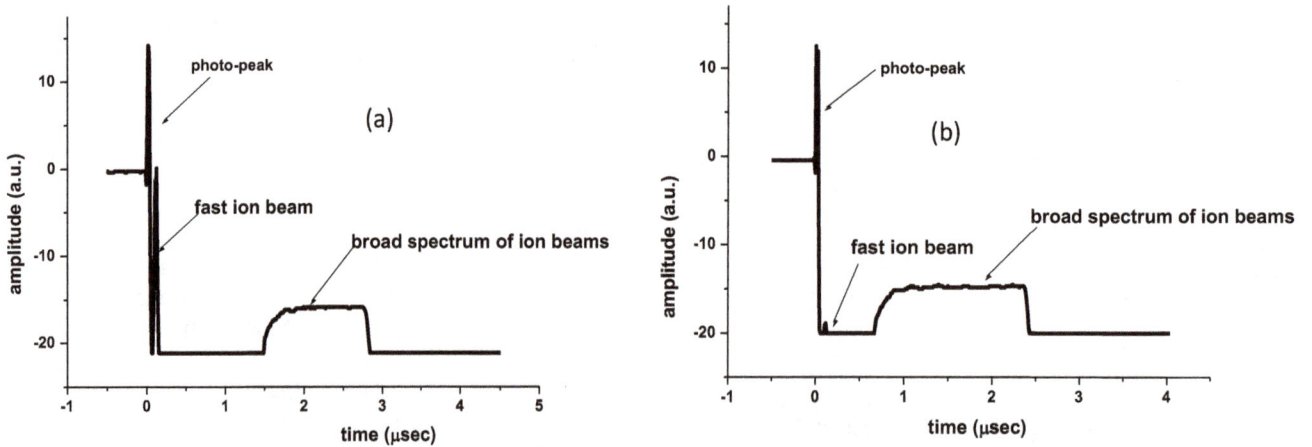

Figure 3. Typical Faraday cup signals obtained from a single discharge left signal (a) indicate good peak of fast ions while right signal shows much smaller fast ions than right one (b).

consequently the best operating pressure for ion detection in this experiment is 0.8 torr. The energy spectrum of helium ion is shown in Figure 5. Data shows two groups of ion spectrum, first group of lower kinetic energy corresponding to helium ion with kinetic energy ranging from 0.3 to 1.0 keV, and second group of higher kinetic value ranging from 15 to 540 keV.

The ion beam of the present image (Figure 6) has a circular cross-sectional area 28 μm^2. The beam contains a large number of fast ions which are distributed uniformly or quasi-uniformly over the beam cross-section. A majority of quasi-uniform ion density, where the density of ions is distributed in an increasing trend toward the beam center.

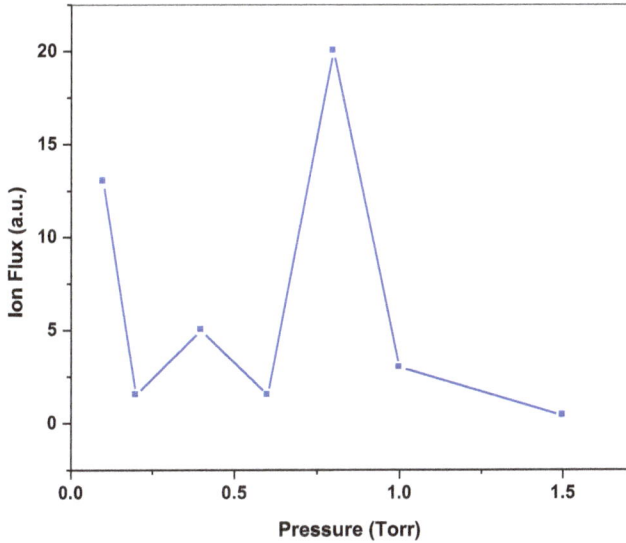

Figure 4. Variation of ion flux with helium gas pressure (torr).

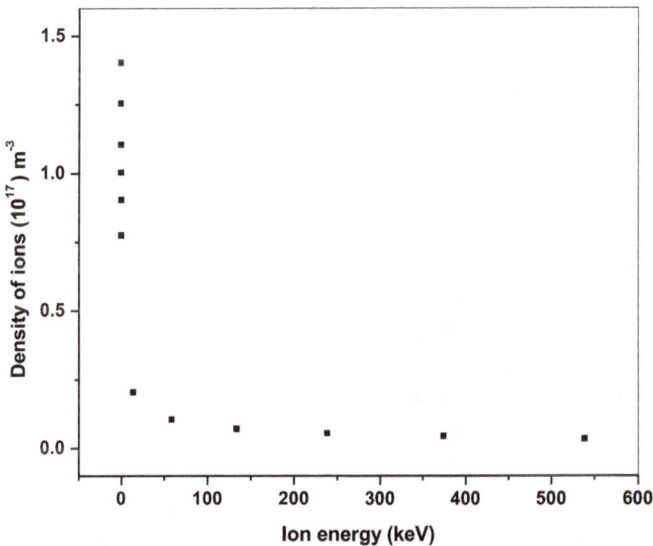

Figure 5. Energy spectrum of ion.

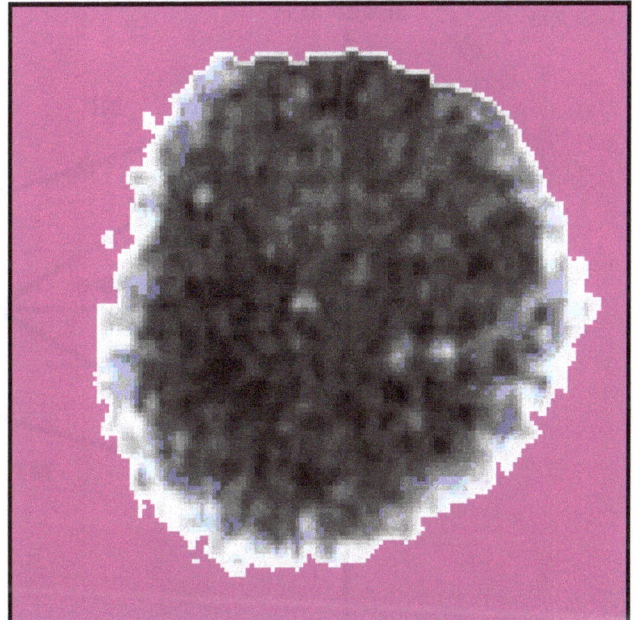

Figure 6. Image of the helium ion tracks (0.8 torr, 15 kV) as obtained from plasma focus discharge with LR-115A film.

which is provided by currents flowing in the direction of rotary motion of the mass of the plasma. The rotational drift velocity is greater for the ions because of their greater mass and hence a net circular current flows will crossed with B_θ to produce a centrifugal acceleration to hold the plasma in the form of an eddy or vortex.

The drift velocity of ions due to viscous forces in rotating plasma provide an azimuthally current density J_θ and an axial magnetic field B_z. The B_z field due to such viscous battery is proportional to the vortices of the fluid. The rotational velocity V_θ is composed of three drifts that is, $E \times B$ drift, diamagnetic drift and centrifugal force drift. The plasma is accelerated not only axially by $J_z \times B_\theta$ force but azimuthally by $J_\theta \times B_z$ force, where J_z is the axial discharge current and B_θ is self-induced azimuthally magnetic field.

Acceleration mechanism

In the radial phase, shock front and the piston starts together at r=a length of the pinch is zero at this time. In any plane the velocity of the shock front is larger than the piston velocity, so the distance between them is a time dependent quantity. The shock front accelerates onto the axis, hitting it; a reflected shock develops and moves in radial outwards (Figure 7). The piston continuous to compress inward until it hits the outgoing reflected shock front. The meeting point between reflected shock and a piston called point of maximum compression of the pinch (minimum radius).

It is possible to estimate the ion flux density of microbeam using a model consisting of a core surrounded by co-central zones (CCZ model) is proposed by EL-Aragi et al. (2007). The registered ion images are obtained for 0.1 J/ 15 kV PF-shots performed at the initial filling gas pressure P = 0.8 Torr helium. The total ion emission (centered on Z-axis) is estimated to be 4.47 × 10^{11} ion/steradian.

The shear in the velocity in conjunction with B_θ will produce centrifugal force which tends to a rotary or vortex pattern. Once the vortex or vortices are setup, the acceleration of plasma is the centrifugal acceleration

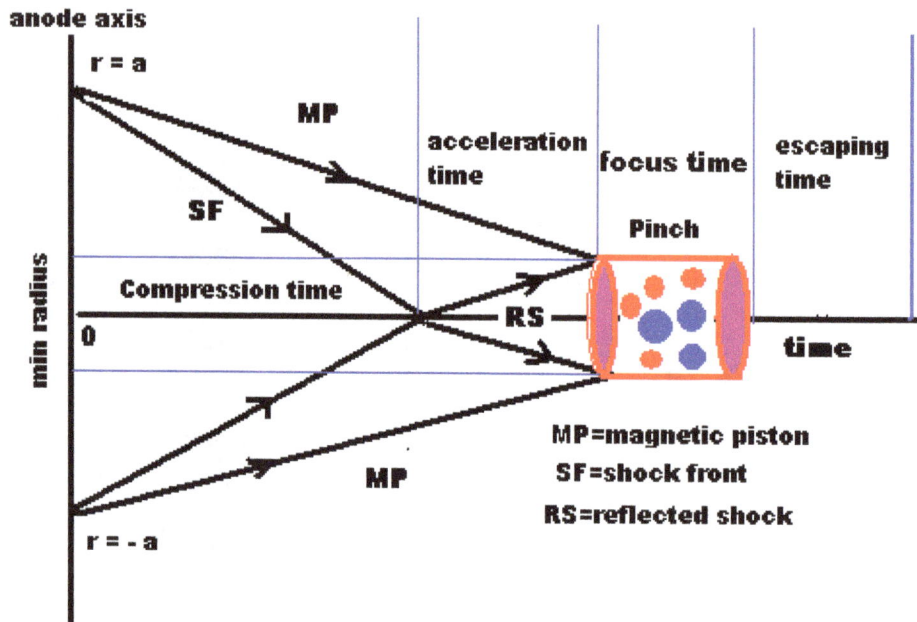

Figure 7. Schematic of radial phase development when magnetic piston separates from.

Compression phase start after transformation of plasma structure to plasma column. This plasma column will compressed adiabatically to form the pinch of final focus. The rapid compression of relatively dense plasma leads to heating plasma to a thermonuclear (fusion) temperature. The plasma collapses toward the center with approximately constant acceleration and flutes which subsequently develop on the outer surface of the plasma are interpreted as Rayleigh-Taylor.

Rayleigh-Taylor instability prevents the compression of the plasma column to get uniformly radial. The magnetic field starts diffusing into the plasma column, leading to an anomalous high plasma resistance as well as increasing the inductance of the system because of increasing the density of plasma column. The sharp change in plasma inductance and high plasma resistance induces high electric field inside the plasma column. This electric field will accelerate the ions and electrons in opposite direction. As result for that, the relative drift between ions and electrons leads to an increasing electron thermal velocity.

If the density clump occurs in plasma, an electric field can cause the ions and electrons to separate, generating another electric field. If there is feedback mechanism that causes the second electric field to enhance the first one, an electric field grows indefinitely and the plasma is unstable. Such instability called drift instability. In some cases, drifts can be self-perpetuating (charge separation leading to a drift, leading to more charge separation and so on), so that plasma instability results.

Generation of high energy particles and radiation in the plasma focus are considered to be an indication of non-thermodynamic equilibrium of the pinch. The presence of a beam of charged particles that forms in the plasma is explained by the appearance of electric fields which can be caused by the development of a Rayleigh-Taylor instability, plasma turbulence, magneto-acoustic wave propagation in the pinch.

In plasma focus, especially in the pinch phase the plasma gets more compressed at the onset of hydrodynamic instabilities ($m = 0$) as shown in Figure 8 (El-Aragi, 2010). The charged particles (ions and electrons) are trapped between two layers of plasma current sheath which acts like two moving magnetic mirrors with radius R and these particles initially have a velocity V_i (Yousefi et al., 2007), which were observed in the plasma column, a ring-shape around the dense plasma column (necking) that it can be attributed to the ion shape in the pinhole image. In fact, it seems that, $m = 0$ instability (necking) cause the ions acceleration with periods of few to tens of nanoseconds.

During compression phase the plasma density is high, then coulomb collisions between electrons and ions will efficiently thermalise the plasma. The dissipated energy ends up distributed equally amongst the particles, forming a Maxwellian distribution. The particles are accelerated in the current sheath medium as result of many collisions which act as a scattering center. Here the charged particles impinging on the oscillating sheath edge suffer a change of a velocity upon reflection back into the bulk plasma. As the sheath moves into the bulk, the reflected particles gain energy, as the sheath moves away, the particles lose energy. As the particle goes deeper into the sheath which acts as a potential barrier, it

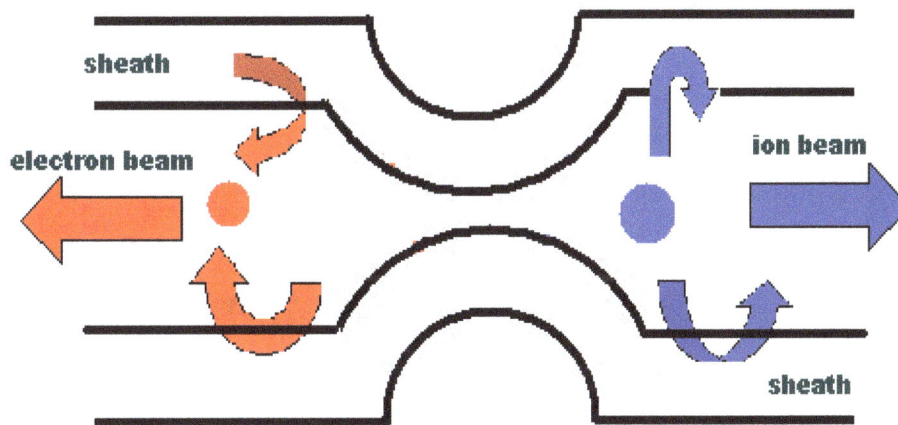

Figure 8. Shows the plasma pinch and emission beams of charged particles onset of instability inducing micro instabilities.

is slowed down by the retarding space charge field. After it has entirely lost its velocity, it turns back and increases its velocity under the action of an accelerating field, returning to the plasma.

Conclusion

Helium ion beam was detected using fast response charge collector system (Faraday cup) and time-integrated solid state nuclear track detector LR-115A. Faraday cup measurements showed that the ion energy spectrum a mixture of faster group of ions that has higher kinetic energy ranging from 15 to 540 keV and slower group that has lower kinetic energy value ranging from 0.3 to 1.0 keV. It is found that the ion flux depend on the filling gas pressure and the maximum ion flux was registered at P = 0.8 Torr of helium gas. The ion flux density has been determined using track etching technique LR-115A. The flux density of ion beam was estimated to be 4.4×10^{11} ion/steradian using CCZ model.

Conflict of Interest

The authors have not declared any conflict of interest.

REFERENCES

Bernstein MJ (1970). Acceleration mechanism for neutron production in plasma focus and Z-pinch discharges. Phys. Fluids. 13:2858–2866.
Deutsch R, Kies W (1988). Ion acceleration and runaway in dynamical pinches. Plasma Phys. Contr. Fusion. 30:263–276.
El-Aragi G (2010). Ion Beam Emission within a Low Energy Focus Plasma (0.1kJ) Operating with hydrogen. J. Zeitschrift fuer Naturforschung 65a.(2010).

EL-Aragi GM, Seddik U, Abd EL-Haliem A (2007). Detection of accelerated particles from pulsed plasma discharge using solid state nuclear track detector, Pramana J. Phys. 68(4):603-609.
Gerdin G, Stygar W, Venneri F (1981). Faraday cup analysis of ion beams produced by a dense plasma focus. J. Appl. Phys. 52:3269-3275.
Jager U, Herold H (1987). Fast ion kinetics and fusion reaction mechanism in the plasma focus. Nucl. Fusion. 27:407–423.
Kelly H, Marquez A (1996). Ion-beam and neutron production in a low-energy plasma focus. Plasma Phys. Contr. Fusion. 38:1931.
Lee S, Saw SH (2013). "Plasma focus ion beam fluence and flux-For various gases". Physics Plasmas. 20(6).
Mather JW (1971). Dense plasma focus, Methods of Experimental Physics Vol 9B, ed H Lovberg and H R Griem (New York: Academic) pp. 187–249.
Pasternak A, Sadowski M (1998). Analysis of ion trajectories within a pinch column of a PF-type discharge. In: Proc ICPP&25th EPS Conf CFPP. Praha, Czech Republic. ECA 22C:2161–2164.
Wong CS, Choi P, Leong WS, Singh J (2002). Generation of High Energy Ion Beams from a Plasma Focus Modified for Low Pressure Operation. Jpn. J. Appl. Phys. 41:3943-3946.
Yousefi HR, Nakata Y, Ito H, Masugata K (2007). Characteristic Observation of the ion beams in the plasma focus device. Plasma Fusion Res. 2:S1084.
Zakaullah M, Ijaz A, Murtaza G, Waheed A (1999). Imaging of fusion reaction zone in plasma focus. Physics of Plasma. 6:3188.

20

Development of a multilayer perceptron (MLP) based neural network controller for grid connected photovoltaic system

A. Ndiaye, L. Thiaw, G. Sow and S. S. Fall

Laboratory of Renewable energy, Polytechnic Higher School, Cheikh Anta Diop University BP 5085, Dakar, Senegal.

This paper focuses on the development of a controller for grid connected photovoltaic energy conversion system. Control design of a single phase inverter interfacing a photovoltaic generator and an electrical grid is performed, based on Artificial Neural Networks. The developed controller is compared with a Proportional Integral (PI) controller through computer simulation. The obtained results show that the neural controller has faster response and lower total harmonic distortion (THD) without overshoots.

Key words: Photovoltaic generator, inverter, maximum power point tracking (MPPT), neural networks.

INTRODUCTION

The main difficulties in the control strategy of real dynamic systems are the non-linearity and strong non-linearity. The lack of right knowledge necessary for the development of the uncertainties. The control of the system requires in general the development of a mathematical model making it possible to establish the transfer function of the system that links the inputs and the outputs. This requires good knowledge of the dynamic and properties of the system. In the non-linear system case, the conventional techniques have often shown their limits mainly when the system to be studied presents mathematical models is somehow the origin of those limits (Mohammed et al., 2007).

Recourse to the control methods based on artificial intelligence has become a necessity. These control methods follow an extraction process of the knowledge of the system to be studied from collected empirical data, so as to be able to react in front of new situations: This strategy is known as intelligent control (Panos et al., 1993).

Artificial neural networks are used in intelligent control due to the fact that they are parsimonious universal approximators (Panos et al., 1993; Rival et al., 1995) and that they have the capacity to adapt to a dynamic evolving through time. Moreover, as multi-input and multi-output systems, they can be used in the frame of the control of the multivariable systems.

A feed forward ANN makes one or more algebraic functions of its inputs, by the composition of the functions made by each one of its neurons (Dreyfus, 2002). These are organized in layers and inter-connected by well-balanced synaptic connections. The supervised training of a neural network consists in modifying the weights to have a given behavior minimizing a cost function often represented by the quadratic error (Panos et al., 1993; Cybenko, 1989).

Several authors have tried to exploit the advantages of neural networks to control a dynamic system (Mahmoud et al., 2012; Zameer and Singh, 2013) precisely, within the field of robotics (Rival et al., 1995; Yildirim, 1997) and for the control of asynchronous motors (Mohammed et al., 2007; Panos et al., 1993; Branštetter and Skotnica, 2000). More details on neural network controllers can be found in Panos et al. (1993), Wishart and Harley (1995), Ronco and Gawthrop (1997), Hagan and Demuth (1996), Wishart and Harley (1995), Ahmed et al. (2008), Tai et al., (1990), Hagan and Demuth (1996), Chen et al. (1997), Norgaard (1996) and Vandoorn et al. (2009) in which a comparative study was made between PI controller, PID controller and a fuzzy logic based controller for an inverter control shows that the PI controller has better performances, though the fuzzy logic based controller is an intelligent one.

In the work presented in this article, the capacities of multi-layer perceptron (MLP) to learn the inverse model of non-linear systems are used to work out the control of a single-phase inverter used as an interface between a photovoltaic generator (PVG) and an electrical grid. The objective is to inject into the grid as much photovoltaic energy as available, with low total harmonic distortion (THD) and good reference signal tracking a characteristic.

METHODOLOGY

Inverter control by using a PI controller

The PI controller is the most used controller in industrial systems. It is easy to implement and it is costs efficient. The control scheme of a grid connected photovoltaic system used in this work is given in Figure 1.

A loop control is elaborated in order to ensure the injection of the maximum available photovoltaic energy into the grid. This loop enables current control to give a reference current determined by the maximum power point tracking system (Figures 1 and 2). In order to determine the controller parameters, the whole system model has been established. The inverter transfer function links inverter output current to the duty cycle. The PI controller parameters can be determined from this transfer function. The input voltage of the inverter is supposed to be constant (ripple are neglected). From Figure 1, Equation (1) can be established.

$$\text{L}_{ac} \frac{di_g}{dt} = \alpha V_{dc} - v_g \tag{1}$$

Where, L = inductor value of the filter; i_g = current injected into the grid; α = duty cycle; V_{dc} = inverter input voltage, and v_g = grid voltage;

Using small signals models, it is possible to write:

$$\alpha = \overline{\alpha} + \widetilde{\alpha}$$
$$I_g = \overline{I} + \widetilde{i}_g$$

Where, $\overline{\alpha}$ is the average value of the duty cycle and $\widetilde{\alpha}$ the duty cycle ripple; $\overline{I}_g$ is the average value of the current and $\widetilde{i}_g$ the current ripple.
Considering that the grid average voltage is null and neglecting its

ripples, Equation (2) can be obtained.

$$\frac{d\widetilde{i}_g}{dt} = \frac{V_{dc}}{L} \widetilde{\alpha} \tag{2}$$

Applying Laplace transform to Equation (2) and considering the control loop represented in Figure 2, we get the open loop transfer function expressed by Equation (3) linking the injected current to the duty cycle.

$$\text{G}_g = \left(k_p + \frac{k_i}{s} \right) \frac{G_{ti}}{v_{ti}} \frac{V_{dc}}{sL} \tag{3}$$

Where, v_{ti} = The magnitude of the carrier, and G_{ti} = gain loop (gain of the current sensor).
Exploiting this transfer function allows the PI coefficients to be determined (Equations 4 and 5).

$$\text{k}_p = \frac{2\pi f_{cL} L v_{ti}}{V_{dc} G_{ti}} \tag{4}$$

$$\text{k}_i = \frac{2\pi f_{cL}}{\tan(p_{hm})} \tag{5}$$

Where, f_{cL} = Cut-off frequency, and p_{hm} = Phase margin.

The PI controller input consists of the error between the current provided by the inverter and its reference. The objective of this control is to correct the current injected into the grid (i_g) so that it follows the reference value (i_{gref}).
This type of controllers is simple but it gives limited performances if the system integrates strongly nonlinear elements such as static inverters. In fact, the determination of the controller parameters can be done through different methods but generally depends on the knowledge of the system to be controlled, and mathematical model of the system is not always available. Equations 4 and 5 show that the parameters of the PI controller (kp and ki) depend on Vcd which is related to meteorological conditions (solar irradiation and temperature). So it is worth adapting this coefficient any time the meteorological conditions change, which seems to be impossible. Therefore an adaptive control has to be set up. This fact has led to carrying out a comparative study of a PI controller and a neural network controller.

Neural network controller for single phase inverter

Principles of artificial neural networks

The ANN network is based on models that try to explain human brain functioning. They are adapted to the treatment in parallel of complex problems such as speech and face recognition, or simulation of nonlinear functions. So they offer a new means of information treatment. In Figure 3, the main elements of an artificial neural are depicted: the input, processing unit and an output.
A formal neuron is characterized by Equations (6) and (7).

$$x_i = f(A_i) \tag{6}$$

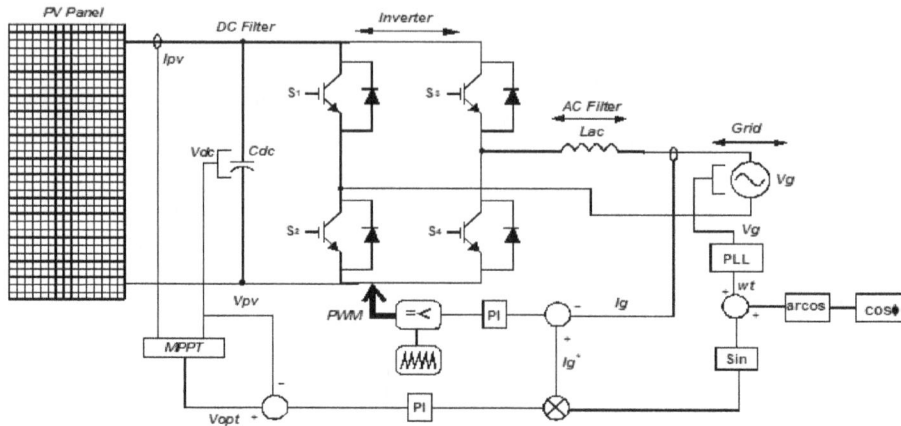

Figure 1. Control loop of a grid connected photovoltaic system.

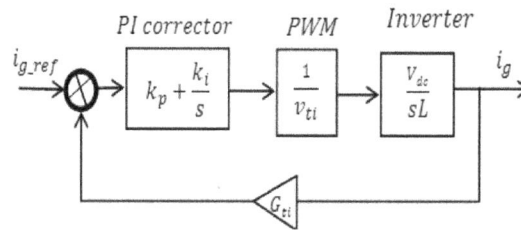

Figure 2. Control loop of the inverter current.

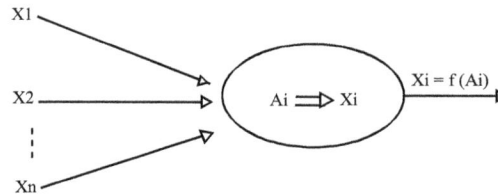

Figure 3. Representation of a formal neuron.

$$A_i = \sum_{i=1}^{N_i} w_{ij} x_i + b_i \qquad (7)$$

With, x_i = State of a neuron j connected to neuron i; A_i = Activity of neuron i; w_{ij} = Weight of the connexion between the neurons j and I, and b_i = Bias.

The MLP network (Figure 4) is a feed forward network that is composed of several layers, each neuron of a layer being totally connected to the neurons of the next layer. The resulting network is able to approximate any nonlinear function.

The error $\delta_{p,k}$ made on the kth output neuron for a sample p is expressed by Equation (8).

$$\delta_{p,k} = O_{p,k} - x_{p,k} \qquad (8)$$

Where, $O_{p,k}$ = Desired output of the neuron k for the sample p, and $x_{p,k}$ = output of the neuron k for the sample p.

As a result, the total error (for all output neurons) is estimated by:

$$e_p = \frac{1}{2} \sum_{i=1}^{N_i} \delta_{p,k}^2 = \frac{1}{2} \sum_{k=1}^{m} \left(O_{p,k} - x_{p,k} \right)^2$$

Where m = number of neurons on the output node.

The synaptic weights are then adjusted so as to reduce the output error for the whole samples of the data base:

$$e = \sum_{p=1}^{N} e_p \qquad (9)$$

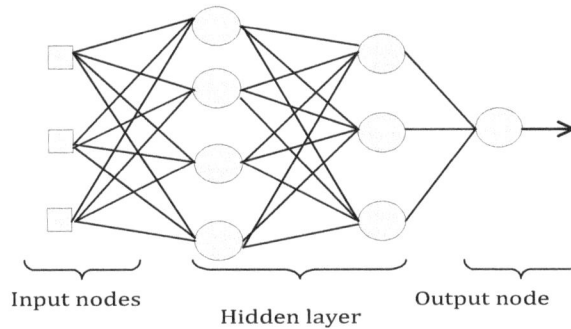

Figure 4. Architecture of an MLP network.

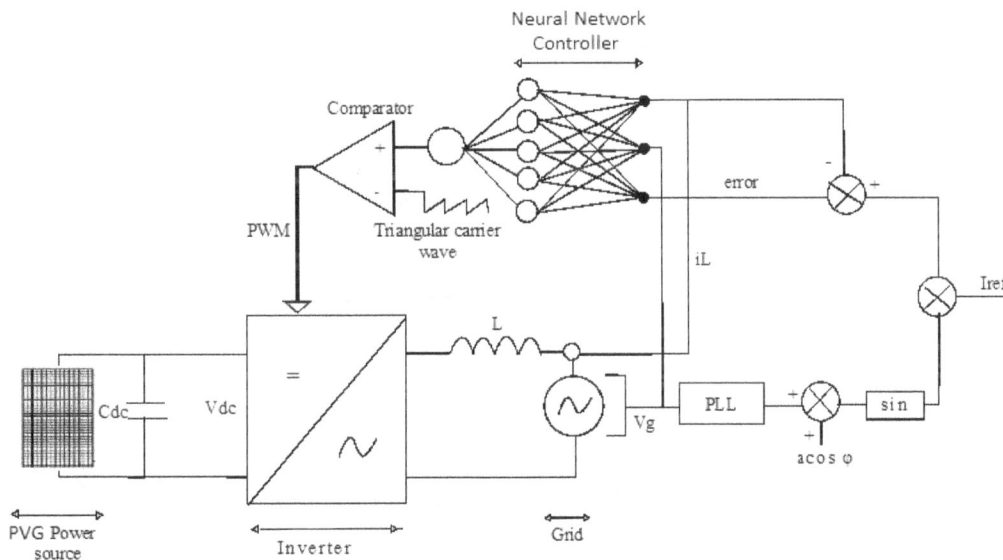

Figure 5. Grid connected photovoltaic system with single phase inverter and neural controller.

Where N designates the size of the database.

The process of the network parameters estimation is called training. The set of parameters that are to be estimated includes all the weights and biases. An algorithm called back propagation is mainly used for the network training. More details on neural networks is given in Ahmed et al. (2008).

Proposed design method of the neural controller

Within the framework of this study, the system to control is a single-phase inverter serving as an interface between a photovoltaic generator and an electrical grid. The structure of the neural controller for photovoltaic energy injection into the grid is represented in Figure 5.

The inputs of the neural controller are the current injected into the grid, the grid voltage and the error between the actual and the reference values of the inverter output current.

Database for the neural controller training is obtainend from the system simulation with several PI controllers, each of witch being determined for a given system operating point, defined by the inverter input DC voltage.

RESULTS AND DISCUSSION

The inverter is designed so that its switches be able to support the maximum current i_{gmax} and the maximum open circuit voltage (Vco) of the photovoltaic generator. Table 1 gives the inverter parameters and those of the photovoltaic generator.

The filter inductor value is determined by Equation (11).

$$L = \frac{V_{dc}}{16 \Delta I_{max} f_s} \tag{11}$$

Where, $Vcd.$ is the inverter input voltage; f_s is the switching frequency, and ΔI_{max} is the maximum value of the output current ripple.

The system is first simulated with the PI controller (Table 2). The injected current and its reference value are presented in Figure 6, whereas Figure 7 shows grid

Table 1. Inverter and photovoltaic generator parameters.

Parameter	Value
DC bus voltage (Vdc = Vopt at 1 kW/m² and 25°C)	800 V
Opened circuit voltage of the PV generator	1000 V
Short circuit current of the PV generator	6.8 A
Filter inductor value (L)	5 mH
ESR value of the inductor	0.2 Ω
Maximum power of the PV generator	4 kW
Grid RMS voltage value (Vgeff)	220 V
Grid frequency (f_o)	50 Hz
Inverter switching frequency (fs)	20 kHz

Table 2. PI controller parameters.

Coefficients	kp	ki
Values	5.23	6.33 104

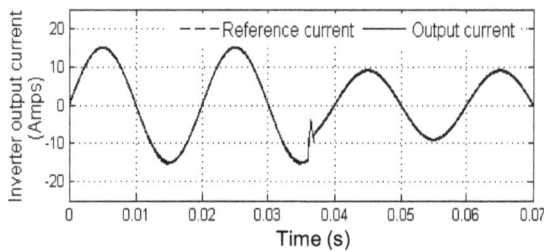

Figure 6. Inverter output current and its reference value when a PI controller is used.

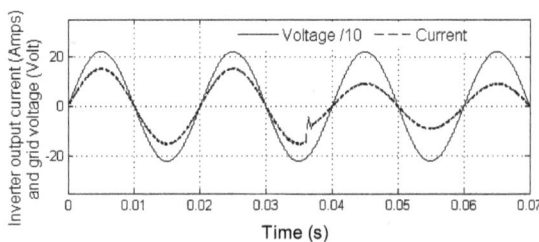

Figure 7. Grid voltage and inverter output current when a PI controller is used.

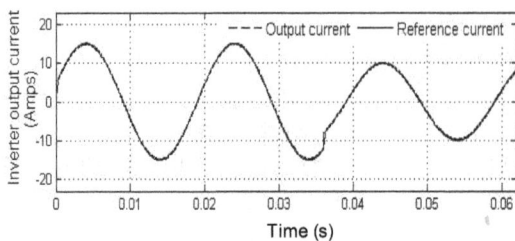

Figure 8. Inverter output current and its reference value when a neural controller is used.

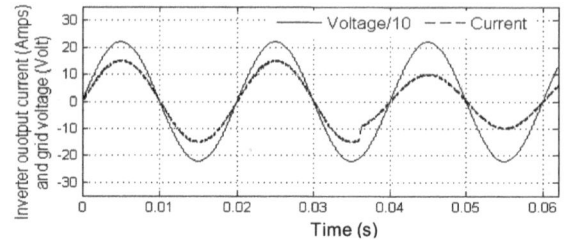

Figure 9. Grid voltage and inverter output current when a neural controller is used.

voltage and injected current for unity power factor. A disturbance consisting of a 33% reduction of reference current magnitude is introduced at t = 36 ms. The PI controller presents a relatively fast reference current tracking but an important overshoot can be noticed. The main drawbacks of this controller is due to the fact that it has to be designed for a given meteorological conditions.

The design of the neural network controller consists of designing several PI controllers for various meteorological conditions. The following values are used for the solar irradiation and the temperature: (0.25 kW/m², 25°C), (0.25 kW/m², 40°C), (0.6 kW/m², 25°C), (0.6 kW/m², 40°C), (1 kW/m², 25°C) and (1 kW/m², 40°C).

Control signals from the PI controllers, grid voltage, inverter output current and its reference value are gathered to form a large database used for the neural controller training.

Figure 8 shows inverter output current and its reference value when neural controller is used for the following meteorological conditions: a solar irradiation of 1 kW/m² and a temperature of 25°C. A disturbance consisting of a 33% reduction of reference current magnitude is introduced at t = 36 ms. The obtained results prove fast tracking capability of the neural controller without overshoots. Grid voltage and injected current for unity power factor are shown in Figure 9.

A comparison study of the two controllers is performed throughout simulation of two cases. In the first case, the simulation is made for the following meteorological conditions: Solar irradiation of 1 kW/m² and temperature of 50°C. The PI controller parameters for these meteorological conditions has resulted in kp=1.16 and ki=7.07 10^3 rad/s.

The total harmonic distortion (THD) of both controllers have been calculated and compared. The obtained results are presented on Figures 10 and 11. They show that the neuronal controller has a THD slightly weaker than the PI controller.

In the second simulation case, the same meteorological conditions were used but a disturbance consisting in a rapid variation of the reference current has been introduced. The simulation results are represented on Figures 12 and in Table 3. These results show that the relative error between the injected current and its

Development of a multilayer perceptron (MLP) based neural network controller for grid connected...

173

Fundamental (50Hz) = 15 , THD= 0.53%

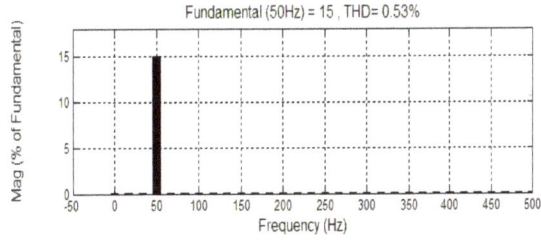

Figure 10. THD obtained with a PI controller for a solar irradiation of 1 kW/m^2 and a temperature of 50°C.

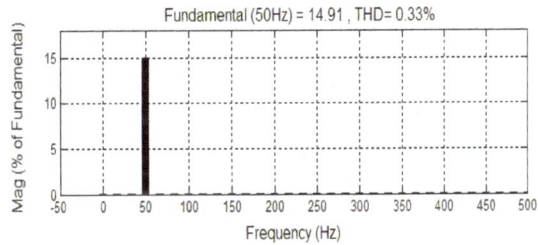

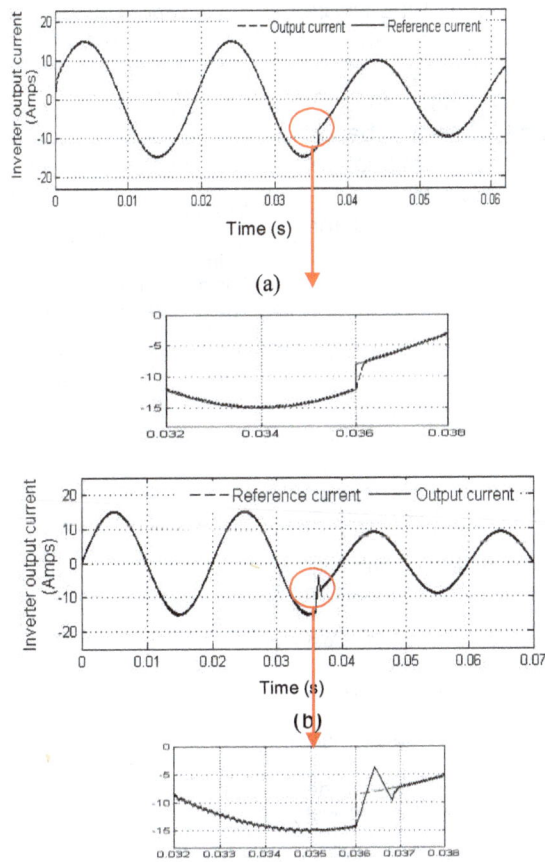

Fundamental (50Hz) = 14.91 , THD= 0.33%

Figure 11. THD obtained with the neural controller for a solar irradiation of 1 kW/m^2 and a temperature of 50°C.

(a)

(b)

Figure 12. Performances of the neuronal controller (a) and PI controller of (b) with disturbance, for an irradiation of 1 kW/m^2 and a temperature of 50°C.

Table 3. Comparison results of the pi and neural controller.

Performance	Neural controller	PI controller
Time response (ms)	0.5	1
Overshoot current (A)	0	5
Total harmonic distortion (%)	0.33	0.53
Magnitude of the fundamental current (A)	14.96	15
Relative current error: 100*(ig-ig_ref) /Ig_refmax	0.67	1.33

reference is weaker for the neural controller, be it the half of the one obtained by PI controller. Moreover, the PI controller has a response time twice greater than that of the neural controller. Unlike the PI controller, the neural controller responds to the disturbance without overshoot. These two controllers provide a fundamental magnitude of about 15 A. Yet, considering the nature of both signals, the neural controller gets closer to the reference, giving its weak THD (Figures 10 and 11).

Conclusion

Development of a MLP based neural controller is presented. The training and validation data of the used neural controller were obtained by simulation of the whole system with several PI controllers calculated for various meteorological conditions. The simulation results show that the neural controller gives better results than a PI controller. The advantage of neural network based controller is that it adapts to the changing of meteorological conditions unlike the PI controller whose performance decreases during a strong variation of the temperature and/or irradiation.

REFERENCES

Ahmed T, Hamza A, Abdel GA (2008). La commande neuronale de la machine à réluctance variable" Rev. Roum. Sci. Techn. – Électrotechn. et Énerg. Bucarest. 53(4):473–482.

Branštetter P, Skotnica M (2000). Application of artificial neural network for speed control of asynchronous motor with vector control, Proceedings of International Conference of Košice, EPE-PEMC, 6-157-6-159.

Chen CT, Chang WD, Hwu J (1997). Direct control of nonlinear dynamical systems using an adaptive single neuron, IEEE Trans on Neural Networks 2(10):33-40.

Cybenko G (1989). Approximation by superposition of a sigmoidal function", Math. In: Control Signals System 2nd ed, pp. 303-314.

Dreyfus G (2002). Réseaux de neurones: méthodologies et applications, editions Eyrolles.

Hagan MT, Demuth HB (1996). Neural network design, Thomson Asia Pte Ltd, 2nd ed.

Mohammed S, Djamel E, Chaouch M, Fayçal K (2007). Commande neuronale inverse des systèmes non linéaires, In 4th International Conference on Computer Integrated Manufacturing CIP, 2007 03-04 November.

Mahmoud AY, Tamer K, Mushtaq N, Mohd AA (2012). An Improved Maximum Power Point Tracking Controller for PV Systems Using Artificial Neural Network, Przegląd Elektrotechniczny (Electrical Review), ISSN 0033-2097, R. 88 NR 3b/2012.

Norgaard M (1996). System identification and control with neural networks", Thesis, Institute of automation, Technical University of Denmark.

Panos J, Antsaklis K, Passino M (1993). Introduction to Intelligent and Autonomous Control, Kluwer Academic Publishers, ISBN: 0-7923-9267-1.

Rival I, Personnaz L, Dreyfus G (1995). Modélisation, classification et commande, Par réseaux de neurones: principes fondamentaux, Méthodologie de conception et illustrations industrielles, Mécanique Industrielle et Matériaux, n°51 (septembre 1998).

Ronco E, Gawthrop PJ (1997). Neural networks for modelling and control, Techncal Report CSC-97008, Center for Systems and Control, Glasgow.

Tai P, Ryaciotaki-Boussalis HA, Tai K (1990). The application of neural networks to control systems: a survey, Signals, systems and Computers, Record Twenty-Fourth Asilomar Conference on Vol.1.

Vandoorn T, Renders B, De Belie F, Meersman B, Vandevelde L (2009). A Voltage-Source Inverter for Microgrid Applications with an Inner Current Control Loop and an Outer Voltage Control Loop, International Conference on Renewable Energies, and Power Quality (ICREPQ09) Valencia (Spain), 15th to 17th April.

Wishart MT, Harley RG (1995). Identification and Control of Induction Machines Using Artificial Neural Networks, IEEE Transaction on Industry Applications, 31:3.

Yildirim S (1997) New neural networks for adaptive control of robot manipulators, Neural Networks, International Conference. 3:1727-1731.

Zameer A, Singh SN (2013). Modeling and Control of Grid Connected Photovoltaic System-A Review, Int.J. Emerging Technol. Adv. Eng. ISSN 2250-2459, ISO 9001:2008. 3(3).

Permissions

All chapters in this book were first published in IJPS, by Academic Journals; hereby published with permission under the Creative Commons Attribution License or equivalent. Every chapter published in this book has been scrutinized by our experts. Their significance has been extensively debated. The topics covered herein carry significant findings which will fuel the growth of the discipline. They may even be implemented as practical applications or may be referred to as a beginning point for another development.

The contributors of this book come from diverse backgrounds, making this book a truly international effort. This book will bring forth new frontiers with its revolutionizing research information and detailed analysis of the nascent developments around the world.

We would like to thank all the contributing authors for lending their expertise to make the book truly unique. They have played a crucial role in the development of this book. Without their invaluable contributions this book wouldn't have been possible. They have made vital efforts to compile up to date information on the varied aspects of this subject to make this book a valuable addition to the collection of many professionals and students.

This book was conceptualized with the vision of imparting up-to-date information and advanced data in this field. To ensure the same, a matchless editorial board was set up. Every individual on the board went through rigorous rounds of assessment to prove their worth. After which they invested a large part of their time researching and compiling the most relevant data for our readers.

The editorial board has been involved in producing this book since its inception. They have spent rigorous hours researching and exploring the diverse topics which have resulted in the successful publishing of this book. They have passed on their knowledge of decades through this book. To expedite this challenging task, the publisher supported the team at every step. A small team of assistant editors was also appointed to further simplify the editing procedure and attain best results for the readers.

Apart from the editorial board, the designing team has also invested a significant amount of their time in understanding the subject and creating the most relevant covers. They scrutinized every image to scout for the most suitable representation of the subject and create an appropriate cover for the book.

The publishing team has been an ardent support to the editorial, designing and production team. Their endless efforts to recruit the best for this project, has resulted in the accomplishment of this book. They are a veteran in the field of academics and their pool of knowledge is as vast as their experience in printing. Their expertise and guidance has proved useful at every step. Their uncompromising quality standards have made this book an exceptional effort. Their encouragement from time to time has been an inspiration for everyone.

The publisher and the editorial board hope that this book will prove to be a valuable piece of knowledge for researchers, students, practitioners and scholars across the globe.

List of Contributors

B. Basirat
Department of Mathematics, Birjand Branch, Islamic Azad University, Birjand, Iran

K. Maleknejad
Department of Mathematics, Karaj Branch, Islamic Azad University, Karaj, Iran

E. Hashemizadeh
Department of Mathematics, Karaj Branch, Islamic Azad University, Karaj, Iran

Stephen Nyende-Byakika
Department of Civil Engineering, Tshwane University of Technology, Pretoria, South Africa

Gaddi Ngirane-Katashaya
Department of Civil Engineering, Makerere University, Kampala, Uganda

Julius M. Ndambuki
Department of Civil Engineering, Tshwane University of Technology, Pretoria, South Africa

José Alfredo Leal–Naranjo
Sección de Estudios de Posgrado e Investigación, Escuela Superior de Ingeniería Mecánica y Eléctrica, Instituto Politécnico Nacional, Unidad Azcapotzalco, Av. de las granjas No. 682 col. Sta. Catarina, delegación Azcapotzalco C.P. 02550, México D.F

Christopher René Torres-San Miguel
Escuela Superior de Ingeniería Mecánica y Eléctrica, Instituto Politécnico Nacional, Unidad Profesional "Adolfo López Mateos", edificio 5, 2do piso, col. Lindavista, delegación Gustavo A. Madero. C.P. 07738, México D.F Instituto Universitario de Investigación del Automóvil, Escuela Técnica Superior de Ingenieros Industriales, Universidad Politécnica de Madrid, Carretera de Valencia, km.7, 28031, Madrid, España

Manuel Faraón Carbajal–Romer
Sección de Estudios de Posgrado e Investigación, Escuela Superior de Ingeniería Mecánica y Eléctrica, Instituto Politécnico Nacional, Unidad Azcapotzalco, Av. de las granjas No. 682 col. Sta. Catarina, delegación Azcapotzalco C.P. 02550, México D.F

Luis Martínez-Sáez
Instituto Universitario de Investigación del Automóvil, Escuela Técnica Superior de Ingenieros Industriales, Universidad Politécnica de Madrid, Carretera de Valencia, km.7, 28031, Madrid, España

S. Prabhu
School of Mechanical Engineering, S.R.M. University, Chennai 603 203, Tamil Nadu, India

B. K. Vinayagam
Department of Mechatronics, S.R.M. University, Chennai 603 203, Tamil Nadu, India

S. Selva Nidhyananthan
Department of ECE, Mepco Schlenk Engineering College, Sivakasi-626005, India

R. Shantha Selva Kumara
Department of ECE, Mepco Schlenk Engineering College, Sivakasi-626005, India

D. S. Roland
Department of ECE, Mepco Schlenk Engineering College, Sivakasi-626005, India

Fatma Özdemir
Department of Mathematics, Faculty of Science and Letters, Istanbul Technical University, 34469 Maslak-Istanbul, Turkey

O. Djebili
Laboratoire Grespi/MAN UFR, Sciences Exactes and Naturelles Moulin de la Housse, BP 1039 51687 REIMS CEDEX 2, France

F. Bolaers
Laboratoire Grespi/MAN UFR, Sciences Exactes and Naturelles Moulin de la Housse, BP 1039 51687 REIMS CEDEX 2, France

A. Laggoun
Department de physique, Faculté des Sciences UMBB Boumerdes, Algéria

J. P. Dron
Laboratoire Grespi/MAN UFR, Sciences Exactes and Naturelles Moulin de la Housse, BP 1039 51687 REIMS CEDEX 2, France

Noradin Ghadimi
Department of Electrical Engineering, Ardabil Branch, Islamic Azad University, Ardabil, Iran

Seyed Mahdi Hatamian
Department of Electrical Engineering, Science and Research Branch, Islamic Azad University, Tehran, Iran

Vahid Ahmadi
Department of Electrical Engineering, Tarbiat Modares University, Tehran, Iran

Elham Darabi
Plasma Physics Research Center, Science and Research Branch, Islamic Azad University, Tehran, Iran

Paras Chawla
Department of Electronics and Communication Engineering, Thapar University, Patiala, Punjab, India

Rajesh Khanna
Department of Electronics and Communication Engineering, Thapar University, Patiala, Punjab, India

Kh. Lotfy
Department of Mathematics, Faculty of Science, Zagazig University, Zagazig P. O. Box 44519 Egyp

Wafaa Hassan
Department of Mathematics and Physics, Faculty of Engineering, Port Said Branch of Suez Canal University, Port Said, Egypt
Department of Mathematics, Faculty of Science and Arts, Al-mithnab, Qassim University, P.O. Box 931, Buridah 51931, Al-mithnab, Kingdom of Saudi Arabia

Mrittunjoy Guha Majumdar
St. Stephen's College, Delhi, India

Y. Song
Savannah College of Art and Design, P. O. Box 3146, Savannah, GA 31402-3146, USA

D. Edwards
Department of Mechanical Engineering, University of Idaho, Moscow, ID 83844-0902, USA

V. S. Manoranjan
Department of Mathematics, Washington State University, Pullman, WA 99164-3113, USA

B. Thole
Ngurdoto Defluoridation Research Station, P. O. Box Usa River, Arusha, Tanzania

W. R. L. Masamba
Harry Oppenheimer Okavango Research Centre, University of Botswana, P/Bag 285, Maun, Botswana

F. W. Mtalo
College of Engineering and Technology, University of Dar Es Salaam, P. O. Box Dar Es Salaam, Tanzania

Tanuj Kumar Garg
Deptartment of Electronics and Communication Engineering, Gurukul Kangri University, Haridwar, India

S. C. Gupt
Deptartment of Electronics and Communication Engineering, DIT, Dehradun, India

S. S. Patnaik
Deptartment of ETV, NITTTR, Sector-26, Chandigarh, India

Vipul Sharma
Deptartment of Electronics and Communication Engineering, Gurukul Kangri University, Haridwar, India

Yung-Hsiang Hung
Department of Industrial Engineering and Management, National Chin-Yi University of Technology, 35, Lane215, Section 1, Chung-Shan Road, TaiPing, TaiChung, 411, Taiwan, R.O.C

Mei-Ling Huang
Department of Industrial Engineering and Management, National Chin-Yi University of Technology, 35, Lane215, Section 1, Chung-Shan Road, TaiPing, TaiChung, 411, Taiwan, R.O.C

Kun-Liang Fanchiang
Department of Industrial Engineering and Management, National Chin-Yi University of Technology, 35, Lane215, Section 1, Chung-Shan Road, TaiPing, TaiChung, 411, Taiwan, R.O.C

M. E. Medhat
Experimental Nuclear Physics Department, Nuclear Research Centre, P. O. 13759, Cairo, Egypt
Institute of High Energy Physics, CAS, Beijing 100049, China

Yifang Wang
Institute of High Energy Physics, CAS, Beijing 100049, China

Ali A. J. Adham
Faculty of Industrial Management, University Malaysia Pahang, Gambang, Pahang, 26300, Malaysia

Ahmad N. N. Kamar
Faculty of Industrial Management, University Malaysia Pahang, Gambang, Pahang, 26300, Malaysia

G. M. El-Aragi
Plasma Physics and Nuclear Fusion Department, Nuclear Research Center, AEA, P. O. Box 13759 Cairo, Egypt

A. Ndiaye
Laboratory of Renewable energy, Polytechnic Higher School, Cheikh Anta Diop University BP 5085, Dakar, Senegal

L. Thiaw
Laboratory of Renewable energy, Polytechnic Higher School, Cheikh Anta Diop University BP 5085, Dakar, Senegal

G. Sow
Laboratory of Renewable energy, Polytechnic Higher School, Cheikh Anta Diop University BP 5085, Dakar, Senegal

S. S. Fall
Laboratory of Renewable energy, Polytechnic Higher School, Cheikh Anta Diop University BP 5085, Dakar, Senegal